The Redshift Controversy

FRONTIERS IN PHYSICS

David Pines, Editor

FRONTIERS IN PHYSICS (*continued*)

David Pines, Editor

M. Gell-Mann and Y. Ne'eman

The Eightfold Way: (A Review—with a Collection of Reprints), 1964

W. A. Harrison *Pseudopotentials in the Theory of Metals, 1966 (2nd printing, 1971)*

R. Hofstadter *Electron Scattering and Nuclear and Nucleon Structure: A Collection of Reprints with an Introduction, 1963*

D. Horn and F. Zachariasen

Hadron Physics at Very High Energies, 1973

S. Ichimaru *Basic Principles of Plasma Physics: A Statistical Approach, 1973*

M. Jacob and G. F. Chew

Strong-Interaction Physics: A Lecture Note Volume, 1964

L. P. Kadanoff and G. Baym

Quantum Statistical Mechanics: Green's Function Methods in Equilibrium and Nonequilibrium Problems, 1962 (2nd printing, 1971)

I. M. Khalatnikov

An Introduction to the Theory of Superfluidity, 1965

J. J. J. Kokkedee

The Quark Model, 1969

A. M. Lane *Nuclear Theory: Pairing Force Correlations to Collective Motion, 1963*

T. Loucks *Augmented Plane Wave Method: A Guide to Performing Electronic Structure Calculations—A Lecture Note and Reprint Volume, 1967*

A. B. Migdal and V. Krainov

Approximation Methods in Quantum Mechanics, 1969

A. B. Migdal *Nuclear Theory: The Quasiparticle Method, 1968*

Y. Ne'eman *Algebraic Theory of Particle Physics: Hadron Dynamics in Terms of Unitary Spin Currents, 1967*

P. Nozières *Theory of Interacting Fermi Systems, 1964*

iii

THE REDSHIFT CONTROVERSY

GEORGE B. FIELD
Center for Astrophysics, Harvard College Observatory
and Smithsonian Astrophysical Observatory, Cambridge, Massachusetts

HALTON ARP
Hale Observatories, Pasadena, California

JOHN N. BAHCALL
Institute for Advanced Study, Princeton, New Jersey

1973
W. A. BENJAMIN, INC.
ADVANCED BOOK PROGRAM
Reading, Massachusetts

London · Amsterdam · Don Mills, Ontario · Sydney · Tokyo

Library of Congress Cataloging in Publication Data

Field, George B   1929-
    The redshift controversy.

    (Frontiers in physics)
    Includes bibliographies.
    1. Redshift.   I. Arp, Halton C., joint author.
II. Bahcall, John N., joint author.   III. Title.
IV. Series.
QB465.F53   523.1'8   74-614
ISBN 0-8053-2512-3(hb)
ISBN 0-8053-2513-1(pb)
ABCDEFGHIJ-HA-787654

Original text reproduced by W. A. Benjamin, Inc., Advanced Book Program, Reading, Massachusetts, from camera-ready copy prepared by the authors.

Printed in the United States of America

CONTENTS

REPRINTS OF PAPERS SELECTED by HALTON ARP 219

COMMENTS ON REPRINTED PAPERS by HALTON ARP 221

EDITOR'S FOREWORD

The problem of communicating in a coherent fashion
the recent developments in the most exciting and
active fields of physics seems particularly pressing
today. The enormous growth in the number of
physicists has tended to make the familiar channels
of communication considerably less effective. It
has become increasingly difficult for experts in a
given field to keep up with the current literature;
the novice can only be confused. What is needed is
both a consistent account of a field and the
presentation of a definite "point of view"
concerning it. Formal monographs cannot meet such
a need in a rapidly developing field, and, perhaps
more important, the review article seems to have

fallen into disfavor. Indeed, it would seem that
the people most actively engaged in developing a
given field are the people least likely to write
at length about it.

FRONTIERS IN PHYSICS has been conceived in an
effort to improve the situation in several ways.
One of these is to take advantage of the fact that
the leading physicists today frequently give a
series of lectures, a graduate seminar, or a
graduate course in the special fields of interest.
Such lectures serve to summarize the present status
of a rapidly developing field and may well constitute
the only coherent account available at the time.
Often, notes on lectures exist (prepared by the
lecturer himself, by graduate students, or by
postdoctoral fellows) and have been distributed in
mimeographed form on a limited basis. One of the
principal purposes of the FRONTIERS IN PHYSICS
Series is to make such notes available to a wider
audience of physicists.

It should be emphasized that lecture notes
are necessarily rough and informal, both in style
and content, and those in the series will prove no

exception. This is as it should be. The point of

the series is to offer new, rapid, more informal,

and it is hoped, more effective ways for physicists

to teach one another. The point is lost if only

elegant notes qualify.

A second way to improve communication in very

active fields of physics is by the publication of

collections of reprints of recent articles. Such

collections are themselves useful to people working

in the field. The value of the reprints would,

however, seem much enhanced if the collection would

be accompanied by an introduction of moderate length,

which would serve to tie the collection together and,

necessarily, constitute a brief survey of the present

status of the field. Again, it is appropriate that

such an introduction be informal, in keeping with

the active character of the field.

A third possibility for the series might be

called an informal monograph, to connote the fact

that it represents an intermediate step between

lecture notes and formal monographs. It would offer

the author an opportunity to present his views of a

field that has developed to the point at which a

summation might prove extraordinarily fruitful, but
for which a formal monograph might not be feasible
or desirable.

Fourth, there are the contemporary classics--
papers or lectures which constitute a particularly
valuable approach to the teaching and learning of
physics today. Here one thinks of fields that lie
at the heart of much of present-day research, but
whose essentials are by now well understood, such
as quantum electrodynamics or magnetic resonance.
In such fields some of the best pedagogical material
is not readily available, either because it consists
of papers long out of print or lectures that have
never been published.

_____

The above words, written in August,1961, seem
equally applicable today (which may tell us something
about developments in communication in physics during
the past decade). The present volume appears to be
especially well-suited to the "Frontiers" format,
dealing as it does with an important area of science
in which there exists a continuing controversy, one

which is best looked at in a bright light rather
than relegated to an obscure corner. The Symposium
of the AAAS, out of which the present volume grew,
represented a planned experiment in communication on
the part of the participants--an experiment designed
to focus attention on the kind of ambiguity which
quite often confronts the scientists working at the
frontier. What is intriguing about the "Great
Redshift Controversy" is that it has continued for
rather longer than such controversies normally do.
Hopefully, the present volume, in encouraging
astronomers and astrophysicists to look carefully
at the basis for assigning redshifts, will play a
role in resolving the controversy. Such a resolution
is devoutly to be wished, because the question is
one of fundamental importance for our understanding
of the universe.

GEORGE B. FIELD

Introduction

INTRODUCTION

By George B. Field
Center for Astrophysics, Harvard College Observatory
and Smithsonian Astrophysical Observatory
Cambridge, Massachusetts

The present book grew out of a Symposium of the American Association for the Advancement of Science, "Redshifts of Galaxies and Quasars," which took place on December 30, 1972, in Washington, D.C.*

In the past few years some astronomers have become increasingly convinced that there is something basically wrong with the conventional picture of the Universe. They question whether the redshifts of all galaxies are really due to the expansion of the Universe, as has been accepted since the 1920's. They believe that at least some redshifts are discordant, in that they cannot be attributed to the

---

*As the A.A.A.S. provided the forum for the debate, we have arranged for any royalties resulting from this book to be placed in a fund to support the activities of the Astronomy Section of the A.A.A.S.

3

expansion. If they are right, modern cosmology is called
into question. Dr. Halton Arp, an astronomer at the Hale
Observatories in Pasadena, has developed this point of view.

Other astronomers have vigorously defended the
conventional interpretation of redshifts. However, there
have not been many opportunities for a direct confrontation
of the two viewpoints in an open meeting. As chairman of
the Astronomy Section of the A.A.A.S. for 1972, I thought it
would be interesting and worthwhile to arrange a debate on
the redshift controversy, and I invited Dr. Arp and Dr. John
Bahcall to participate. Dr. Arp is well known for his work
on globular clusters, stellar photometry, and unusual galaxies.
Dr. Bahcall, a theoretical astrophysicist at the Institute for
Advanced Study in Princeton, has an extensive background in
neutrino astronomy, stellar structure, and the physics of
quasars. Recently both Dr. Arp and Dr. Bahcall have been
working actively on extragalactic problems. Both Dr. Arp
and Dr. Bahcall have in the past been awarded the Warner
Prize for excellence in research by the American Astronomical
Society.

While a number of possible theoretical explanations of
anomalous redshifts have been proposed, the participants
agreed that there is little profit in debating them as long
as the interpretation of the observations is disputed. We

therefore agreed to focus the debate on observational

evidence. Each participant spoke for 45 minutes and had

opportunity for rebuttal. The floor was then opened for

what turned out to be a lively discussion.

The present volume contains the material presented at

the debate, as well as reprints selected by the participants

to illustrate their points.* Our purpose in publishing this

volume is, first of all, to stimulate further research in

this area. If Dr. Arp's interpretation is correct, a

revolution in astronomy, and perhaps in physics, is in the

offing. If, on the other hand, Dr. Bahcall's analysis is

correct, then scientific progress would be served by a

disproof, which may be stimulated by discussion of this kind.

Our other goal is to illustrate for students of

astronomy the ambiguity which sometimes besets astronomical

research. While students sometimes get the impression that

questions have simple yes-or-no answers, working astronomers

know that one can often reply only "probably" or "perhaps" to

current questions. The redshift controversy should be

understandable to students with relatively little background

and we hope this book will be useful to them. We believe

_____

*We agreed that the literature used would be limited to that
available to both participants before January 1, 1973.

that the addition of selected preprints will aid in
understanding the background of the controversy. In addition,
students as well as experienced professional astronomers will
enjoy reading some of the basic papers in extragalactic
astronomy.

It may be helpful to review the concepts employed in the
discussion. Both participants use well-established techniques
of astronomy, according to which the distances of the stars
can be determined, and their distribution within our Galaxy
(the Milky Way) found. The Galaxy contains about 100 billion
stars in a disk about 60,000 light years in diameter and 1000
light years thick. (A light year is 10 trillion kilometers.)

Stellar atmospheres absorb light at discrete wavelengths,
which are just those where various gases absorb in the
laboratory. The wavelengths of these so-called absorption
lines are shifted by small amounts in the spectrum of every
star. These shifts, which are observed to be proportional to
the wavelengths of the lines, are usually interpreted as
Doppler shifts, that is, as due to the motion of the star
along the line of sight. The Doppler-shift phenomenon is well
established in the laboratory, and is incorporated into the
special theory of relativity. The amount of the shift, which
is the difference between the observed wavelength λ_{OBS} and

the laboratory wavelength λ_{LAB} of the line, is proportional
to λ_{LAB} and to the velocity V of the light source away from
the observer (as long as the velocity is much smaller than
the speed of light, c = 300,000 km/sec). Thus, the
fractional wavelength shift, denoted by z (a dimensionless
quantity), is

$$z = \frac{\lambda_{OBS} \; \lambda_{LAB}}{\lambda_{LAB}} = \frac{V}{c}$$

Sometimes wavelength shifts are expressed in terms of
the "symbolic velocity" V = cz, expressed in km/sec. In a
case where the shift is really due to motion, and z is much
less than unity, V so defined is also the true velocity of
the source. However, if z is large, relativity gives
corrections to the simple formula, so cz does not equal the
true velocity. Note that λ_{OBS} may be larger or smaller than
λ_{LAB}, depending on whether V is positive (recession) or
negative (approach). If it is larger, the wavelength shifts
toward the red end of the spectrum, and one speaks of a
redshift.

Almost all wavelength shifts observed in the Galaxy are
Doppler shifts, as proved, for example, by double stars, where
the spectrum of each of the two stars periodically shifts
back and forth as the star alternately approaches and recedes

from the observer. The spectra of unusually compact stars
like white dwarfs are redshifted because of another phenomenon,
called gravitational redshift, predicted by the general
theory of relativity to occur in strong gravitational fields.

From our location within the Galaxy, we can see many
other galaxies external to our own, each composed of billions
of stars. Distances to the nearest galaxies are found to be
of the order of a million light years, by observing certain
types of stars within them and comparing their apparent
brightnesses with these of known distance within the Galaxy,
using the inverse-square law. Brightnesses of stars and
galaxies are reckoned on a logarithmic magnitude scale such
that each factor of 10 decrease in brightness corresponds to
an increase of 2.5 magnitudes. Thus, Sirius, the brightest
star in the sky, is −1.4 magnitude, while the Andromeda
Nebula, the nearest large galaxy, is +3.5 magnitude, or about
100 times fainter. The apparent faintness of Andromeda (even
though it contains billions of stars) is due to its much
greater distance (2,000,000 light years, compared to 9 light
years for Sirius).

Galaxies tend to occur in groups, ranging from about a
dozen galaxies (like the Local Group in which the Galaxy and
Andromeda are located) to rich clusters containing thousands
of galaxies (like Virgo or Coma). In the case of rich

clusters, there is no doubt that the galaxies are physically located in a small region, because so many galaxies appear in a small area of sky; there are 2500 galaxies within a radius of 6° around the center of the Virgo cluster.* The reality of small groups is less certain, because the number of galaxies in the apparent group is not much larger than is expected from an accidental alignment of galaxies having different distances. This uncertainty plays a role in the present discussion because Dr. Bahcall emphasizes that the apparent proximity of various objects on the sky does not necessarily establish their physical association.

Galaxies more than a few tens of millions of light years from the Earth are so distant that individual stars cannot be resolved, and distances must be based on some other method. One method is to compare the apparent magnitudes of the galaxies themselves with those of closer galaxies whose distances are known. However, the intrinsic luminosities of galaxies in general differ considerably from one another, so this method is not very reliable in any particular case. Thus, any given galaxy may be intrinsically luminous and relatively distant, or equally well, intrinsically faint and relatively close.

*Distances on the sky are reckoned in degrees and minutes of arc; one degree (1°) equals sixty minutes (60').

In the 1920's it was found that the spectral lines of
galaxies (which resemble those of stars) are shifted by large
amounts. As the shifts are proportional to wavelength, they
were interpreted as Doppler shifts; Dr. Bahcall discusses the
tests of this hypothesis. As more galaxies were observed,
it developed that nearly all shifts were toward the red,
indicating that most galaxies are receding from us. Thus
was born the concept of the expanding Universe. About the
same time it was shown that such an expansion is consistent
with the predictions of the general theory of relativity, and
a specific prediction, that the redshifts of galaxies should
be proportional to their distances (the Hubble Law of the
Redshift) was verified by plotting the observed redshifts
against the distances determined as explained above. For
galaxies of the same kind, which are believed to have the
same luminosity, relative distances can be found from their
apparent magnitudes. These relative distances are observed to
be accurately proportional to the observed redshifts. Hence
it is possible to assign a distance to any distant galaxy
from its redshift alone, which is easily and accurately
observed. As Dr. Arp points out, however, there is no
independent proof that this is a valid procedure for all
galaxies, in particular, for those kinds of galaxies for
which it has not yet been tested. In other words, it is hard

to prove that any single galaxy is at the distance indicated by its redshift (sometimes called its redshift distance or cosmological distance).

Dr. Arp believes he has found objects which are physically associated with other objects of different redshift. Since the distances of the objects are the same (it is claimed), at least one redshift would be discordant, in the sense that it does not conform to the Hubble Law. The objects showing this effect tend to be strange in other respects as well, showing morphological peculiarities or radio emission. The strangest extragalactic objects of all, quasars (quasistellar objects) look like individual points of light at the telescope, as do stars, and often have intense radio emission. Their redshifts are large, ranging up to 3.4, corresponding to 91% of the speed of light. If they are at their cosmological distances, they are intrinsically a hundred times more luminous than normal galaxies, leading to the speculation that quasar redshifts are discordant, so that quasars are not at their cosmological distances.

Dr. Bahcall, on the other hand, has taken the lead in two efforts to demonstrate the validity of the Hubble Law by trying to find quasars in clusters of galaxies having known redshifts, and by constructing the analogue of a Hubble diagram for the brightest quasars.

The redshift controversy is particularly sharp for quasars, for quasar redshifts are the largest known. Both this fact and the other puzzling properties of quasars have led to the speculation that new physical principles may be required to explain them.

In this book we address the observational evidence on discordant redshifts. Before adopting the view that current physics is wrong, the reader should decide for himself whether he thinks discordant redshifts are a proven phenomenon.

HALTON ARP

Evidence for Discordant Redshifts

EVIDENCE FOR DISCORDANT REDSHIFTS

By Halton Arp
Hale Observatories
Carnegie Institution of Washington
California Institute of Technology

Because it is believed that we live in an expanding universe, distant galaxies are expected to be receding from us with large velocities. Recession velocities are the only known way to shift spectral lines of whole stellar assemblages to the red (Doppler shifts). Therefore it is assumed that every highly redshifted galaxy observed is out at a large distance where the universe is expanding with appreciable velocity. If a galaxy with a high redshift could be demonstrated to be at a close distance where expansion velocity could not be responsible, then there would be no other available explanation for its redshift. We would be faced with a critical paradox in terms of physics theory. We could call this an example of a "discordant" redshift.

In the following paper, examples of discordant redshifts are presented. In each case the proof of a redshift anomaly stands or falls on the certainty with which the distance to the object can be estimated. Therefore I would like first to make some comments on distance criteria.

DISTANCES IN THE UNIVERSE

Large distances in astronomy are very difficult to
measure directly. For example, trigonometric parallaxes,
using the earth's orbit around the sun as a base line, are
not accurate beyond about 20 light years. Within our own
Galaxy, which is approximately 60,000 light years in diameter, the distances to the brightest stars and star clusters have to be calibrated by means of statistical motions
or by relating them to more common stars like those in the
solar neighborhood whose luminosities we know. The nearest major galaxy, the Andromeda Nebula, is about 2 million
light years away. At that distance only the brightest objects
in a galaxy, such as novae, globular clusters, Cepheid variables, and brightest stars, whose luminosities we think we
know, can be used to estimate the distance. But such distances are cosmically small, and out beyond the Local
Group of galaxies is where the overwhelming population of
the universe resides. At distances from the order of 10^8
light years to the edge of the observable universe at 10^{10}
light years, there is no direct way of measuring distance
with the uncertain exception of supernovae.

What can be done in the face of this difficulty? Can
we devise some method for getting at least a crude estimate
of distances of remote galaxies? Astronomers in the 1920s
thought they had found the answer when they discovered that,
on the average, the fainter the galaxy they observed the
higher was its redshift. It was natural to believe that the
apparently fainter galaxies were the more distant and to assume that the redshift was caused by an expanding universe
in which the expansion away from any observer is greater
the greater the distance to which he observes.

Because there is no other parameter besides redshift that is easily observable in a faint, featureless galaxy,
the custom of assigning the distance to such a galaxy according to the size of its redshift has become established.
If a galaxy has a faint apparent magnitude for its redshift, we
say it is underluminous or a dwarf, and the reverse if it is
apparently bright for its measured redshift. I wish to emphasize that there is no way of ever producing any discordance with the redshift-distance relation for even one single
object when operating from the base of current assumptions.
This is true because no matter where a galaxy point falls in
the redshift-apparent magnitude diagram its position can be
explained in terms of high or low intrinsic luminosity. For
example, the quasars fall generally above the Hubble line in
the redshift-apparent magnitude diagram, but they are not
concluded to have excess redshift—they are instead said to
have excess luminosity.

If, on the other hand, we wish to test the hypothesis
that redshift always is caused by distance, we must find
some different method of measuring distance. There are
only two methods of measuring distances directly. One is
by means of clustering or grouping. If we see a group of
galaxies clustered together on the sky, we may conclude that
they form a physical group at the same distance from us.
This is essentially a statistical criterion of distance. That
is, we must be able to show that for any objects assigned to
this cluster the chance is small for them to be background
or foreground objects accidentally projected into the apparent area of the cluster. In the past, of course, any discor-
dantly high redshifts measured in a cluster were simply assumed to be background galaxies without any further investigation.

The second direct method of measuring distances is to observe two or more galaxies either connected together with a luminous filament or interacting gravitationally. In either case we are sure the galaxies are essentially at the same distance. The latter is perhaps the most direct possible kind of observation and is capable of establishing from a single case a violation of the redshift-distance relation. It must be remembered that any number of cases may obey the redshift-distance relation, but only one well-documented case is necessary to establish the existence of a new phenomenon—namely, a noncosmological redshift in a galaxy.

The conventional viewpoint assumes that all galaxy redshifts are due only to Doppler velocities of recession. Therefore, if we can produce just one example of a redshift difference that cannot be explained as a velocity difference, then we have broken the assumption on which the redshift-distance relation is always applied to derive distances. In this eventuality it would then become necessary to reexamine each category of the different kinds of galaxies in order to see whether current distance assignments would need to be revised.

The explanation for any such noncosmological redshifts is not readily available from current physics. Because the discordant redshifts are overwhelmingly redshifts and not blue shifts, peculiar Doppler velocities cannot be invoked. This is because in the case of high random velocities as many approach (blue shift) anomalies as recession anomalies should be observed. Gravitational redshifts require too much mass and too abrupt local gradients of field strength to be reconciled with observations of diffuse galaxies, even if complicated models could be made to work for the more compact quasars.

In the following paper I present recent evidence that directly associates high-redshift objects with low-redshift objects. Throughout I try to form a related picture of what kinds and classes of objects exhibit these anomalous redshifts. For anyone wishing to delve into the details of any case, I also try to reference the papers that refer to any discussion or debate, pro or con, of the result of its significance.

QUASARS

Properly, quasars are objects that look like unresolved stars but have redshifts which place them outside our own Galaxy. Because their redshifts are so large—often approaching large fractions of the velocity of light—their distances are, on conventional interpretation, assumed to be also enormous—approaching large fractions of the radius of the universe. This would make the quasars by far the most luminous objects known. Since some of them vary their brightness in time scales as short as a month, the dimensions of the variable portion can be no larger than about a light month, and the huge luminosity of many times the brightest galaxy has to emerge from the relatively tiny volume. There are well-known physical difficulties in getting the photons out through such a high energy density region.[1] Considerations of these enormous luminosities and exceedingly high energy densities originally led to doubts as to whether the quasars were really at the large distances indicated by their redshifts.

EVIDENCE FOR QUASARS AT CONVENTIONAL REDSHIFT
DISTANCES. The only direct way in which we could prove
that quasars were really out at the distances indicated by
the conventional interpretation of their redshifts would be to
find them physically associated with other objects which we
have reason to believe are, in fact, at great distances from
us. The only other objects that we believe we know anything about at such distances are the brightest galaxies in
clusters of galaxies. The lowest quasar redshifts start at
about $z = 0.1$. Galaxies can be photographed with the largest
optical telescopes down to apparent magnitudes which probably have redshifts of the order of $z = 0.5$ to $z = 0.8$. It is
only in this overlap region then that a cluster, if it were
associated with a quasar, could be observed around the
quasar.

Of the roughly 43 quasars between $z = 0.1$ and $z =
0.5$, only three cases have been put forward as representing
associations of quasars with distant clusters of galaxies.
The quasars Pks 2251+11 ($z = 0.31$), 3C 323 ($z = 0.26$), and
RN8 ($z = 0.184$) seem to be bona fide quasars insofar as
they are unresolved star images with telescopic resolution
of about one arc second. In each case there are some fuzzy
images nearby that have approximately the same redshift as
the quasar.

In the first case of Pks 2251+11 there is at most a
very poor group of faint galaxy-like images.[2] Only about 7
candidate objects are visible. This is a far cry from a
cluster of galaxies, the poorest of which is called richness
class zero and which Abell defines as 30 galaxies in a range
2 magnitudes fainter than the third brightest galaxy. In
fact, Arp[3] showed a deep photograph of the field around
Pks 2251+11 in order to illustrate there was no conspicuous

grouping of galaxies in the general field of the quasar. (In general, we also must be careful to realize that every faint fuzzy image which is observed may not necessarily be an intrinsically luminous, distant galaxy. The few spectra obtained of objects around quasars, insofar as anything can be seen in the spectra, are said to resemble normal galaxies at large distances, but the resolution of the telescope is not good enough to prove whether the morphology of these objects is like that of normal, nearby galaxies. Dwarfs and shreds and pieces of galaxies, of course, can also have galaxy-like spectra but do not need to be of very high luminosity.)

In the second case of a quasar possibly associated with a cluster, 3C 323. 1, the cluster of galaxies is fairly rich but the quasar lies quite far from the center of the cluster.[4] It is claimed that the cluster extends a long, low-density bridge of galaxies toward the quasar, but the number of galaxies or faint images in this supposed outer isophote is relatively small.

The best example seems to be the third, RN8.[5] In that case the cluster is rich and the candidate quasar is near the center. Three of four faint images measured in the cluster have essentially the same redshift as the quasar. One difficulty with this case, however, is that the quasar itself may not be the source of the radio emission (which is also called 3C 61. 1). That radio emission is a wide pair of lobes[6] centered more on the cluster generally than on the blue stellar object specifically. The blue stellar object itself is only about $B = 18$ mag, and at such a low redshift comes out to be a very low-luminosity quasar. In fact, it is not certain how this object is physically different from a blue compact galaxy or N galaxy, objects that it would not be very surprising to find in a cluster.

In addition, for a redshift of $z = 0.18$ the brightest galaxies in a rich cluster should be brighter than 17 mag, and the ones in this cluster seem considerably fainter. A photograph of this quasar and the surrounding galaxies, taken with the 200-inch telescope, is shown in Figure 1.

If we look, instead, at the clearest and most famous prototypes of quasars like 3C 273 ($z = 0.16$) and 3C 48 ($z = 0.37$), quasars that are much more luminous than bright galaxies if they are at their redshift distance, then we see absolutely no evidence of clusters or groups of companion galaxies. Particularly an example like 3C 273, which is more than 5 mag brighter than RN8 at roughly equivalent redshift, should have companion cluster galaxies nearby as bright as $m_{vis} = 16$ mag and have appreciable apparent angular diameters. The deepest photographs with the the 200-inch telescope around 3C 273 and 3C 48, down to nearly 24th magnitude, reveal no surrounding clusters or groups of galaxies.[7]

In a broader sense, Bahcall[8] looked at all quasars with redshifts less than $z = 0.2$ and found none, even within 2 radii of Abell clusters, which clusters include all classes of richness, zero or greater, out to that redshift. If most distant matter is not aggregated into clusters of galaxies, then perhaps quasars could be out at these distances and not appear in the clusters. But there are indications that most, if not all, galaxies do occur in clusters or groups[9], and in this case the distant quasars would be indicated as occurring in the empty spaces between the clusters. Vorontsov-Velyaminov suggested at one time that distant massive quasars might actually represent the origin of clusters and would evolve into clusters of galaxies. The observational

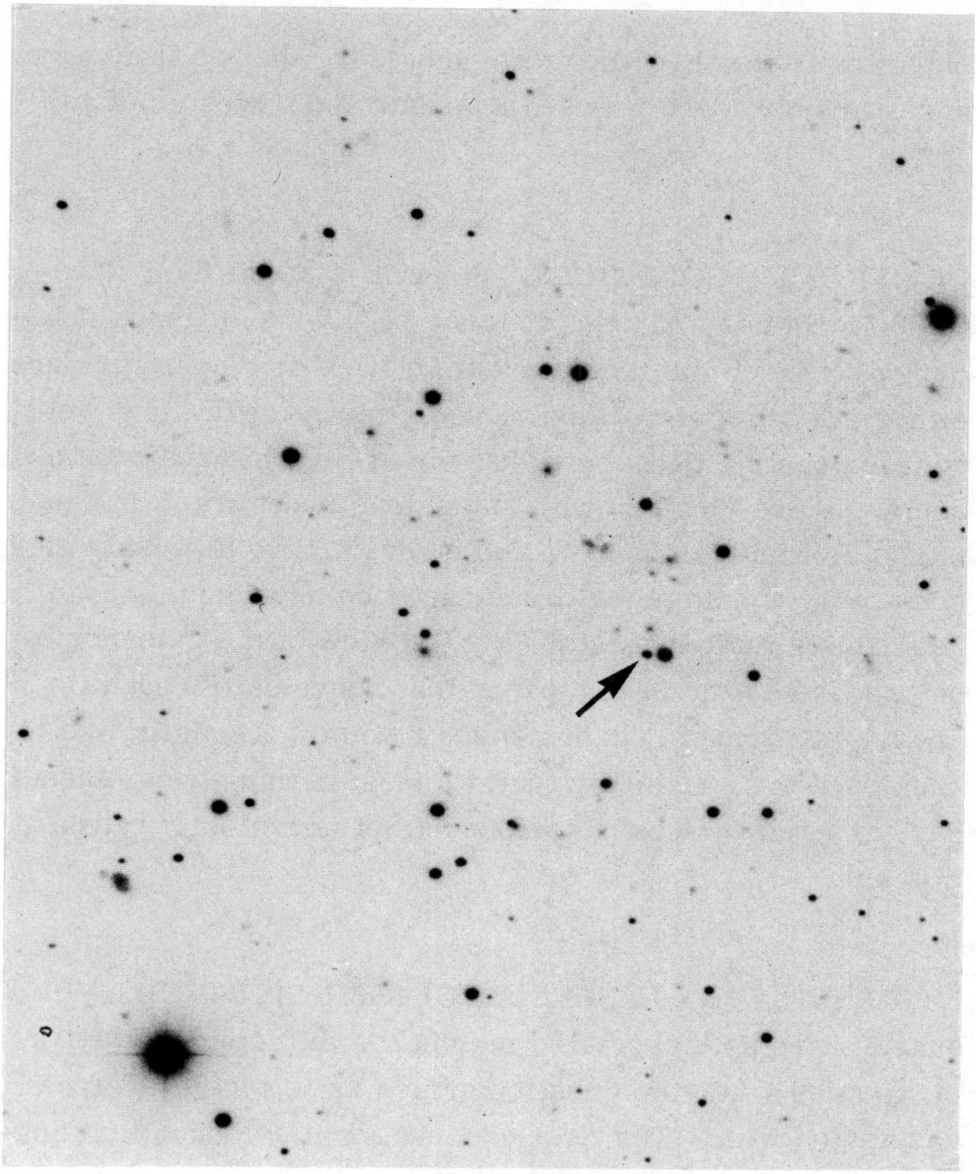

Figure 1

The quasar RN8 is marked by an arrow. Three of the galaxies nearby have redshifts close to that of the quasar ($z = 0.18$). Redshifts by Miller, Robinson, and Wampler with 120-inch Lick telescope; photograph by Arp with 200-inch Palomar telescope.

objection to this hypothesis is, however, that no transition cases are observed between a quasar and a cluster of galaxies.

COMPROMISE SOLUTIONS. Since it is so difficult to establish with certainty the distance of any one quasar, it is obviously much more difficult to prove that *all* quasars are either at cosmological distances or more local. It is, of course possible that some quasars are out at great distances and some are close by with anomalous redshifts. G. Burbidge, Rowan-Robinson[10], and others[11] have discussed this possibility and some evidence has been advanced that, in fact, this situation exists. But, in view of the evidence to be produced later in this paper that nonvelocity redshifts are a quite general phenomenon, I would personally like to keep open the possibility that all quasars and young objects have greater or lesser components of nonvelocity redshift in them.

EVIDENCE FOR QUASARS AT CLOSER DISTANCES. If quasars are not associated in general with distant cluster of galaxies, then the question arises as to what they are associated with. The first observational evidence that quasars were not at cosmological distances came from the discovery that radio sources tended to pair across nearby peculiar and disrupted galaxies.[12] Some of these paired radio sources turned out to be high-redshift galaxies and quasars, which then had to be at the close distance of the low-redshift central galaxy. Although these associations were originally selected on the basis of the alignment of the nearby radio sources, the associations were independently tested by another property—the average closeness of those radio sources to the peculiar galaxies. The latter test, by means of just the distance between the radio sources and the peculiars, showed that the association could be accidental only in one out of a hundred cases.[13] Two other papers[14] were published that were critical of the associations of high-redshift objects with peculiar galaxies, and one in rebuttal.[15]

Recently it has become clear that there exists a class of quasars with flat radio spectra and small apparent angular dimensions. There is no unanimity among authors, but there does seem to be more suspicion of the validity of the redshift distance for this particular class of quasars than for the more extended variety.[10,16] It is noteworthy, then, that among the radio sources originally associated by Arp[12] with peculiar galaxies there were seven quasars, and all of these seven were of small apparent angular diameter. The 3C R (*Third Cambridge Catalogue Revised*) contains 16 quasars with $\theta < 3$ arc sec and 24 with $\theta > 3$ arc sec. Rowan-Robinson[10] computes the chance of picking all seven quasars accidentally from the small-diameter group is only 0.0016. Despite the fact that we have little knowledge of what the physical significance of these observed properties are, it is nevertheless impressive to find a property—six years after the original associations were made—that indicates there was less than one chance in 500 of having accidentally associated the original set of quasars with the relatively nearby peculiar galaxies.

After studying the nature of radio sources around peculiar galaxies, the next investigation addressed itself directly to the crucial question of the quasars alone. In the latter investigation the problem was inverted, and it was asked what kind of galaxies fell in the neighborhood of this homogeneous sample of quasars. It was possible to show that these quasars fell closer, on the average, to bright galaxies than would a randomly distributed set of quasars. For randomly distributed quasars, the average distance to a galaxy was $10°.7$ (on the sky); whereas, for real quasars that had brightness $8.0 < m_{pg} < 10.6$ mag, the average distance to galaxies turned out to be only $5°.8$.[17] These same quasars were shown *also* to have a tendency to be aligned across these bright galaxies. (Recently a paper has been written analyzing a quantity related to the dispersion of objects on a sphere.[11]   It is difficult to see how this quantity relates to the distribution of bright galaxies. Nevertheless, that analysis disagrees with the association of faint QSRs in general with bright galaxies, but supports the association of a certain class of QSR—flat spectral index with small angular diameter—with the original Arp peculiar galaxies.)

Some critics objected that in the first investigation high-redshift objects had been associated with peculiar galaxies and in the second investigation with a different set of objects, namely, with bright but presumably normal galaxies. But it is clear on studying galaxy distribution on the sky that the peculiar galaxies are strongly clustered and usually in the vicinity of the large normal galaxies. It appears that in both cases the high-redshift objects were being associated with the same things, namely, the centers of nearby clusters and groups of galaxies. This point again becomes important further along when we cite investigations of the neighborhoods of large, nearby spiral galaxies.

It is important to note that *different* sets of objects led to the *same* conclusion, namely, that high-redshift objects were associated with nearby low-redshift objects. Moreover, tests with independent properties, such as average separation of objects as well as angular alignments, led to the same conclusion. The situation then must be faced that if one of these analyses or properties accidentally gave an indication of an association, then the other tests should certainly *not* have also given a confirmation. The fact that independent tests lead to the same conclusion requires that the separate improbabilities be multiplied, giving an extremely small probability that the association could be spurious. In the final section, further independent tests will be cited that again lead to this same conclusion.

But, returning to the subject of quasars and bright galaxies, about a year later other investigators reported that out of a homogeneous sample of quasars and galaxies four high-redshift quasars fell within 7 arc min of bright low-redshift spiral galaxies—a result that had only one chance in 250 of being an accident.[18] Analyses of a fainter set of quasars showed no positive correlation with bright galaxies;[19] but additional bright quasars were found very close to some fainter galaxies[20,21] (Figure 2, this paper), which made the overall chance accidental association now of the order of one in 10^{-5}. This time the improbability calculation of close correlation between some quasars and galaxies in the sample was criticized because it was said that the sample tested was the sample in which the correlation was first noticed. This of course is not true, because all the previous analyses by Arp had shown that quasars were in fact associated with just such galaxies or groups containing such galaxies, and the Burbidge, Burbidge, Solomon, and Strittmatter[18] paper was simply just another test of that association with a different set of data. In fact,

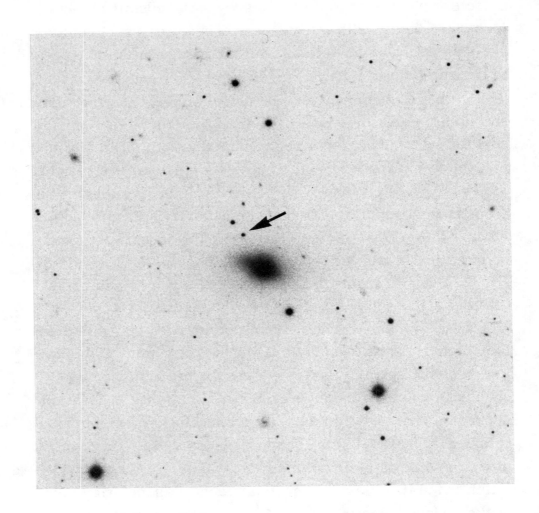

Figure 2   The quasar 3C 455 is marked by an arrow. The nearby galaxy is NGC 7413 and has much lower redshift. Note the peculiar extension of the galaxy to the northeast, in the general direction of the quasar. Photograph is a 25-min exposure on an Eastman Kodak 103a-J plate taken with the 200-inch telescope. Initially, before accurate radio positions were obtained, the radio source was identified with the galaxy.

Arp had shown in a paper that was accepted only in the Soviet journal *Astrofizika*,[22] that of high-redshift objects associated with bright galaxies, the high-redshift *galaxies* were preferentially associated with elliptical galaxies and that *quasars* were preferentially associated with spiral galaxies. The four galaxies falling so close to the quasars in the B^2S^2 paper were all spiral galaxies. It is significant to note that the association of quasars with peculiar galaxies and with spirals—which in most recent interpretations now also appear to be nonpermanent ejection phenomena[23,24] — associates the quasars that are estimated to have short lifetimes with galaxies and disturbances in galaxies that also have relatively short lifetimes.

During the 1971 study by Arp of the distribution of quasars and bright galaxies on the sky, a line of four quasars was discovered on the sky. It turned out that this line originated exactly in the second brightest, drastically exploding galaxy in the sky.[17] This was obvious evidence for ejection of quasars from the centers of nearby galaxies and supported the conclusion from the original paper[12] that, like other kinds of radio sources, quasars are also, in fact, ejected from galaxies. The properties of these particular quasars were such as to make their mutual association with the exploding galaxy have less than about one chance in a million of being accidental.[25] It was claimed by Barnothy[26] that several different straight lines could be drawn through quasars in the area, but he had not restricted himself to a homogeneous, complete sample of radio quasars and had allowed very wide acceptance limits for a "straight" line. Overall, this case must be viewed as the "experimentum crucis" because of its almost vanishing possibility of only accidentally associating quasars with a nearby galaxy.

Although in most cases it is difficult to form an opinion as to whether a given quasar is associated with a large nearby galaxy or with a smaller peculiar or companion galaxy in the neighborhood, it appears possible to get a general idea of the distances of different kinds of quasars. The brightest apparent-magnitude quasars, which usually have redshifts in the range $z = 0.2$ to $z = 0.5$, and also the highest redshift quasars, with $z = 1.8$ or greater, seem to fall preferentially in the south galactic hemisphere.[11,27]

That, of course, is the direction of the Andromeda Nebula. M31 is the dominant member of the Local Group of galaxies of which our own Milky Way is a member. The distribution of the bright apparent-magnitude quasars, from $40°$ to $60°$ around the position of M31 on the sky, is such that this category of quasars must be related to M31[11,25] and the Local Group of galaxies. The remaining quasars, those with redshifts between about $z = 0.5$ and $z = 1.8$, must then fall in the remainder of the supercluster of galaxies of which our own Local Group is just a small part. This is confirmed by the fact that the distribution of these latter kinds of quasars is richer in the northern galactic hemisphere, which is just the direction of the center of the local supercluster. In fact, the local supercluster center and anticenter are one of the most important concepts in understanding the distance and distribution of objects in space around us. It is known that the number of bright galaxies is far greater in the north galactic hemisphere, toward the supercluster center. It is now also known that the quasars with the largest apparent radio diameters show an excess in this direction.[28] As mentioned, it is readily ascertainable that the optical quasars are more numerous in this direction.[17] Recently, asymmetries in the radio-source counts have been shown to exist between the north and south galactic hemispheres.[29] On conventional redshift assumptions, these latter counts are supposed to reflect conditions out toward the edge of the universe, but, like quasars, they instead show relationships to the relatively local supercluster.

In the following section we will discuss the problem of what really comprises the environs of clusters and groups of galaxies. We will see that there is evidence for each group of galaxies to contain a range of objects from quasars through compact galaxies, and intrinsically high-redshift peculiars, as well as the more diffuse and normal types of galaxies of which the clusters and groups of galaxies are now supposed to exclusively consist.

GALAXIES AND DISCORDANT REDSHIFTS

Since the quasars have such generally large redshifts, demonstrating that they fall within the bounds of the local supercluster (about 100 Mpc), demonstrates immediately the existence of discordant redshifts. But quasars are not unique objects completely unrelated to any other kinds of extragalactic objects. Most astronomers admit that there is a continuity of observed physical characteristics going from quasars progressively through compact galaxies, N galaxies, Seyfert objects, peculiar galaxies, and through to normal galaxies. If quasars have discordant redshifts, then we would predict some of these related objects also to have discordant redshifts. Of course if we can demonstrate discordant redshifts in any nonquasars, we have arrived just as surely at the earlier-mentioned

physical paradox.   In this section we go on to show a number of discordant redshifts in galaxies in both statistical and
individual cases.

THE VIRGO CLUSTER.   The Virgo Cluster is certainly the
best known and richest nearby cluster of galaxies; consequently it has been studied intensively.   As long ago as 1962,
Holmberg[30] in a careful analysis of the available data showed
that among accepted cluster members there was a significant systematic trend for the fainter galaxies to have higher
redshifts.   G. de Vaucouleurs then discussed this result
and concluded that a systematic spectroscopic error in the
measurement of the redshifts could not account for the effect.   Most recently, de Vaucouleurs,[31] who more than anyone else has expert knowledge of this cluster from having
classified and studied its galaxies, reported that the average redshift of its elliptical members was $\bar{z}$ = 900 km s^{-1}
and that the average redshift then increased *steadily* for
nonelliptical members until it reaches the end of the Hubble
sequence at Sc spirals, which have just about 900 km s^{-1}
greater redshift, or a mean value of $\bar{z}$ = 1800 km s^{-1}.   Such
a redshift discrepancy is far too large to be a systematic
error of any kind.   The only other explanation, that the
spirals represent a more distant cluster behind the ellipticals that is receding with greater cosmological recession
velocity, is not tenable because there are various subclusters
of ellipticals and spirals which are coincident within the
Virgo Cluster.   Each one of these would have to be aligned
accidentally in front of a concentration of more distant spirals.   Of course, neither would we expect a continuous
gradation of mean redshift for the different morphological

classes.

Very recently Tammann[32] has written a paper on the
Virgo cluster claiming there are no systematic redshift differences between morphological classes. Despite the fact
that he has included galaxies from outside the Virgo Cluster
and has averaged unlike morphological types together, however, the systematic redshift differences still seem very
apparent to me and in the same sense as derived by other
authors.

THE COMA CLUSTER. The second best-known cluster of
galaxies is the more compact Coma Cluster. It contains
almost entirely E and S0 (also called lenticular) galaxies.
(An S0 is more flattened or disk-shaped than an E, but like
an E does not contain any conspicuous young stars or dust
and gas.) The core of the Coma Cluster is so dense that
there is no question of any significant contamination by
background field galaxies. And yet the mean of the redshifts
for the non-E galaxies is about 700 km s^{-1} greater than that
for the E galaxies.[33] Again the difference is so large as to
rule out any systematic measuring error, and it is conceptually inadmissable to consider all the non-E galaxies having
a systematic velocity of recession away from the E galaxies.
It is also important to note that all the Coma galaxies were
observed for redshift and type-classified *before* the redshift
difference between the types was noticed, so there could be
no prior bias from the classification standpoint.

CLUSTERS, GROUPS, AND ASSOCIATIONS IN GENERAL.
It has been shown that in other groups and clusters of

galaxies the fainter galaxies tend to have somewhat higher
redshifts.[34] Recently the Finnish astronomer Jaakkola[35]
has investigated all the major clusters, groups, and associations of spiral, elliptical, and other kinds of galaxies
with each other. He finds that, overall, the redshifts of
spirals with the greatest population of young stars (Sc
spirals) average about 200 km s^{-1} more than elliptical and
Sa galaxies that are at the same distance. We will see as
this section progresses that the discordant redshifts generally apply to companion galaxies, that is, fainter or
younger companions physically associated with large bright
galaxies. E galaxies are generally acknowledged to be the
oldest dynamically relaxed, old-star populated galaxies
that we know. Spiral galaxies are naturally in more transient stages than E galaxies, contain more young stars, and
are dynamically unrelaxed compared to the E's. So the results of all the tests described here do point to the same
general conclusion of correlation of excess redshift with
youth of the galaxy, and the different tests are independent
ways of testing the same general result of discordant redshifts. Tests like Jaakkola's are particularly powerful because they encompass all the current data in one section of
the field.

COMPANIONS TO NEARBY GALAXIES. The result that it
was the fainter galaxies that tended to have the excess redshifts led Arp[34] to compare redshifts of well-known, traditionally accepted companion galaxies like those around M31,
those around M81, and companions linked on the end of spiral arms. The result was that although small, the average
excess redshift of the companions of around +70 km s^{-1} was

significant in terms of the scatter of the redshift differences. This result has been debated in the literature.[36]   The interpretation of the smallness of the excess redshift was that these companions were very closely the same epoch of formation or the same kind of material as the main galaxy.   It is also true that much higher-redshift companions would not have been traditionally accepted as valid companions.   For example, NGC 404 is not usually listed as a companion to M31 although there is good evidence that it is a companion and its excess redshift is only +237 km s^{-1}.   In any case, even a small systematic excess redshift for the companions discussed above is significant because these galaxies have redshifts that are generally accurately measured, and the companions are firmly established to be at the same distance.   It would be unacceptable for the companions to be systematically receding, even slowly, away from the main galaxy just along our line of sight in every case.

More peculiar and younger companion galaxies will be discussed in the following sections.   There the redshift discrepancies are much larger, and in those cases the whole crux of the argument swings instead on whether we can prove them to be physical companions of the central galaxy.

HOLMBERG'S COMPANIONS OF MORE DISTANT GALAXIES

The most comprehensive study of the neighborhood of spiral galaxies was made by Holmberg.[37] By studying *Palomar Observatory-National Geographic Society Sky Survey* prints of moderately bright spiral galaxies, he was able to show that a number of faint galaxies out to about 10

radii from the central galaxies were physically associated.
By inspecting control fields away from the galaxies on the
same prints, it was possible to show there was an absolute
excess of small galaxies around this set of spirals, and it
was possible also to show these excess galaxies were concentrated in a cone along the extended minor axis of the
central galaxy. This latter is very powerful evidence for
an ejection origin of the companions from the central galaxy.
But the importance of this result from the standpoint of the
subject of discordant redshifts is that these fainter galaxies
were shown with statistical validity to be associated with
the central galaxy. Arp has been measuring redshifts of
those companions in fields where over 50% are statistically
associated. Although the program is far from finished because of lack of telescope observing time, essentially all
the companion redshifts measured so far are much in excess
of the redshift of the central galaxy. Actually an experienced
observer would predict these higher redshifts by just visual
inspection of the faintness, relatively high-surface brightness, and general appearance of most of the companions.

Holmberg is noted as a careful and accurate worker.
The work has never been challenged in the literature. Yet,
because, I suppose, the majority of the companions are obviously high-redshift systems, the result has been generally
ignored.

Holmberg did not attempt to classify the morphological types of the companion galaxies in the fields of the bright
spirals. But, in the work described below, it now seems
possible to isolate particular kinds of morphological types
that are characteristically companions and not faint background galaxies.

INTERACTING DOUBLES IN THE NEIGHBORHOOD OF SPIRAL GALAXIES

Working from a complete list of 32 spiral galaxies between apparent magnitude 10.0 and 11.2, Arp showed that interacting double galaxies were commonly encountered within about 50 arc min of these spirals.[38] By comparing control fields on the *Sky Atlas* prints he was able to show that these interacting doubles were many times more numerous around the spirals and, hence, that the majority were physically associated. Again spectra have not been obtained for all, but enough redshifts have been derived to show that essentially all the peculiar companions are above the $z = 5000$ km s^{-1} range, whereas the central galaxies are of the order of $z = 1000$ km s^{-1} or less. This, independently, is the same kind of result that Holmberg obtained. The difference in the Arp case is that the companions are a subset of peculiar galaxies that are much less apt to be found in the general background field. Their extremely unstable forms insure that they are young in some sense, and therefore makes them candidates for exhibiting higher excess redshifts.

MULTIPLE INTERACTING COMPANIONS

A more special and rarer case of interacting double galaxies is the case of interacting multiple galaxies. Two examples would be Seyfert's Sextet and Stephan's Quintet. There are only about six outstanding examples of such kinds of systems. It has been shown that all of these fall close to large spiral or peculiar galaxies.[38] In all but one of the cases, the redshift discrepancies are known to be more than 5000 km s^{-1}. As mentioned in the introduction, just proving that one object, or in this case one set of objects, is at the same distance as another set of much lower or higher redshifts is enough to establish the redshift discrepancy. In addition to the strength for the case of the multiple interacting systems alone, however, the result has been independently confirmed by the results on double interacting systems. And, in turn, both results have been confirmed by the Holmberg companion results.

STEPHAN'S QUINTET

To take a study in depth of just one of the systems mentioned above, the results on Stephan's Quintet may be reported. There the identical size and appearance of the H II regions in two of the component galaxies, NGC 7320 and NGC 7318, show that from this criterion they must be at the same distance (see Figure 3). This is despite the redshift of NGC 7320 being z = 800 km s^{-1} and the redshift of NGC 7318 being z = 5700 and 6700 km s^{-1}. But, the even stronger evidence for these two galaxies being at the same distance is the evidence from optical features of actual physical interaction of the two galaxies and probable radio evidence for the same interaction.

Stephan's Quintet falls only 30 arc min away from the large spiral NGC 7331 (see Figure 4). It is suggested by the presence of radio sources and luminous filaments in the area that Stephan's Quintet is a multiply interacting companion actually associated with NGC 7331. Just a few arc minutes to the east of NGC 7331 three other,

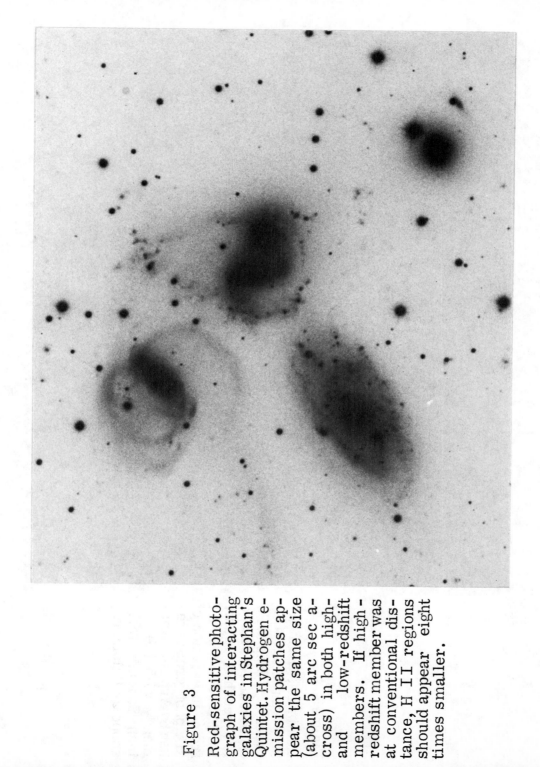

Figure 3
Red-sensitive photo-
graph of interacting
galaxies in Stephan's
Quintet. Hydrogen e-
mission patches ap-
pear the same size
(about 5 arc sec a-
cross) in both high-
and low-redshift
members. If high-
redshift member was
at conventional dis-
tance, H II regions
should appear eight
times smaller.

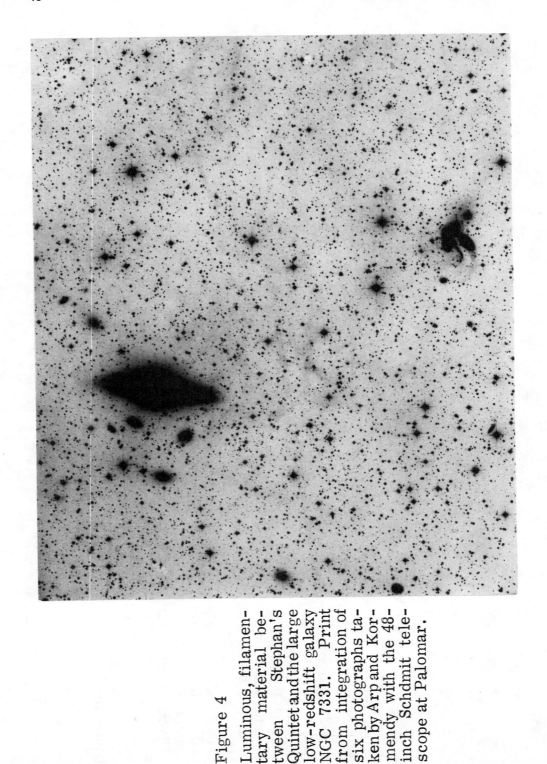

Figure 4

Luminous, filamentary material between Stephan's Quintet and the large low-redshift galaxy NGC 7331. Print from integration of six photographs taken by Arp and Kormendy with the 48-inch Schdmit telescope at Palomar.

moderately large companions appear statistically to be also associated despite their redshifts of $z = 5000$ to 6000 km s^{-1}.

DISCORDANT REDSHIFTS IN CHAINS OF GALAXIES

The existence of chains of galaxies seems to be a fundamental but puzzling phenomenon. It has been suggested that they arise from ejection processes from central galaxies.[39] This would link them to what has been suggested as another fundamental process in galaxies, namely, the ejection of radio sources, quasars, compact galaxies, and multiple interacting galaxies. But, regardless of the origin of these chains, their existence enables us to use them as a criterion of objects all at the same distance. There are four chains I would particularly like to mention:

1. Cen A chain. This stretches about $18°$ across the sky and is centered on the giant active and peculiar elliptical galaxy NGC 5128 ($z = 400$ km s^{-1}). At least seven bright galaxies and six radio sources are apparently involved. The galaxy IC 4296, which is itself a triple radio source, is indicated to be a member of this chain despite the fact that its redshift is $z = 3700$ km s^{-1} and highest remaining member of the chain is $z = 1140$ km s^{-1}.[39]

2. NGC 4736 chain. The large spiral in the chain has a redshift of $z = 300$ km s^{-1}. Four fainter, but still quite bright galaxies stretch in a line to the west. They are all double, peculiar interacting galaxies and have redshifts that vary between $z = 500$ and 800 km s^{-1}.

3. NGC 4038/9 is an interacting double galaxy of large apparent diameter with a redshift of $z = 1650$ km s^{-1}. Two out of the three galaxies extending in a line to the

southwest are peculiar interacting doubles with redshifts of about $z = 1900$ km s^{-1}. The peculiar interacting double along the same line to the northeast has a redshift of $z = 14,500$ km s^{-1}.

4. VV 172. This very close chain of five galaxies was observed spectroscopically by Sargent[40] (see reference 25 for photograph). Four were found to have redshifts around $z = 16,000$ km s^{-1}, but the fifth one, slightly smaller and more compact, had a redshift of $z = 37,000$ km s^{-1}.

It also has been shown (R. Weymann, private communication) that the discordant galaxy is bluer than the other galaxies in the chain. If it were really a background galaxy, then it would have to be a very high-luminosity spiral-like galaxy, which would make the chance of accidental projection in just this position even more unlikely.

If we were to find only one example of a chain containing a discordant redshift, we might be able to argue that this was the result of some extraordinary coincidence where a background galaxy just happened to project into the right position in a foreground group. But the importance of the above chains is to demonstrate that even though such exactly aligned chains are rare, they often contain discordant redshift galaxies. If someone wished to demonstrate that these cases were only accidental, he would have to produce hundreds to thousands of examples of similar chains where the discordant redshift position was empty.

The case with the multiple interacting galaxies of the previous section is similar in the sense that those systems are also very rare but a large percentage contain discordant redshits. What the chains and multiple interacting systems have in common, however, is that they are dynamically short-lived. This again indicates that the youth of the objects is associated with the probability of the appearance of discordant redshifts.

DISCORDANT REDSHIFT SYSTEMS CONNECTED BY LUMINOUS FILAMENTS

Another kind of interaction between galaxies which can immediately establish a common distance is one in which the two galaxies are linked together by a luminous filament. One of the first cases of discordant redshifts in such a system was discovered by Zwicky.[41] In that one, two galaxies of about $z = 7000$ km s^{-1} were linked together by a filament that led to a third galaxy that had a redshift of only a few hundred km s^{-1}.

More recently, further examples of connected galaxies with discordant redshifts have been discovered. Some of these are listed below:

1. NGC 772 is a disturbed Sb spiral with a redshift of about $z = 2400$ km s^{-1}. It has a long filament coming off to a small galaxy, which itself is an interacting double and which has a redshift of about $z = 20,000$ km s^{-1}.[42] There are other small peculiar and presumably high-redshift galaxies immediately in the vicinity of this very disrupted central galaxy.

2. NGC 7603 is an extremely disturbed Seyfert galaxy (see Figure 5). A single luminous filament curves out of it and terminates on a smaller, peculiar companion galaxy.[43] The redshift of the central galaxy is $z = 8700$ km s^{-1}, and the companion has a redshift of 16,800 km s^{-1}. The redshift difference is too great to interpret as a velocity of ejection or collisional encounter of the companion

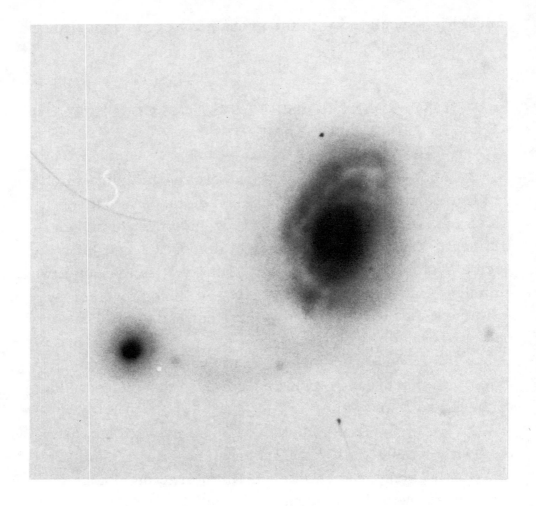

Figure 5   The Seyfert galaxy NGC 7603 with redshift $z =$ 8800 km s^{-1} is shown connected to a peculiar companion with redshift $z = 16,900$ km s^{-1}. The exposure is for three hours on a IIIa-J plate with the 200-inch telescope. The plate is printed lightly here to show the extreme disturbance in the inner regions in the larger galaxy. See original reference for stronger registration of peculiar luminous filaments that link companion galaxy to main galaxy.

because in that case the companion would be escaping too
rapidly to gravitationally perturb out the connecting filament.
In any event, all the cases we have shown yield the companions with higher redshifts and, as we have explained before,
ejection or orbital velocities should furnish as many blue as
redshifts.

This one case could be taken as a crucial test of the
reality of discordant redshifts. I think it is clear that we
can say that if this companion, with an absorption-line spectrum, is not an accidental projection of a background galaxy, then it must represent a galaxy which somehow gener-
ates a nonvelocity redshift throughout the stars of which it
is presumably composed. So the vital question to be answered is: Could the companion be a background galaxy?
The answer seems to be clearly no on a whole series of
counts. First, there is the very disturbed nature of the
parent galaxy. If the companion is not in some way responsible, then what is? Second, there is the luminous filament
which shows much better in red-sensitive photographs. Such
features almost always show better in blue photographs.
Therefore, the luminous filament itself is marked as a very
unusual feature. Third, there is a second, fainter luminous
filament coming out of the galaxy and terminating in the
same spot as the companion. So the position of the companion on the sky is marked from the low-redshift object by a
second unusual feature. Fourth, the companion is morphologically peculiar. It has a sharp central disk with a very
faint halo. I would estimate roughly that not one galaxy in
1000 looks like this particular companion. So what would
the chances be if a background galaxy were to fall in that
position that it would be so peculiar? Fifth, and most important of all, a narrow rim of the faint halo brightens

perceptibly just where the arm from the central galaxy
touches it.   This peculiar behavior established an interaction
between the arm and the companion and ensures that they
are at the same distance.   The figure included in this article
to show this behavior was a special print that was automa-
tically dodged in order to show the inner disturbed regions
of the central galaxy and the peculiarity and interaction of
the companion, as well as showing at the same time the
faintest parts of the connecting filament.

3.   NGC 4319 is another peculiar spiral galaxy that
is close to an object which technically must be called a
quasar (Markarian 205).   Markarian 205 appears as a stellar object on normal high-resolution photographs and has a
redshift of $z = 21,000$ km s^{-1}.   The galaxy has a redshift of
about $z = 1800$ km s^{-1}.   Very deep photography on special
plates by Arp showed a luminous connection between the
galaxy and the quasar.[44].   Later Lynds and Millikan[45] confirmed this connection on the same kind of plates with a
different telescope.   The latter authors argued, however,
that the connection was not real but was instead a blending
effect between the outer edges of the quasar and the galaxy.
But, just in case it was real, they also argued it was an
accidentally superposed background galaxy that gave the
appearance of a bridge.

Neither argument seems valid to me, however, because (1) inspection of the photograph (see Figure 6) readily
reveals that the connection is not cusp-shaped, as would be
expected from blending of images, but straight-sided and
narrow, and (2) the connection is low-surface brightness
and not saturated, as inspection of normal background galaxies in other parts of the plate readily reveals.

Arp showed also a photograph taken in the red where

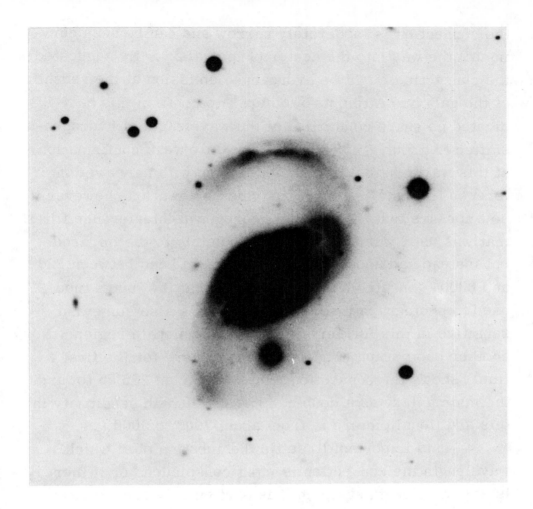

Figure 6   Photograph of the spiral galaxy NGC 4319 with a redshift of $z = 1800$ km s^{-1} and the quasar Markarian 205, which is 42 arc seconds south and has a redshift of $z = 21,000$ km s^{-1}. The exposure is 4 hours on a baked IIIa-J plate with the 200-inch telescope. A narrow straight-sided connection is seen between the high redshift object and the galaxy. Note how the arms of the spiral galaxy are detached and disturbed, giving support to the conclusion that some explosive or fission process has ejected the quasar from the center of the larger galaxy.

the connection is moderately narrow and can be seen curv-
ing all the way into the center of the galaxy. He suggested
the connection might be hydrogen-α emission at the redshift
of the galaxy. Attempts by others[46] to confirm this have indicated no sharp connection at the wavelength of hydrogen-α
shifted to about z = 2000 km s^{-1}. Only weak confirmation
at this wavelength was obtained by Arp with narrower interference filters. In rechecking this observation, however, it
now appears that the original hydrogen-α interference filter
that was used was not blocked completely to the infrared,
but instead had about a 45% transmission band between 8200
and 8300 Å (angstroms). It turns out that the image tube
used to photograph through the filter was also somewhat
sensitive in this further-to-the-red wavelength region. Now
another photograph (Figure 7), shown here for the first
time, shows the connection from Markarian 205 up toward
the main galaxy as a double. The wavelength sensitivity in
this additional picture is from about 7000 to 9000 Å.

It is important to settle the question of at which
wavelength the connection is most conspicuous or if there
is any wavelength at which it is particularly conspicuous.
For example, from the information on hand at this moment,
the connection could be hydrogen-α even more redshifted
than Markarian 205, or it could be some kind of relatively
cool infrared feature. More observations are obviously
needed on this important object in order to tell us something about the very interesting question of what is the physical nature of the connection. But the existence of the connection seems very well established. For example, Figures
7 and 8 show that the Lynds-Millikan connection falls exactly
on the connection as recorded by Arp with the image-tube
photographs in red wavelengths. Just the existence of the

connection is of first order importance because it clearly established the existence of the redshift discordance between the quasar and the galaxy to which it is connected. The fact that there is *some kind* of connection between Markarian 205 and NGC 4319 is attested to by four different photographs, and it seems to me that this evidence simply must be faced.

Another interesting feature of this system is a faint starlike image very close to Markarian 205 which appears to be connected into the image of the quasar by a thin luminous filament. Weedman and T. Adams said they could see this on their good-seeing plates and Arp was later able to see it on his. This may be the same unusual close doubleness phenomenon found in compact galaxies in general[42] and in the high-redshift systems on the ends of the arms in both NGC 772 and NGC 4151 (following). Second, we can comment that the spiral galaxy NGC 4319 seems to be in an unusually disturbed and unstable condition even for a spiral. The arms are unsymmetrical and become detached, particularly the northern arm, as they approach the inner disk of the galaxy.

Recent photometric isophotes of photographic plates of NGC 4319 with the GALAXY machine (courtesy N. M. Pratt) shows some of the inner isophotes of NGC 4319 to have a distortion in the direction of Markarian 205.

Finally, we can cite the radio analysis of the area[43], which shows an optically unidentified radio source in a symmetrical position on the side of NGC 4319 away from Markarian 205.

4. NGC 4151 is one of the six original Seyfert galaxies. Kraft and Anderson[47] have shown there is mass ejection from it. Deep photographs on the 48-inch show a long

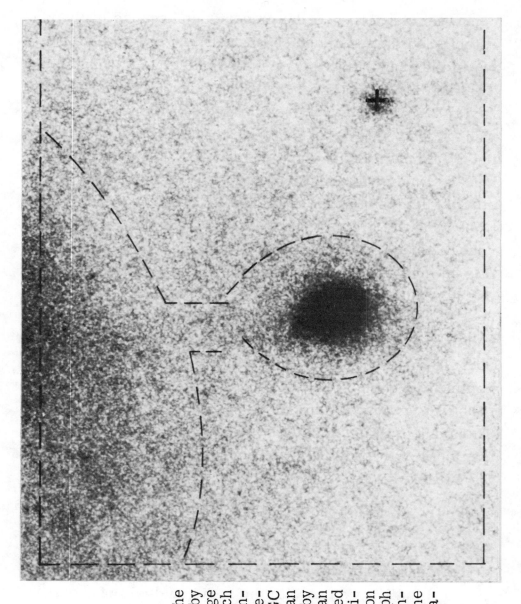

Figure 7

Photograph in the red and infrared by Arp with the image tube on the 200-inch telescope. Isodensity tracing of region between NGC 4319 and Markarian 205 as recorded by Lynds and Millikan is shown as dashed outline. The luminous connection on the Arp photograph runs directly coincident with the Lynds–Millikan feature.

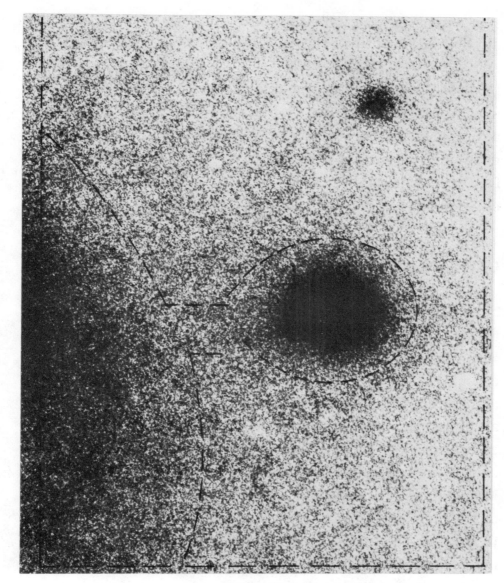

Figure 8

Another photograph, in a different red and infrared region, by Arp with the image tube on the 200-inch telescope. Connection between NGC 4319 and Markarian 205 now looks double, but still directly within the isophotes from the Lynds-Millikan photograph.

thin extension of a spiral arm to a nearby small spiral galaxy. The redshift of the small spiral is z = 6700 km s^{-1}
and of NGC 4151 is z = 900 km s^{-1}. As the faint arm from
NGC 4151 enters the companion, there is a small luminous
knot of material present in the small spiral, indicating an
actual physical interaction. The companion in turn has an
even smaller galaxy attached to it by a luminous jet, quali-
fying this companion as one of the previously-mentioned
category of interacting double galaxies.

CERTAIN TIGHT GROUPINGS OF BRIGHT GALAXIES

Two systems are now discussed which fall so close
together on the sky and involve such relatively bright galaxies that there is very small chance that they could be accidental.

The first shown below in the diagram (Figure 9) is
an isolated triplet of galaxies. All catalogued NGC and IC
galaxies in the large areas shown are plotted. The bright
central galaxy is NGC 7585 at a redshift of z = 3500 km s^{-1}.
Morphologically it appears to resemble a galaxy composed
of Population II stars (relatively old stars), but its form
is very disrupted. The next brightest galaxy, the companion southwest, proves to be a high-surface brightness,
symmetrical spiral with tightly wound arms. It is appreciably fainter in apparent magnitude and has a redshift of
z = 3700 km s^{-1} (redshifts from de Vaucouleurs and de
Vaucouleurs[48]). The faintest of the three, the northern
companion, turns out to be a strongly interacting double
galaxy, just the kind that were established in the study referred to earlier in this paper as being typical high-redshift

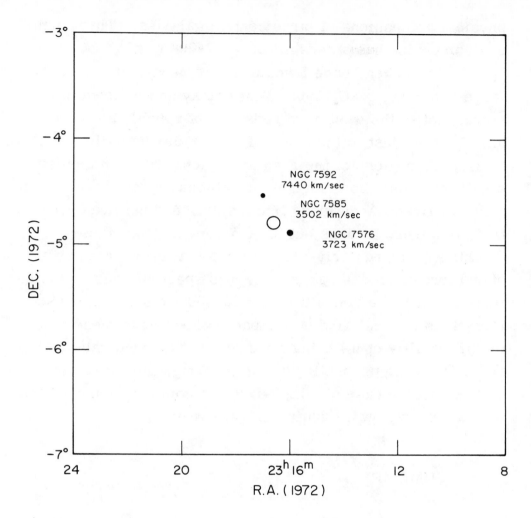

Figure 9   Diagram of a complete region in the sky in
            which all catalogued NGC and IC galaxies are
            plotted. The only galaxies in the region are the
            tight group of three shown. The two faint gal-
            axies that pair across the brighter galaxy both
            have larger redshifts, and the northern one,
            which is double and very peculiar, has a much
            larger redshift than the central galaxy.

physical companions of large central galaxies. This northern companion has a redshift of $z = 7400$ km s^{-1}.

The second case I would like to report is that of the
large E galaxy, NGC 1199. That galaxy and the peculiar
spiral just to the west have redshifts of roughly $z = 2500$
km s^{-1}. But just to the south of NGC 1199, actually in the
faint outer regions of the E galaxy, along with some other
small irregular, low-surface brightness objects is an object
which at first glance looks like a star. At high magnifica-
tion, however, it turns out to be a very high-surface
brightness, irregularly-shaped lump that is bisected by a
sharp irregular absorption line. Its spectrum has emission
lines and it has a redshift of $z = 13,000$ km s^{-1}. Because it
is such an unusual kind of compact galaxy, it is clear that
the probability of its being a random background galaxy is
very low. The probability is very strong, therefore, that
this is another case of a high-redshift companion associated
with a nearby low-redshift central galaxy.

SUMMARY

We have found that in every assemblage of galaxies
that has been close enough to show faint members, the fainter
members, on the average, have higher redshifts, and some
fainter members in particular have much higher redshifts.
Every test we have been able to apply gives this same answer. In general, the fainter members tend to be of higher
surface brightness and more disturbed and are suspected to
be younger, and in particular those that are believed to be
the youngest show the most discordant redshifts. It cannot
be stressed too strongly, however, that these discordant

redshifts are not discovered in just one or two isolated cases that have no relation to each other. But in every case we can test—large clusters, groups, companions to nearby galaxies, companions to middle-distance galaxies, companions linked by luminous filaments, galaxies interacting gravitationally, chains of galaxies—in every conceivable case, we come out with the same answer: the same discordant redshifts for the same general class of younger, fainter galaxies. This evidence, taken together with the same kind of evidence for the quasars—which are a kind of extremely young and, if this evidence is correct, intrinsically faint companion—forms a coherent picture of the kind of galaxies that have excess intrinsic redshifts.

If we take the foregoing as a working hypothesis, we must then perform the reexamination of the distances of each category of different kinds of galaxies referred to in the opening sections.   If we do this, we realize that our own Local Group of galaxies and nearby groups probably contain some compact objects and interacting objects of high intrinsic redshift which we have hitherto placed much farther out in distance. This reevaluation of local cluster space will have two salutory effects. First, it will give our Local Group of galaxies a normal complement of these kinds of peculiar objects which they are now lacking. Second, it will take peculiar objects like quasars and bright compact galaxies like 3C 120, which under present assumptions just exist isolated in space, and put them in association with other galaxies and groups of galaxies.

In conclusion, it should be remarked that if the material discussed in the preceding pages is wrong it cannot be just slightly wrong or slightly misinterpreted. If this evidence is wrong, it must be massively wrong, all down the line in case after case. And if it is to be so proved wrong, it does no good to simply say in each case that the evidence could be wrong—it will have to be demonstrated to be wrong by the same careful observing, amassing, and presentation of the evidence that has been put forth to establish the case as it stands now for discordant redshifts.

REFERENCES

1. F. Hoyle, G. R. Burbidge, and W. L. W. Sargent, *Nature*, 209, 751, 1966; also, L. Woltjer, *Astrophys. J.*, 146, 597, 1966.

2. J. E. Gunn, *Astrophys. J.*, 164, L113, 1971; L. B. Robinson and E. J. Wampler, *Ibid.*, 171, L83, 1972.

3. H. Arp, *Astrophys. J.*, 162, 811, 1970.

4. A. Oemler, J. E. Gunn, and J. B. Oke, *Astrophys. J.*, 176, L47, 1972.

5. J. S. Miller, L. Robinson, and E. J. Wampler, *Astrophys. J.*, 179, L83, 1973.

6. G. H. McDonald, S. Kenderdine, and A. C. Neville, *Mon. Not. Roy. Astron. Soc.*, 138, 259, 1968.

7. A. R. Sandage and W. C. Miller, *Astrophys. J.*, 144, L1238, 1966; A. R. Sandage, *Intern. Astron. Union Draft Reports*, Prague, p. 584, 1967.

8. John N. Bahcall, *Astrophys. J.*, 158, L87, 1969.

9. G. de Vaucouleurs, in *Stars and Stellar Systems*, Vol. 9, *Galaxies and the Universe*, G. P. Kuiper and B. M. Middlehurst, eds. (Chicago: University of Chicago Press, 1973), Chap. 17 (in press).

10. M. Rowan-Robinson, *Nature*, 236, 112, 1972, and *Astron. Astrophys.*, 23, 331, 1973.

11. S. H. Plageman, Dunsink Observatory, unpublished.

12. H. Arp, *Science*, 151, 1214, 1966, and *Astrophys. J.*, 148, 321, 1967.

13. D. Lynden-Bell, R. D. Cannon, M. V. Penston, and V. C. A. Rothonan, *Nature*, 211, 838, 1966; R. V.

Wagoner, *Ibid.*, 214, 766, 1967.

14. H. van der Laan and F. Bash, *Astrophys.J.*, 152, 621, 1968; D. J. Holdern, *Mon.Not.Roy.Astron.Soc.*, 133, 225, 1966.

15. H. Arp, *Astrophys.J.*, 152, 633, 1968.

16. L. Woltjer, Sixth Texas Symposium on Relativistic Astrophysics, 1972, New York City.

17. H. Arp, *Astron.J.*, 75, 1, 1970.

18. E. M. Burbidge, G. R. Burbidge, P. M. Solomon, and P. A. Strittmatter, *Astrophys.J.*, 170, 233, 1971.

19. J. N. Bahcall, C. F. McKee, and N. A. Bahcall, *Astrophys.J.(Lett.)*, 10, 147, 1972; also C. Hazard and P. A. Strittmatter, *Ibid*, 11, 77, 1972.

20. H. Arp, E. M. Burbidge, C. D. Mackay, and P. A. Strittmatter, *Astrophys.J.*, 171, L41, 1972.

21. F. Hoyle, Russell Lecture, American Astronomical Society Meeting, Seattle, April 1972.

22. H. Arp, *Astrofizika*, 4, 59, 1968.

23. H. Arp, *Sky and Telescope*, 38, 385, 1969.

24. P. C. van der Kruit, J. H. Oort, and D. S. Mathewson, *Astron. Astrophys.*, 21, 169, 1972.

25. H. Arp, *Science*, 174, 1189, 1971.

26. M. F. Barnothy and J. M. Barnothy, *Bull.Am.Astron. Soc.*, 4, 239, 1972.

27. C. R. Lynds and D. Wills, *Astrophys.J.*, 172, 531, 1972.

28. G. K. Miley, *Mon.Not.Roy.Astron.Soc.*, 152, 477, 1971.

29. A. Yahil, Institute of Advanced Studies, Princeton, preprint, 1972.

30. E. Holmberg, *Astron.J.*, 66, 620, 1961.

31. G. de Vaucouleurs, *Nature*, 236, 166, 1972.

32. G. A. Tammann, *Astron.Astrophys.*, 21, 355, 1972.

33. W. G. Tifft, *Astrophys.J.*, 175, 613, 1972.

34. H. Arp, *Nature*, 225, 1033, 1970.

35. T. Jaakola, *Nature*, 234, 534, 1971, and editorial, p. 505.

36. B. M. Lewis, *Nature (Physical Science)*, 230, 13,1971, and reply by H. Arp, *Ibid.*, 231, 103, 1971.

37. E. Holmberg, *Uppsala Astron. Medd.*, No. 166, 1969.

38. H. Arp, invited paper, American Astronomical Meeting. Seattle, Washington, April 1972; abstract, *Bull. Am. Astron. Soc.*, 4, 397, 1972.

39. H. Arp, *Pub. Astron. Soc. Pacific*, 80, 473, 1968.

40. W. L. W. Sargent, *Astrophys. J.*, 153, L135, 1968.

41. F. Zwicky, *Astronomie*, 72, 285, 1958; *Ergeb. Exakt. Naturwiss.*, 39, 354, 1956.

42. H. Arp, *Astrophys. Lett.*, 5, 257, 1970; *External Galaxies and Quasi-Stellar Objects, Intern. Astron. Union Symposium No.* 44, D. S. Evans, ed. (Dordrecht: Reidel Publishing Co., 1972).

43. H. Arp, *Astrophys. Lett.*, 7, 221, 1971.

44. H. Arp, *Astrophys. Lett.*, 9, 1, 1971.

45. R. Lynds and A. G. Millikan, *Astrophys. J.*, 176, L5, 1972.

46. T. F. Adams and R. J. Weymann, *Astrophys. Lett.*, 12, 143, 1972; H. C. Ford and H. W. Epps, *Ibid.*, 12, 139, 1972.

47. K. S. Anderson and R. P. Kraft, *Astrophys. J.*, 165, L3, 1971.

48. G. de Vaucouleurs and A. de Vaucouleurs, *Reference Catalog of Bright Galaxies* (Austin, Texas: University of Texas Press, 1964).

JOHN R. BAHCALL

Redshifts as Distance Indicators

REDSHIFTS AS DISTANCE INDICATORS[*]

by

John N. Bahcall
The Institute for Advanced Study
Princeton, New Jersey

I.   INTRODUCTION

The usual interpretation of redshifts as distance indicators underlies practically all our understanding of extragalactic astronomy and astrophysics. Much of what both laymen and scientists believe about the large scale structure of the universe is derived from the assumption that the redshifts of galaxies indicate distance. It is therefore appropriate that this discussion take place under the aegis of the American Association for the Advancement of Science, which has always concerned itself with important scientific issues that are also of interest to non-scientists.

I am grateful to Professor Field for organizing this discussion which forces us to define and confront the scientific issues. I hope that our efforts will contribute

---

[*] To Neta

to an increased understanding of the questions in dispute and ultimately to their resolution.   Of course the reason that our discussion takes place now is that Dr. Arp and others (especially G. R. Burbidge, E. M. Burbidge, and F. Hoyle) have in recent years vigorously challenged the conventional interpretation of redshifts.   Their ideas have been given a great deal of attention both in the popular press and in scientific journals.   It seems to me appropriate, in the face of this fully articulated challenge, that the classical reasons for believing in the conventional interpretation, augmented by recent evidence wherever possible, be put forward and that the arguments supporting the new challenges be examined critically.

I shall begin by reviewing the evidence that supports the usual interpretation of redshifts and then analyze in detail some of the outstanding examples of what Arp calls "discordant redshifts."   It will not be possible of course to discuss in depth all of the many examples that have been put forward; I shall instead concentrate on those examples which have been advanced with the greatest fanfare and assurance and which have therefore attracted the most careful observational checks and theoretical analysis.   I shall suggest, as we go along, observational tests that can be made of both the conventional and the unconventional interpretations.

II.   PREDICTIONS OF THE USUAL INTERPRETATION

II-1.   FLASHLIGHT ANALOGY

Much of the observational evidence that supports the conventional interpretation of redshifts as distance indicators can be understood in terms of an analogy. Suppose one collects a large bag containing flashlights of different sizes and with different powers of illumination. Now let us select from the bag only the brightest flashlights, all of which come from the same manufacturing company and can therefore be assumed to have approximately the same illuminating power. If we place these brightest flashlights at different distances from us, then the apparent brightness we see from an individual flashlight will be a measure of the distance of that flashlight from us (apparent brightness $\propto 1/(\text{distance})^2$). This example is illustrative of the technique developed especially by Hubble, Humason, and Sandage of determining the distances of the brightest galaxies (the brightest flashlights in the above analogy) in rich clusters of galaxies. They found that distances determined in this way are, to high accuracy, linearly proportional to the measured redshifts of the brightest galaxies (the so-called giant ellipticals).

Suppose instead that one selects flashlights of different intrinsic brightnesses and distributes them randomly at different distances from us. Then one will expect only an average correlation of apparent brightness (as seen by us) with distance. For example, some of the apparently bright objects may be very far away if they happen to be the intrinsically brightest objects. We shall see in Section II-3 that for ordinary galaxies the redshift is a good distance indicator, although if galaxies of all types are considered the correlation between redshift and apparent magnitude necessarily has a lot of scatter in it.

One may use another test to determine distance if one selects a sample of flashlights (or astronomical objects) all of which are about the same size. Place the equal-sized flashlights at different distances from us. Then the angular size (apparent size) of each of the flashlights is a measure of its distance from us (apparent size $\propto$ 1/distance). The analogous relation for galaxies is discussed in Section II-5.

We shall now review some of the most important observational results that exemplify the lines of argument presented above. The six tests we discuss are summarized, along with possible new tests, in Section II-7.

II-2. THE HUBBLE DIAGRAM

A great breakthrough in extragalactic astronomy was effected in 1929 when Edwin Hubble showed in a brief paper,[1] which is reprinted in the present volume, that the measured redshifts were proportional to known distances for the galaxies for which good distance estimates could then be made. These distance estimates were established by comparing the apparent brightness of individual stars in the distant galaxies with their known intrinsic brightness (primarily determined by measurements made on stars in our own galaxy). The correlation that Hubble found is shown in Fig. I, which is taken from his original paper. Note the approximately linear dependence of measured velocity [≡ (wavelength shift/laboratory wavelength) × velocity of light] on distance. The common notation in which redshifts are expressed in terms of velocities reflects the conventional interpretation that the observed large redshifts are Doppler shifts caused by the expansion of the universe.

Hubble put forward the linear relation between redshift and distance as a tentative hypothesis and immediately set about testing this relation on fainter objects. He encouraged his associate at Mount Wilson, Milton Humason,

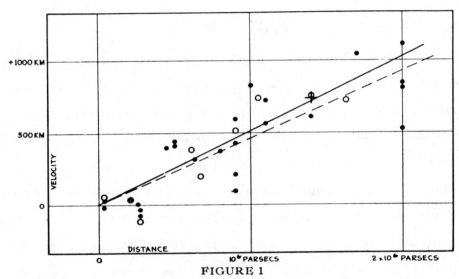

FIGURE 1

Velocity-Distance Relation among Extra-Galactic Nebulae.

Radial velocities, corrected for solar motion, are plotted against distances estimated from involved stars and mean luminosities of nebulae in a cluster. The black discs and full line represent the solution for solar motion using the nebulae individually; the circles and broken line represent the solution combining the nebulae into groups; the cross represents the mean velocity corresponding to the mean distance of 22 nebulae whose distances could not be estimated individually.

FIGURE I

to take spectra of the brightest galaxies in groups and clusters in order to extend the tests to the largest red-shifts then measurable. In fact, Hubble's 1929 paper is preceded in the same journal by a short note by Humason[2] in which he reported the then very large velocity of 3,779 km/sec for the galaxy NGC 7619, which is a member of a group of galaxies. Humason's result for NGC 7619 was consistent with what one would expect from an extrapolation of Hubble's linear relation.

The tests of the linear relation between redshift and distance were carried out to the limit of the Mount Wilson 100" telescope; the spectroscopic results were reported by Humason[3] in a brief paper that is reprinted here. By the most careful observations, Humason was able to extend the redshift measurements to 19,700 km/sec (for the brightest galaxy in the Leo cluster--which required a 13 hour exposure). The analysis of the new spectroscopic data was described in a joint paper in the same issue of the journal by Hubble and Humason.[4] We show in Fig. II the relation they found between measured redshift (expressed as a velocity) and apparent brightness for the galaxies in groups or clusters. The velocity-distance relation obtained by these authors when they combined the then most recent work with Hubble's earlier study[1] is shown in Fig. III. Note that the original relation remains valid even though there is a twenty-fold increase in the range of velocities and distances that are covered.

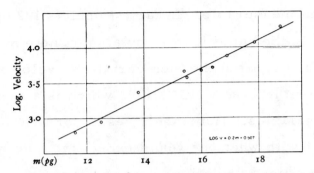

Correlation between the quantities actually observed in deriving the velocity-distance relation. Each point represents the mean of the logarithms of the observed red-shifts (expressed on a scale of velocities) for a cluster or group of nebulae, as a function of the mean or most frequent apparent photographic magnitude.

FIGURE II

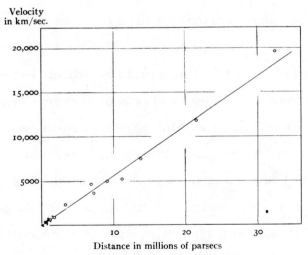

The velocity-distance relation. The circles represent mean values for clusters or groups of nebulae. The dots near the origin represent individual nebulae, which, together with the groups indicated by the lowest two circles, were used in the first formulation of the velocity-distance relation.

FIGURE III

The correlation between redshift and apparent magnitude (i. e. , the logarithm of apparent brightness) has become known as the Hubble diagram (see also the 1928 paper by H. P. Robertson[5]). It has been the focal point of extensive investigations by many observers over the past forty years. Some of the most recent results are shown in Fig. IV, which exhibits Figs. 2-4 of a recent paper by Sandage[6] on the Hubble Diagram. It is an awesome and sobering fact that the tiny black box in the lower left-hand corner represents all the data that were available to Hubble in 1929 when he suggested the linear relation between redshift and distance!

II-3. AVERAGE GALAXIES

In his epochal 1929 paper, Hubble considered galaxies of all types for which distances could be established. Attention has been focused in recent years on the brightest galaxies in rich clusters because their redshifts can be measured at the greatest distances and because they define an accurate relation between redshift and apparent brightness. We expect, using the flashlight analogy, that if one plots the apparent brightness versus distance for all galaxies there will be much more scatter in the diagram. Some apparently faint galaxies will be nearby, because they correspond to intrinsically faint objects, and some apparently bright galaxies will be far away if they are among the intrinsically brightest sources. Nevertheless one would expect for the galaxies (as for the flashlights)

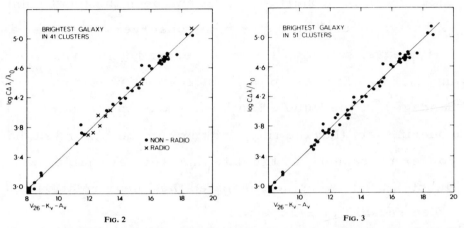

FIG. 2

FIG. 3

FIG. 2.—The Hubble diagram for first-ranked galaxies in 41 clusters from the data of table 2. *Abscissa*, the corrected V_c magnitude; *ordinate*, the logarithmic redshift. The box in the lower left is the approximate interval within which Hubble established the redshift-distance relation in 1929. A line of slope 5, required by all homogeneous models in the $z \rightarrow 0$ limit, is fitted to the data in zero point only. It is equation (4) of the text.

FIG. 3.—Same as fig. 1, with data from table 3 added to those of table 2.

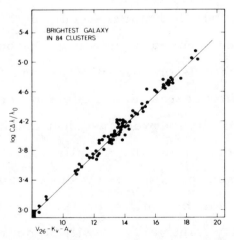

FIG. 4.—Same as fig. 1 for the combined data of tables 2, 3, and 4

FIGURE VI

that on the average the fainter objects are further away.

The correlation of apparent brightness with redshift is given in Fig. V for all types of galaxies; the redshift-magnitude relation shown is taken from the classical paper of Humason, Mayall, and Sandage.[7]   The observed relation is just what one would expect from the flashlight analogy if galaxy redshifts are good distance indicators.   The most important inference to be drawn from Fig. V is that galaxy redshifts seem to be good distance indicators for average galaxies, not just for the brightest ellipticals.

II-4.   MEAN REDSHIFTS AND DISPERSION IN REDSHIFTS

Humason also discussed briefly, in 1931,[3] another critical test of the usual interpretation of redshifts. He had obtained spectra for a number of galaxies in each of several clusters and noted that the mean redshift for the cluster galaxies was much larger than the dispersion in redshifts.   This is just what was expected if one assumed that the mean cluster redshift gave the distance to the cluster and that the dispersion about the mean was caused by individual galaxy motions within the cluster.   This interpretation has received much additional observational support as more redshifts have become available for galaxies in clusters (see, e.g., the discussion in Section III and references 7-9).

An even simpler test of the conventional interpretation of redshifts has been made by Page[10,11] in connection with

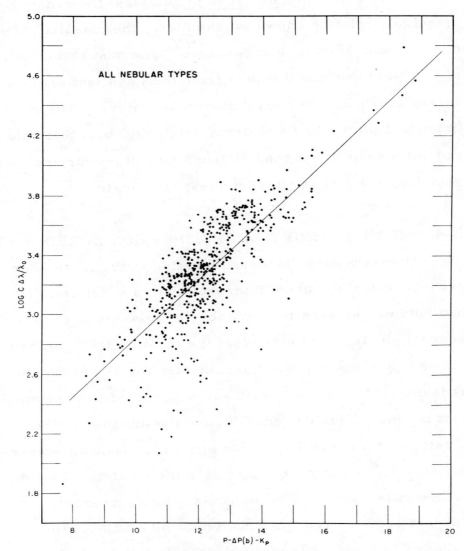

Figure 10. The redshift-magnitude relation for 474 field nebulae of all nebular types.

FIGURE V

his studies of pairs of close galaxies. Page found, as
expected on the conventional interpretation, that the mean
redshift of close pairs was much larger than the difference
between the two redshifts of the pair. This result is of
course consistent with the usual interpretation in which
the mean redshift of the pair is indicative of their average
distance and the difference between individual redshifts is
caused by their comparatively small relative velocities.
It is important to note that Page's results refer to double
galaxies of all types: spirals, ellipticals, and irregulars.

II-5. ANGULAR DIAMETERS VERSUS REDSHIFT

There is another set of tests that is obvious from
the flashlight analogy discussed earlier, namely, the
relation between apparent size and redshift. If redshifts
are good distance indicators, then one expects for a given
class of objects (same intrinsic size) that the angular
diameter, θ , is proportional to $(\text{redshift})^{-1}$ (at very large
redshifts, there are corrections to this relation that de-
pend on the cosmological model used;[12] these corrections
are unimportant for our purposes). More explicitly,

$$\theta \ \propto \ (\text{distance})^{-1} \quad \text{(fixed class of objects)} \quad \text{(1a)}$$

$$\propto \ (\text{redshift})^{-1} \quad \begin{array}{l}\text{(if redshift proportional} \quad \text{(1b)}\\ \text{to distance)}\end{array}$$

The first part of the above relation, equation (1a),
was established by Hubble in the famous paper[13] in which

he also systematically described for the first time his phenomenological classification of galaxies. Hubble showed that for a fixed class the apparent angular dimensions represented, in a statistical sense, the relative distances. Fig. VI shows some results from this pioneering study. The quantities actually plotted are apparent magnitude and angular size, but since apparent luminosity $\propto$ (distance)$^{-2}$ (see Sec. II-2 and -3) the approximate straight line relation of Fig. VI is equivalent to equation (1a).

The second part of equation (1) has been tested extensively only for giant ellipticals. Sandage[12] has obtained a relatively homogeneous set of photographic plates of giant galaxies in clusters. Some of his results are shown in Fig. VII. We see that the estimated angular diameters, θ, are proportional to (redshift)$^{-1}$, as is expected if redshifts are proportional to distance.

An independent test of the relation angular diameter $\propto$ (redshift)$^{-1}$ has been initiated by N. Bahcall.[14,15] She has measured the angular diameters of the cores of five rich clusters of galaxies and has shown that the core diameters of the clusters are approximately constant over a range in redshifts of a factor of 16 (redshifts from 0.022 to 0.38) if equation (1b) is satisfied. Bahcall's method must be tested on more clusters before it can be regarded as conclusive but the preliminary results are suggestive.

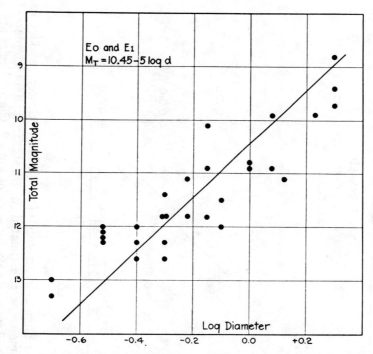

FIG. 2.—Relation between luminosity and diameter among nebulae at the beginning of the sequence of types—Eo and E1 nebulae.

FIGURE IV

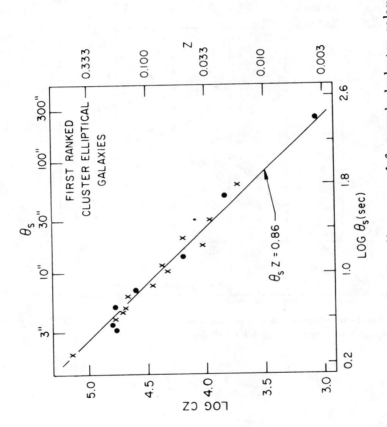

FIG. 1.—Correlation of estimated angular diameters of first-ranked cluster galaxies (uncorrected isophotal) with redshift. Angular diameters were estimated (in arc sec) from 200-inch plates (103a-D + GG11). Ordinate is log redshift, expressed on a velocity scale of km s^{-1}. Closed circles are from plates exposed on the same night and developed homogeneously. Crosses are from plates taken at other times. Line is $\theta_s \propto z^{-1}$. Data are from table 1.

FIGURE VII

II-6. RADIO AND OPTICAL REDSHIFTS

There is an extremely precise test of the conventional interpretation of galaxy redshifts which consists of comparing the redshifts measured at radio and at optical wavelengths.   According to the conventional (Doppler) interpretation of redshift, the measured shift $(\Delta\lambda/\lambda)$ must be independent of wavelength.   Roberts[16] has summarized the observations which show that redshifts of the neutral hydrogen 21-cm line determined by radio techniques are in excellent agreement with redshifts measured with optical lines (3500 Å to 6000 Å).   The results are illustrated in Fig. VIII.   We see that the agreement is better than one part in a hundred over a range of 6,000 km/sec and a wavelength factor of a factor of 5×10^5.   Note also that the 130 galaxies included in Fig. VIII include galaxies of all types.

II-7. SUMMARY OF PREDICTIONS AND FURTHER TESTS

The hypothesis that redshifts are good distance indicators has survived six observational tests since the proposal was first clearly stated by Hubble in 1929.   Also of great importance is the fact that the specific form of Hubble's law, redshift $\propto$ distance, has a very simple theoretical interpretation.   Hubble's law is predicted by all cosmologies that assume the universe is expanding and is (at least locally) homogeneous and isotropic.

We now summarize the observational confirmations.

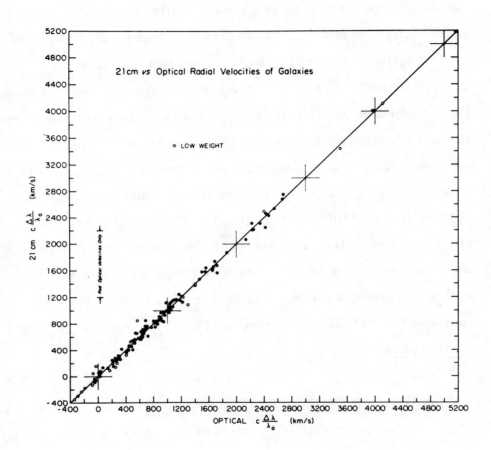

21cm *vs* Optical Radial Velocities of Galaxies

○ LOW WEIGHT

FIGURE VIII

First, the original relation between redshift and apparent brightness was tested and found valid for the brightest galaxies over a redshift range that is more than one hundred times larger than the range originally available to Hubble. Second, the average apparent brightness of all galaxies irrespective of type decreases like $(redshift)^{-2}$, with a scatter about the mean relation that can be understood in terms of the range in intrinsic luminosities. Third, the mean redshift of double galaxies was found to be much larger than the difference in redshifts between the two members of each such pair. Fourth, the mean redshift of galaxies in a rich cluster was found to be much larger than the dispersion in redshifts of individual galaxies in the cluster. Fifth, the apparent angular diameter of the brightest galaxies in rich clusters decreases as $(redshift)^{-1}$. Sixth, redshifts that are determined by radio and optical techniques agree to high accuracy.

Other quantitative tests of the usual interpretation of redshift could be carried out. Further redshift studies of galaxies in distant clusters, as well as in smaller groups, could test the prediction that the mean redshift of the cluster or group must be large compared to the dispersion of redshifts. The relation between redshift and the angular size of the core of rich clusters of galaxies found by N. Bahcall[14,15] can be tested on a larger number of clusters. One can also test the predicted equality of the mean redshifts of different types of galaxies in groups or

clusters.[*] Careful estimates of the background are required for redshift tests involving groups or clusters,
i. e. , one must determine the likelihood that some candidates for group or cluster membership may not be
members but may be accidentally superimposed on the
area of interest. Finally, we note that the data on double
galaxies could be greatly increased using the faster techniques for obtaining spectra that have recently become
available and the results could be organized by galactic
type to investigate the limits on discordant redshifts as a
function of type.

---

[*] This test has been carried out for the Virgo and Coma
clusters with results that are consistent with the usual
interpretation of redshifts (see Section III-6).

III. EXAMPLES OF ALLEGED DISCORDANT REDSHIFTS

III-1. GENERAL REMARKS

Before we discuss specific cases of "discordant redshifts," I would like to make a few general remarks. None of the examples of alleged discordant redshifts is unambiguously in contradiction to the usual interpretation of redshifts; the examples are at best "suggestive." Moreover many of the heuristic arguments that have been advanced are not well defined ahead of time and they do not exhibit an obvious logic that interested observers can use to make predictions.

Why are discordant redshifts discussed in terms of examples and the conventional interpretation in terms of quantitative predictions? The basic reason is that the conventional interpretation is a specific hypothesis: the observed redshifts of galaxies are caused by the familiar Doppler effect. On this basis, the large redshifts of galaxies are interpreted as evidence for the expansion of the universe; small differences in redshifts of galaxies that are supposedly close together (e. g. , double galaxies or galaxies in clusters or groups) are interpreted in terms of dynamical motions.   This interpretation gives a partial but consistent picture of the universe in agreement with the laws of physics as we know them from terrestrial experiments.

The interpretation of Arp and his associates is open-ended.   There is no known hypothesis besides Doppler shifts that is both consistent with the laws of physics and able to explain the large redshifts of galaxies (the typical gravitational redshift of a galaxy is only of the order of $GM_{galaxy}/(R_{galaxy}c^2) \sim 10^{-7}$).   If discordant redshifts truly exist, then the known laws of physics do not apply to some galaxies.

The world of physics is four-dimensional, with three dimensions of space and one of time.   But we can obtain by astronomical observations only two-dimensional snapshots of distant galaxies.   The third space dimension, representing distance from us, is not directly observable.   All that we can measure directly are angular separations on the sky; two galaxies that are accidentally close together in angular separation may in reality be very far apart.   We have no way of knowing their actual separation unless there is some independent distance indicator such as the redshift of spectral lines.   The skies when photographed with large telescopes reveal so many individual objects on any photographic plate that one can find almost any configuration one wants if one just hunts: even stars arranged as four-leaf clovers.

The times required for events to occur on galactic scales may be many millions or even billions of years. Thus we cannot make use of the temporal sequence of events (e. g. , two galaxies moving around their common center of mass) for deciding if two faint galaxies are at the same distance. In physics, one discovers and tests new laws with controlled experiments; in astronomy, one must make do with observations of the heavens in two dimensions at one specific time. We shall see in what follows how the limits on astronomical observations continually beset an observer when he begins hunting for "discordant redshifts."

III-2. CONNECTING FILAMENTS

Arp has stressed on a number of occasions the importance of luminous filaments that he believes have been shown photographically to connect pairs of objects. These claims have been publicized by, among many others, Physics Today (February 1972) and have been cited by several authors as decisive evidence for discordant red-shifts. The argument is simple: If two objects are connected by a filament they must be at the same distance and hence, if their redshifts are different, redshifts cannot be good distance indicators. In my opinion, the argument is wrong for two reasons: (1) in some cases, the alleged filament does not exist; and (2) filaments may appear to connect objects that are really at very different distances because both objects are seen only in projection on the sky. I shall discuss the two most famous cases of alleged connecting filaments: M 205 + NGC 4319, and NGC 7603 plus "companion" galaxy.

MARKARIAN 205 AND NGC 4319. The most extensively investigated example of an alleged connection between objects with greatly different redshifts involves Markarian 205 (a compact luminous galaxy[17-20]) and an apparently normal spiral galaxy, NGC 4319. Weedman[17] first drew special attention to this pair of objects; he stressed that M 205 lies within the apparent area included by the outermost spiral arm of NGC 4319. Weedman suggested that M 205 and NGC 4319 were physically associated because of their _apparently_ close proximity; the projected position of M 205 appears to be within a spiral arm of NGC 4319. Weedman argued[17] that the usual redshift-distance relation may not hold for M 205 and NGC 4319 because accidental superposition is an unlikely explanation for the observed configuration. His method for arriving at this conclusion was not stated explicitly, but he did give a hint which, as we shall see in Section III-3, indicates that he used an incorrect basis for estimating the probabilities.

Arp[21] attempted to prove that M 205 and NGC 4319

were at the same distance by finding photographic evidence of their interaction. His first attempts in June 1970 were unsuccessful.

However, Arp persisted in his efforts to demonstrate a luminous connection between M 205 and NGC 4319 and was convinced[21] that such a connection was visible on a special 4-hour exposure he obtained with the Hale 200" telescope on the fine-grained IIIa-J emulsion in March 1971. He also claimed to have even clearer evidence for a luminous connection that could be seen on plates taken primarily in the hydrogen $H\alpha$ emission line. This apparent connection is in fact very striking in one of Arp's photographs (his Fig. 3).[21]

Lynds and Millikan,[18] in a paper reprinted in this volume, tested Arp's claims using new photographic material and laboratory experiments and were unable to confirm his results. They noted that the supposed connection on the continuum IIIa-J plate could be explained by a spurious photographic effect which they demonstrated with laboratory experiments. They showed with photographs of laboratory objects that an apparent connection may be caused by the close proximity in two dimensions of separated objects. They also took a deep IIIa-J photograph of M 205 and NGC 4319 which gave the same impression of a connection as did Arp's photograph. However, Lynds and Millikan measured the equal-brightness contours around both M 205 and NGC 4319. They showed that within the accuracy of the measurements the entire impression of a connecting feature could be explained by the overlap of the two separate brightness distributions, as in the laboratory experiments. They also noted the possibility that there might be a small contribution from a faint distant galaxy (there are many in the field) that happens by chance to lie in the region between NGC 4319 and M 205.

The connecting feature in Hα that Arp[21] claimed to see (and apparently illustrated so clearly in his Fig. 3) has not been found by other observers. Lynds and Millikan[18] were unable to find evidence of a connecting feature in Hα; they used photographic material that was at least as good as Arp's[21] as well as spectroscopic plates. Ford and Epps[19] obtained an Hα photograph that was slightly superior to Arp's in information content and found no evidence for the feature claimed by Arp. Similar results were reported by Adams and Weymann.[20] Abstracts of these important papers by Adams and Weymann and Ford and Epps are reprinted in the present volume.

The moral of the story of the apparent connection between NGC 4319 and M 205 is clear: Seek and ye shall find, but beware of what you find if you have to work very hard to see something you wanted to find.

NGC 7603 AND "COMPANION GALAXY." The relation between NGC 7603 and an apparently nearby companion galaxy has been described by Arp[22] as "a crucial test of the reality of discordant redshifts." The redshift of

NGC 7603 is[23] z = 0.029 and that of the fainter companion
is z = 0.056. Arp[23] claims to have demonstrated a luminous connection between the two galaxies that is slightly
brighter where the extended spiral arm from the main
galaxy reaches the companion. He has concluded from
this evidence that "a major part of the redshift difference
between NGC 7603 and its companion is caused by some
effect other than velocity difference."

This example merits detailed investigation by other
observers in view of the importance attached to it by Arp
and his supporters and in view of the failure of other investigators to confirm Arp's conclusions in the analogous
case of NGC 4319 and M 205. One can in fact make a
prediction using the conventional interpretation of redshifts
as distance indicators. Since NGC 7603 and its companion
have such different redshifts, a detailed intensity tracing
should show that Arp's suggested brightening of the companion in the direction of N G C 7 6 0 3 can be explained
quantitatively by the superposition of intensity profiles
from the two objects (as Lynds and Millikan[18] have demonstrated for NGC 4319 and M 205). I have the impression,
when inspecting Arp's published[23] photograph of this pair,
that the spiral arm of NGC 7603 actually extends beyond its
apparent companion. One would expect on the basis of the
usual interpretation of redshifts that the intensity profile
of any extension would be a smooth extrapolation of the
part of the arm that is interior to the companion, a

prediction that also could be tested by a detailed intensity tracing.

III-3. ELEMENTARY STATISTICS OF APPARENT CONNECTIONS*

REMARKS. The way Arp carries out his observational programs, <u>searching</u> <u>for</u> <u>peculiarities</u> <u>that</u> <u>are</u> <u>not</u> <u>clearly</u> <u>specified</u> <u>before</u> <u>the</u> <u>observations</u>, actually prevents one from using the argument that a particular observed configuration is too unlikely to be due just to chance. The reader can see easily that such arguments when applied <u>a posteriori</u> may be misleading. Suppose you take a detailed large-scale photograph of Times Square that shows at one time in two dimensions all the people in the Square. The <u>a priori</u> probability that all the people have by chance the observed angular separations and apparent configurations seen on the photograph is "negligible"; the number of alternative possibilities is very large. However, the <u>a posteriori</u> probability for the observed configuration is unity, just because it is the observed configuration!

------

* We shall not discuss apparent chains of galaxies in which one or two alleged members have redshifts very different from the others. However, most of our general considerations (see below) apply equally well to apparent chains and the reader may find it instructive to apply formula (2) to alleged examples of chains. Other arguments have been given by Barnothy and Barnothy[24] against the claimed significance of the most famous supposed chain of quasars and galaxies.

Any complex situation when described in sufficient detail will have too small an _a priori_ probability to happen by chance, but a unit probability of having happened if the calculation is performed after the fact.

EXPECTED NUMBER OF ACCIDENTAL SUPERPOSITIONS.

a. _General Expression_. We now describe a method for estimating the number of accidental superpositions that suggest physical connections. Let N be the probable number of accidental configurations of an "unusual kind" that might be expected. For example, N may be the number of spiral galaxies that have apparent companions accidentally superimposed on the end of spiral arms and that appear, as do NGC 7603 and its companion, in Arp's[25] Atlas of Peculiar Galaxies. We may write symbolically

$$N = CABO, \qquad (2)$$

where the four factors are explained below. We denote by C the total number of cases that have been examined; e.g., the number of galaxies whose vicinity has been searched for faint companions. The area of interest around any particular object is denoted by A; i.e., in order to be considered a "close companion" an object must fall within the nearest A square degrees around, e.g., NGC 7603. We denote the number of background objects per square degree by B. This may be, for example, the number of galaxies like the companion of NGC 7603 down to a fixed

limiting brightness.   Finally, O is a catch-all parameter
designed to describe special features which might suggest
physical association; e. g. , O might be the fraction of area
around the parent object which exhibits protuberances (if
connecting filaments are claimed as evidence of physical
association).

None of the quantities C, A, B, or O were specified
by Arp before he focused attention[23] on NGC 7603 and its
"companion" or by Weedman[17] before he drew attention to
the apparent pair M 205 and NGC 4319. Any claim to have
discovered a causal relation on the basis of a posteriori
statistics involving these objects is subject to the same
criticism that applies if one infers causal relations between
random people seen in a single snapshot of Times Square.

b.   Numerical Estimates. It is still of interest to estimate the expected number, N, of accidental close pairs
like NGC 7603 and its "companion" on the assumption that
the conventional interpretation of redshifts is correct.
(The reader may find it instructive to do for himself the
analogous calculation involving M 205 and NGC 4319.) We
begin by trying to estimate the quantity C. Arp[25] does
not say how many galaxies were examined in order to find
the close pairs in his catalogue; he does say that he examined galaxies in the catalogue of Vorontsov-Velyaminov[26]
and personal lists compiled by a number of well-known
astronomers (A. G. Wilson, E. Herzog, T. Page, C. A.

Wirtanen, W. W. Morgan, F. Zwicky, C. Kowal and G. Reaves) through years of study of various sets of photographic plates. The number of galaxies that these hard-working astronomers may have examined in the course of their collective researches is of course unknown and unknowable! We may make a crude guess, probably conservative, at the total number of galaxies that were inspected and set C equal to $2.6 \times 10^{+3}$, the number of galaxies listed in the Reference Catalogue of Bright Galaxies compiled by de Vaucouleurs and de Vaucouleurs.[27]

How close must a companion object be before the closeness is "suggestive" of physical association? In order to estimate the number of accidental cases that are expected, one must fix $A = \pi r^2_{suggestive}$, where $r_{suggestive}$ is the maximum separation suggestive of physical association. For the special case of NGC 7603 and its companion, we may take 1 arc minute as the minimum value of $r_{suggestive}$ because this is the actual separation of the centers of the two galaxies. Thus a lower limit on A is 10^{-3} square degrees.*

---

* In the statistical consideration at which he hinted, Weedman[17] apparently concluded that accidental superposition was an unlikely explanation for the observed configuration of M 205 and NGC 4319 because galaxies of the same type and brightness as M 205 are rare. The rarity of objects such as M 205 is of course not in question. The quantity that must be calculated is the probability of finding in the vicinity of any object, rare or common, an apparent companion in a "suggestive" position.

The kind of objects that would have been considered as close companions and the magnitude limit to which one would note companions is of course unknown.   We may make a crude stab at guessing the background by setting B equal to 80 objects per square degree, which is the number of galaxies brighter than 19th magnitude as determined in the famous Lick survey.[28]

Finally, we must estimate the orientation factor O. I guess from Arp's Fig. 2[23] that of the order of one-sixth of the total area surrounding NGC 7603 contains "interesting" protuberances, i. e., $O \sim 1/6$.

Collecting together all of the factors guessed at above, we find:

$$
\begin{aligned}
N_{accidental} \quad &\sim \quad 3 \times 10^{3} \times 10^{-3} \times 80 \times 1/6 \\
&\sim \quad 40 \ (\underline{but} \ \underline{highly} \ \underline{uncertain}) \\
&\quad \underline{in \ Arp's \ catalogue.}[25]
\end{aligned}
\tag{3}
$$

It should be stressed that the above estimate refers to the number of accidental close pairs like NGC 7603 and its companion that are expected to be present in Arp's catalogue.   The number is of course highly uncertain (e. g., it is proportional to $r^{2}_{suggestive}$) but is in satisfactory agreement with the actual number, 64, of such cases listed by Arp.[25]   <u>It is possible to explain all of the close companions in Arp's catalogue on the basis of accidental superposition.</u>

III-4. STEPHAN'S QUARTET

The five galaxies NGC 7317, 7318a, 7318b, 7319, and 7320 form an apparent group known sometimes as Stephan's Quintet. This group has been the subject of many detailed studies.[29-33] Arp[34] has cited "the Quintet" as an outstanding example of discordant redshifts caused by the ejection of small compact galaxies from larger galaxies. All of the galaxies lie within an area of about ten square minutes of arc (see especially the photographs published in references 30 and 33). The average radial velocity of the first four members of the group is about 6400 km/sec[7] and the differences from the mean are very small (± 500 km/sec) compared to the average velocity. This agreement in redshift, along with their close juxtaposition on the sky, justifies the usual assumption that NGC 7317, 7318a, 7318b, and 7319 are at the same distance and form a physically associated group which we call Stephan's Quartet.[31, 32]

The situation is different with the fifth galaxy, NGC 7320. This galaxy has[30] a redshift of only 800 km/sec, much less than its four apparent neighbors. The conventional interpretation is that NGC 7320 is a foreground galaxy that is accidentally superimposed on the area near the physical group of Stephan's Quartet. The conventional interpretation is supported by independent evidence derived from (1) optical photography[31]; (2) optical spectroscopy[33]; and (3) radio astronomy.[32] Arp[34] has argued instead that all <u>five</u> of the galaxies are at the same distance and that some galaxies were ejected from others with non-Doppler redshifts. We review the evidence below.

OPTICAL PHOTOGRAPHY. Tammann[31] has pointed out that NGC 7320 is clearly resolved into individual stars on Arp's excellent 200-inch photograph,[25] but the visual appearance of the four members of Stephan's Quartet is very different. Thus Tammann concludes on the basis of optical appearance that the four members of Stephan's Quartet (which all have similar redshifts) are at the same distance and NGC 7320 (with a smaller redshift) is much closer. Tammann also points out in his paper, which is reprinted in the present volume, that there is evidence for gravitational interaction between NGC 7318a, 7318b, and 7319 because their optical shapes are strongly distorted. He suggests that there is no evidence for such distortion in the image of NGC 7320, a fact that is again most easily interpreted by assuming that NGC 7320 is not at the same distance as the members of Stephan's Quartet. Tammann also suggested (following van den Bergh[35]) that NGC 7320 is at the same distance as NGC 7331, a bright spiral galaxy that is less than half a degree in angular separation from NGC 7320. Both of these galaxies have practically the same small redshift ($\sim$ 800 km/sec). He noted that the Quartet lies in an extended region that includes an unusually high surface density of galaxies; this region was identified by Zwicky and Kowal[36] as an open cluster of

galaxies.   Tammann suggested that NGC 7320 and 7331
(with small redshifts) are members of a close nearby
group that is seen in projection against a more distant
cluster (with larger redshifts) that contains Stephan's
Quartet.   This suggestion is based on the conventional
interpretation of redshifts as distance indicators and is
subject to observational test (see below) by measuring
the redshifts of many galaxies in the field.

OPTICAL SPECTROSCOPY. Lynds[33] measured redshifts
for a number of other galaxies in an extended field including Stephan's Quartet and NGC 7331. He showed, in a
paper that is reprinted in this volume, that the galaxies
in the extended field broke up into two groups on the basis
of their redshifts. One group, with smaller redshifts
($\sim$ 1000 km/sec), included NGC 7320 as well as NGC 7331
and NGC 7343. The second group is actually a rather extended cluster with redshifts $\sim$ 7000 km/sec that includes
Stephan's Quartet and at least 17 other anonymous galaxies
in the field. The beautiful results of Lynds confirm the
suggestion of superimposed groups made by van den Bergh[35]
and Tammann.[31]

RADIO ASTRONOMY. Allen[32] reported radio measurements that enabled him to estimate the distance to NGC 7320
by a method that is independent of the radial velocity of the
galaxy. He measured the radio power received from
NGC 7320 in the neutral hydrogen 21-cm line; this

measurement gave a value for the mass, M_H, of neutral hydrogen gas which depends on the square of the distance, D, to the galaxy. He also measured the width of the velocity profile from the 21-cm line, which gave a value for the total mass, M_T, of the galaxy. Finally, he used a measurement of the apparent optical brightness of NGC 7320, which gave a value for the optical luminosity, L_{opt}, that depends on D^2. Allen formed the ratios $M_H/(M_T - M_H)$ and M_T/L_{opt}, both of which depend on D, and compared the ratios for NGC 7320 with those of other galaxies of the same type whose distances had been independently measured or determined from the conventional interpretation of redshifts as distance indicators. Allen showed that a comparison of the observed ratios led to the conclusion that NGC 7320 is probably a nearby dwarf spiral galaxy that is seen superimposed on the more distant cluster containing Stephan's Quartet, in agreement with the conclusions of Tammann[31] and Lynds.[33]

ARP'S INTERPRETATION. Arp[34] has suggested as an alternative interpretation that the five galaxies forming Stephan's Quintet are ejected from NGC 7331. He asserts that the conventional interpretation implies two distinct groups of galaxies, both with the same redshift, but one confined to the area near NGC 7331 and one confined to the area near Stephan's Quartet. He then claims that this configuration is too unlikely to be acceptable. This argument does not apply to the work described above by

Lynds[33] and Tammann[31] since both of these investigators
(along with van den Bergh) suggested that the cluster containing Stephan's Quartet and the apparent companions of
NGC 7331 extended throughout the field under discussion.

Burbidge and Burbidge[29, 30] estimated, by an ex
post facto statistical calculation, that the chance of
NGC 7320 being accidentally superimposed on Stephan's
Quartet is only about one in a thousand. Their estimate
neglects the much higher than normal density of galaxies
in the area around NGC 7320; this high density of background galaxies is due to the cluster containing Stephan's
Quartet. Moreover, the estimate of Burbidge and Burbidge
is proportional to $r_{suggestive}^2$ (see Section II-3) and they
chose the minimum possible value for $r_{suggestive}$ (equal
to the observed separation, 2 arc minutes, of NGC 7320
and Stephan's Quartet).

Arp has offered two other arguments in support of
his proposal that all of the galaxies in Stephan's Quintet
are at the same distance and that NGC 7320 was ejected
from NGC 7331. He has claimed that the sizes of the H II
regions in NGC 7320 and in NGC 7318 is the same, which
he says proves that the two galaxies are at the same
distance. This argument is invalid because NGC 7320 is[32]
a dwarf spiral and NGC 7318 is a more normal spiral.
Galaxy mass and type influence the size of H II regions and
one cannot use the apparent size of H II regions for different
kinds of galaxies as a distance indicator. Arp has also

argued[34, 37] that there are connecting optical or radio features between NGC 7331 and Stephan's Quintet, showing that all of the galaxies are at the same distance. This argument is not convincing since all that has been shown is that there may be evidence for connecting features between NGC 7331 and NGC 7320 (which have the same redshifts); there is no evidence that the claimed connecting features involve the (large-redshift) members of Stephan's Quartet.

III-5. CLUSTERS AND GROUPS OF GALAXIES

Arp and his associates have frequently drawn attention to evidence for discordant redshifts in clusters and groups of galaxies. We give below a critical review of this evidence in the best-investigated cases.

VIRGO CLUSTER. The Virgo Cluster is the richest nearby cluster of galaxies and consequently has been intensively studied since the earliest days of extragalactic astronomy.[1, 4, 38] More recently, several authors have concluded that the spiral galaxies in the Virgo Cluster have a significantly larger ($\sim$ 500 km/sec) redshift than do the elliptical and lenticular galaxies.[39-41] On the other hand, Kowal[42] has argued that all the galaxies in the Virgo Cluster are at the same distance on the basis of the apparent brightness of the supernovae that are observed in both spiral and elliptical galaxies.

This combination of results, if correct, is in conflict

with the conventional interpretation of redshifts as distance indicators; two sets of galaxies, spirals on the one hand and ellipticals and lenticulars on the other hand, are alleged to be at the same distance but to have significantly different redshifts.   De Vaucouleurs[43] and Arp[22] have suggested on the basis of the above-mentioned evidence that the redshift difference between spirals and ellipticals is to be interpreted as a non-velocity redshift.

Because of this apparent conflict with the conventional interpretation of redshifts, Tammann[9] has re-examined the redshift distribution for probable members of the Virgo Cluster.   He made use of redshifts for 122 galaxies; this is many more than were available at the time de Vaucouleurs[39] and Holmberg[41] concluded that the spirals and ellipticals had significantly different redshifts.   Tammann[9] also corrected the redshifts for a small systematic error which was discovered by M. Roberts[16] using redshifts obtained with radio techniques. A careful analysis of all the available data led Tammann[9] to the conclusion that there is no statistically significant radial-velocity difference between spiral and elliptical-lenticular galaxies in the Virgo Cluster.

COMA CLUSTER.   The Coma Cluster of Galaxies is the nearest of the symmetric regular clusters of galaxies and, like the Virgo Cluster, has been intensively investigated (for references see the papers by Rood et al.[8] and

N. Bahcall[15]). Tifft[44] has stated that the elliptical and
non-elliptical galaxies in the Coma Cluster have significantly different redshifts and, along with Arp, has claimed
that this provides evidence for non-velocity redshifts.

The conclusions of Tifft and Arp are not supported
by an analysis of the recent systematic observations reported by Rood, Page, Kinter, and King.[8] Using the data
in Table 1 of the paper by the latter authors, I find that the
elliptical galaxies have a mean redshift of 6827 km/sec
(42 objects) and the non-ellipticals have a mean redshift
of 7007 km/sec (53 objects); the standard deviation of the
mean in both cases is of the order of 10^2 km/sec. In
calculating the above averages, I have followed Rood et al.[8]
in omitting, as foreground objects, galaxies with redshifts
less than 3000 km/sec and the three doubtful members with
redshifts ~ 4500 km/sec at distances from the center of
more than 100 minutes of arc.

The difference (180 km/sec) between the means
depends most strongly on where one chooses the cutoff in
acceptable velocity for members of the cluster. For example, if one omits the two objects with the largest velocities (both non-ellipticals with redshifts of 9358 and 9398
km/sec, respectively) the average velocity of the non-
ellipticals decreases by 94 km/sec to 6913 km/sec. If
one omits the two ellipticals with the smallest velocities,
4665 km/sec and 4945 km/sec, the average redshift of the
ellipticals increases by 101 km/sec to 6928 km/sec. Thus

one <u>can</u> <u>make</u> <u>the</u> <u>velocity</u> <u>means</u> <u>agree</u> <u>or</u> <u>disagree</u> <u>by</u>
<u>slightly</u> <u>altering</u> <u>the</u> <u>acceptance</u> <u>criteria</u> <u>for</u> <u>cluster</u> <u>mem-</u>
<u>bership.</u>

The above discussion shows that if one is interested
in comparing the average redshifts of different types of
galaxies in the same cluster one must carefully evaluate
the uncertainty introduced by foreground or background
objects. Unfortunately, estimates of this effect will depend on the model assumed for the velocity and spatial
distribution of galaxies in the cluster.

There is another effect reported by Tifft[44] which
has been extensively discussed. Tifft claims that the
galaxies in the Coma Cluster fall into redshift-magnitude
"bands", i. e. , that the excess redshifts (about the cluster
mean) are correlated with apparent magnitude. This
argument is an example of <u>ex</u> <u>post</u> <u>facto</u> statistics; i. e. ,
looking for unspecified patterns in limited amounts of
data (cf. Section III). The reader can verify for himself
that the patterns claimed by Tifft are less prominent when
the data corresponding to the part of the cluster known as
Coma β[8,44] is included. It will be interesting to see
whether or not the bands disappear entirely when the
results are available from current observational programs
to determine more redshifts.

REDSHIFTS OF COMPANION GALAXIES. Arp[45] has
claimed that the dominant, larger galaxy in three groups

of galaxies has a smaller redshift (by $\sim$ 70 km/sec) than its companion galaxies and has concluded that this is strong evidence for non-velocity interpretations of red-shifts. His claims are open to criticism on several grounds.

Arp says that he limits himself to the three groups (M 31, M 81, and NGC 5128) because these are the only cases in which the companions "are unquestionably ac-cepted as belonging to the larger galaxies." On the other hand, Lewis[46] argues that there are other associations that are at least as well established and in which the effect claimed by Arp is not statistically significant. The claim by Arp has no predictive power if the statement is limited to the cases in which the alleged relation was discovered.

Simkin,[47] in a paper that is reprinted in the present volume, has drawn attention to the fact that systematic errors may arise between the measured redshifts of companions and dominant galaxies because different spectral lines are used to measure the redshifts (for earlier references see the papers by Page[11] and de-Vaucouleurs and de Vaucouleurs[48]). She suggests that the effects Arp attributes to non-velocity redshifts are perhaps due to nothing more than "exotic errors of measurement."

It seems to me that a study as complete as possible of one group on Arp's list of three would be very informa-tive and might resolve the points in dispute. Such a study should examine all lines of evidence for membership of each alleged companion in the group and determine if any other nearby galaxies might be companions (for appropriate lines of investigation, see references 31-33). Whenever possible, both radio and optical redshifts should be obtained. The effect of using different optical lines for determining redshifts should be evaluated[47] by using a spectograph which responds linearly to intensities, and the uncertainties associated with the choice of lines and their method of measurement should be evaluated explicitly for the chosen group.

III-6. NGC 7741 AND ITS COMPANION STARS

A posteriori arguments about special cases can be very misleading. I give below a fictional report (not intended to be realistic) by an imaginary Dr. X illustrating how such reasoning can lead to invalid conclusions. The example concerns the apparent relation between the spiral galaxy NGC 7741 and two bright stars in our own galaxy that are seen in projection near the tip of one of the spiral arms in NGC 7741.*

REPORT OF DR. X. "Figure IX shows a 30-minute photograph of NGC 7741 and its companion stars taken by Hubble[29] on October 30, 1950 with the Hale 200" telescope

---

* I am grateful to my wife for bringing this example to my attention.

FIGURE IX

on a 103a-0 plate.   This photograph provides one of the
most powerful pieces of direct evidence for the ejection
hypothesis; i. e. , the hypothesis that stars in our own
Galaxy are the result of explosions in what have usually
been called "other galaxies."   The most clearly defined
spiral arm of NGC 7741 points exactly to the bright star
at its tip.   This exact alignment of bright star with an
extended, almost straight spiral arm is obviously not the
result of chance.   This one case can be taken as a definite
proof of the existence in nature of discordant redshifts;
the redshift of NGC 7741 is 729 km/sec while the redshift
of its bright stellar companion is practically zero.   Clearly
we can say that if the companion star, which has a normal
absorption-line spectrum, is not an accidental projection,
then it must represent the result of a violent process which
is associated with the large non-velocity redshift of
NGC 7741.   It may have something to do with the possibility
that barred spirals can explosively fragment only along
certain preferred directions.   It may also be related to
the formation of the so-called quasars; i. e. , stars which
have abnormally large redshifts.

The vital question to be answered is: Could the
bright star possibly be a foreground object?   The answer
seems to be clearly "no" on a whole series of counts.
First, there is the very disturbed nature of the parent
spiral galaxy, which has a highly unusual morphological
configuration.   Not only does the extended straight spiral

arm point exactly to the bright star, but also the southern continuation of the same spiral arm can be traced back to the bright companion. Moreover, there is a faint luminous filament, part of another spiral arm, which is directed at a fainter star just south of the bright star. Note that the center of the bar of NGC 7741 has a faint bulge which almost certainly reflects the strong tidal interaction of the companion star on the spiral galaxy. If the companion stars are not in some way responsible for the peculiarities, then what could be? Second, the companion stars of NGC 7741 form a close interacting double, which is characteristic of many examples of discordant redshifts. Third, and most conclusively, the overall shape of the spiral galaxy is in the form of a distorted z (the symbol by which redshift is universally denoted)! The precise z-shape is most clearly seen on the original blue-sensitive photograph that is contained in the archives of Hale Observatory but definite indication of its existence may also be inferred from the reproduction of Fig. IX. I would guess roughly that not one galaxy in 1000 looks like this particular spiral. I do not see how anyone can doubt the significance of this exceptional spiral galaxy with two obviously related stellar companions that have a much different redshift than the parent galaxy!"

IV. QUASARS

IV-1. EVIDENCE FOR COSMOLOGICAL HYPOTHESIS

The most controversial subject in extragalactic astronomy in recent years has been the significance of the large redshifts detected in quasar spectra. I think the reason is that quasars were discovered much more recently than galaxies; there has been only a short time (~ 10 years) since their discovery for the evidence to accumulate regarding the origin of quasar redshifts.

The standard arguments supporting the conventional interpretation of the redshifts of quasars as distance indicators have been summarized in excellent articles by Sandage[50,51] and by Schmidt.[52,53] The paper by M. Schmidt entitled, "Quasars and Cosmology," the Halley Lecture for 1971, is reprinted in the present volume.

GENERAL ARGUMENTS. I will summarize some general considerations which seem to me to support the conventional interpretation.

The so-called Cosmological Hypothesis for the origin of the redshifts is very specific and makes at least three negative predictions that have been verified. These predictions include[53] the absence of blueshifts, the absence of proper motions, and the lack of demonstrable physical associations with smaller redshift objects. The fact that the Cosmological Hypothesis, which assumes a specific (Doppler) mechanism for the origin of the redshifts, has survived a decade of vigorous attack also suggests that it may be correct.

Many authors have pointed out the striking similarity between various compact galaxies and quasars and have stressed the continuity in physical properties between extended galaxies, compact galaxies, and quasars that results when observations are interpreted using redshifts as distance indicators for all the systems. Bahcall has even suggested[54] that quasars may not be a physically distinct class of objects and that future improved observations will show that all of the smaller redshift quasars will be resolvable optically into a galaxy plus a bright stellar nucleus. Important (but still partial) confirmation of this hypothesis has recently been provided by Kristian[55] and by Sandage.[56] If the general point of view that quasars are only a special kind or stage of galaxies is correct, the arguments presented in Section II supporting the conventional interpretation of galaxy redshifts may also support the

IV.   QUASARS

IV-1. EVIDENCE FOR COSMOLOGICAL HYPOTHESIS

The most controversial subject in extragalactic astronomy in recent years has been the significance of the large redshifts detected in quasar spectra. I think the reason is that quasars were discovered much more recently than galaxies; there has been only a short time (~ 10 years) since their discovery for the evidence to accumulate regarding the origin of quasar redshifts.

The standard arguments supporting the conventional interpretation of the redshifts of quasars as distance indicators have been summarized in excellent articles by Sandage[50, 51] and by Schmidt.[52, 53] The paper by M. Schmidt entitled, "Quasars and Cosmology," the Halley Lecture for 1971, is reprinted in the present volume.

GENERAL ARGUMENTS. I will summarize some general considerations which seem to me to support the conventional interpretation.

The so-called Cosmological Hypothesis for the origin of the redshifts is very specific and makes at least three negative predictions that have been verified. These predictions include[53] the absence of blueshifts, the absence of proper motions, and the lack of demonstrable physical associations with smaller redshift objects. The fact that the Cosmological Hypothesis, which assumes a specific (Doppler) mechanism for the origin of the redshifts, has survived a decade of vigorous attack also suggests that it may be correct.

Many authors have pointed out the striking similarity between various compact galaxies and quasars and have stressed the continuity in physical properties between extended galaxies, compact galaxies, and quasars that results when observations are interpreted using redshifts as distance indicators for all the systems. Bahcall has even suggested[54] that quasars may not be a physically distinct class of objects and that future improved observations will show that all of the smaller redshift quasars will be resolvable optically into a galaxy plus a bright stellar nucleus. Important (but still partial) confirmation of this hypothesis has recently been provided by Kristian[55] and by Sandage.[56] If the general point of view that quasars are only a special kind or stage of galaxies is correct, the arguments presented in Section II supporting the conventional interpretation of galaxy redshifts may also support the conventional interpretation of quasar redshifts. A particularly significant step in this connection has recently been provided by Sandage[57] who has shown that the compact N-galaxies (possibly the intermediate stage between quasars and Seyfert galaxies) obey the classical Hubble relation between redshift and apparent magnitude.

Another general consideration is that there are a plethora of theoretical explanations for quasars that are based on the known laws of physics and the usual interpretation of redshifts (these theories include galaxies forming or collapsing; black holes at the centers of galaxies accreting matter; many small pulsars; one big pulsar; etc.). I do not know of a theoretical model that both provides an alternative (non-Doppler) explanation of quasar redshifts and also is consistent with their other observational properties.

Finally, I should caution against a kind of argument that has been used by some opponents of the Cosmological Hypothesis. The argument may be summarized as follows: "A distinguished astrophysicist has been unable to derive, using the Cosmological Hypothesis, an explanation for some phenomenon observed in quasars; for example, rapid time variation or total energy output. Hence, the Cosmological Hypothesis is wrong." A more likely conclusion from this state of affairs is that the phenomenon is very complicated.

POSITIVE EVIDENCE. We discuss two lines of current
investigation that provide observational evidence supporting the Cosmological Hypothesis for quasars. The first
type of investigation concerns the association of quasars
with groups and clusters of galaxies.[58-63] The general
idea has been to search in the vicinity of relatively small-
redshift quasars for groups or clusters of galaxies that
have the same redshift. The paper by Gunn[61] is one of
the most important papers in this series of studies and is
reprinted in the present volume. More recently, N. Bahcall,
J. Bahcall, and M. Schmidt[64] have initiated a new program
to search optically for previously unknown quasars within
the areas of known clusters of galaxies. This method offers
the unbiased possibility of finding either evidence supporting the Cosmological Hypothesis (by discovering a small-
redshift quasar in a cluster of galaxies with the same
redshift) or evidence in conflict with the Cosmological
Hypothesis (by finding an excess of large-redshift quasars
in clusters of small redshift).

There is a second line of investigation that is analogous to the Hubble diagram for the brightest galaxies: the
Hubble diagram for the brightest quasars. Since there are
no known clusters of quasars, one selects the brightest
quasars in a manner that is most analogous to the procedure
used in the galaxy Hubble diagram, i.e., one selects the
intrinsically brightest quasar within each of several redshift groups that contain equal numbers of catalogued

quasars. Bahcall and Hills have described,[65] in a paper that is reprinted in the present volume, a systematic method for carrying out this procedure and for correcting for the selection effects arising from the fact that intrinsically faint quasars can be seen only at relatively small distances. They have also estimated the effect of the dispersion in intrinsic quasar luminosities. A recent summary[66] exhibits explicitly the close analogy of the steps followed in this work with the steps followed by Sandage in constructing the Hubble diagram of the brightest galaxies. The results of Bahcall and Hills are shown in Fig. X. The Hubble diagram for the brightest quasars is in excellent agreement with what is expected on the basis of the Cosmological Hypothesis (the theoretical line in Fig. X). This relation should be tested further, just as was done for the Hubble diagram for galaxies, in the next few years when many more quasar redshifts and magnitudes will become available.

Observations of the angular separation of the radio components of quasars that are double sources can be interpreted consistent with the Cosmological Hypothesis but the interpretation depends strongly on the theoretical model used to describe the evolution of an individual source.[67, 68] The situation is similar with respect to the observed Faraday depolarization of quasars.[69]

COMPROMISE SOLUTIONS. A number of authors have

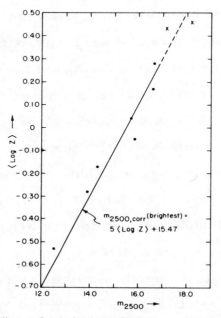

FIG. 1.—The Hubble diagram for the brightest of 105 quasars. The circles represent the intrinsically brightest quasars (for $q_0 = +1$) in successive redshift bins containing 15 objects each; the brightest objects are, respectively, 3C 273, 3C 232, PKS 1354+19, PKS 1252+11, PKS 1127−14, 3C 298, and TON 1530. The two crosses represent the large-redshift quasars PHL 957 ($z = 2.69$) and PHL 938 ($z = 2.88$); they were not included in the analysis. The parameters used are $m_v(\max) = 19.5$ mag, bin size = 15, and log $F_{\min} = 22.4$ clusters. The plotted points include K- and B-corrections. The straight line in figure 1 corresponds to objects ~ 5 mag brighter than the brightest cluster galaxies (i.e., log $F = 24.2$).

FIGURE   X

suggested that there are two kinds of quasars. For the first type, the conventional interpretation of redshifts is assumed valid and these quasars are said to be much brighter than normal galaxies. The second type of quasar is said to be at the same distance as, but much fainter than, the normal galaxies with which it is supposed to be associated. This proposal seems to me to yield an unpleasant theoretical solution. It would require separate theories to explain the "local" and the "distant" quasars and would create a <u>new</u> theoretical problem: why do quasars with apparently similar observed properties have entirely different luminosities and origins of their redshifts?

IV-2. QUASARS AND GALAXIES

Arp and his associates[70, 71] have argued that quasar redshifts are not indicative of distance because some quasars are observed to be in configurations that suggest physical association with selected samples of galaxies. The apparent spatial correlations were discovered by searching <u>a posteriori</u> for suggestive relations among quasars and galaxies. It is not surprising that some unlikely configurations were found since improbable relations are observed whenever large numbers of objects appear on a two-dimensional photograph (whether the objects be grains of salt on a table or people in Times Square). Moreover, the statistical significance of Arp's claims have been tested by searching for similar relations among catalogues of unphysical objects whose positions were produced by a computer. In a paper that is reprinted in this volume, van der Laan and Bash[72] showed that equally large statistical fluctuations could be found among randomly positioned objects. This question is still being debated since, like many questions regarding post facto statistics, there is considerable freedom for interpretation.

The most famous example of an alleged association between bright galaxies and quasars was adduced by E. M. Burbidge, G. R. Burbidge, P. Solomon, and P. Strittmater.[73] These authors pointed out that four out of a sample of 47 radio-emitting quasars are apparently closer (in two dimensions) to objects in the Reference Catalog of Bright Galaxies[27] than would have been expected. Burbidge et al. argued[73] that the four quasars are so close to their apparent neighbors that the probability that this was a chance occurrence was less than 5×10^{-3} if the typical region of interest around a quasar is about 15 arc minutes in radius. Of course, their probability estimate depends upon the chosen value of $r^2_{suggestive}$ (cf. Section II); had they chosen a typical value of $r_{suggestive} = 4^{\circ}$, as might have been inferred a priori from Arp's[70, 71] earlier claims, the number of apparent pairs would not have appeared surprising.[*]

---

[*] Arp[22] has claimed that the results of Burbidge et al.[73] are a confirmation of his earlier studies. This is not correct; there would not have been a statistically significant effect if Burbidge et al.[73] had chosen the value of $r_{suggestive}$ ($\sim 4^{\circ}$) consistent with Arp's prior claims of correlations.

Bahcall, McKee, and Bahcall,[74] in a paper reprinted in the present volume, tested with other catalogs of quasars the claim[73] that some quasars are closer to bright galaxies than would be expected by chance. The Bahcalls and McKee made use of a much larger sample of 222 radio-emitting quasars (from four different radio catalogs) and also included 166 optically-discovered quasars (from the Bologna catalogue); the positions of the quasars were compared with those of bright galaxies in the same catalog as was used by Burbidge, Burbidge, Solomon, and Strittmater. The results with the larger sample were found to be consistent with a completely random correlation of the positions of bright galaxies and quasars.[74] Similar results were obtained by Hazard and Sanitt,[75] in a paper that is also reprinted in the present volume; these latter authors extended the tests to include fainter galaxies and other quasars.

The results obtained with the larger samples[74, 75] suggest that a statistical fluctuation in the smaller sample was responsible for the originally claimed[73] association of quasars and bright galaxies.

V. SUMMARY

We reviewed in Section II the six successful tests of
the usual interpretation of the redshifts of galaxies that
have been carried out; further indicated tests of the conventional interpretation are possible and important. In
Sections III and IV, we discussed alleged examples of
discordant redshifts and showed how one can estimate
probabilities that the suggestive configurations are accidental superpositions. We found unreliable the a posteriori
statistical arguments that are frequently used to try to prove
physical association between objects that appear close together on the sky. We did conclude that it is frequently
possible to test observationally the claimed examples
either by critically examining the data or by obtaining
more data. Sometimes it is possible to test the claims by
generalizing the statistical peculiarity so that it applies to
objects not included in the original investigation. I hope
that discussion regarding the redshift controversy will

concentrate in the future, not on special examples (whose significance is hard to evaluate), but on observational tests of well-defined predictions for a statistically significant sample of objects.

REFERENCES

1.   E. Hubble, Proc. Natl. Acad. Sci. U.S. 15, 168 (1929).

2.   M. L. Humason, Proc. Natl. Acad. Sci. U.S. 15, 167 (1929).

3.   M. L. Humason, Astrophys. J. 74, 35 (1931).

4.   E. Hubble and M. L. Humason, Astrophys. J. 74, 43 (1931).

5.   H. P. Robertson, Phil. Mag. 5, 385 (1928).

6.   A. Sandage, Astrophys. J. 178, 1 (1972).

7.   M. L. Humason, N. U. Mayall, and A. Sandage, Astron. J. 61, 97 (1956).

8.   H. J. Rood, T. L. Page, E. C. Kintner, and I. R. King, Astrophys. J. 175, 627 (1972).

9.   G. A. Tammann, Astron. and Astrophys. 21, 355 (1972).

10. T. Page, Astrophys. J. 116, 63 (1952).

11. T. Page, Astrophys. J. 159, 791 (1970).

12. A. Sandage, Astrophys. J. 173, 485 (1972).

13. E. Hubble, Astrophys. J. 64, 321 (1926).

14. N. A. Bahcall, Astrophys. J. 180, 699 (1973).

15. N. A. Bahcall, Astrophys. J. 183 (August 1, 1973).

16. M. Roberts, "The Gaseous Content of Galaxies, in External Galaxies and Quasi-Stellar Sources, IAU Symposium 44 (ed. by D. S. Evans, Springer-Verlag New York, Inc./New York, 1972).

17. D. W. Weedman, Astrophys. J. $\underline{161}$, L113 (1970).

18. R. Lynds and A. G. Millikan, Astrophys. J. $\underline{176}$, L5 (1972).

19. H. C. Ford and H. W. Epps, Astrophys. Letters $\underline{12}$, 139 (1972).

20. T. F. Adams and R. J. Weymann, Astrophys. Letters $\underline{12}$, 143 (1972).

21. H. Arp, Astrophys. J. $\underline{161}$, L113 (1971).

22. H. Arp, invited talk at the 139th meeting of the American Association for the Advancement of Science, Washington, D. C. , on the topic "Evidence for Discordant Redshifts," December (1972).

23. H. Arp, Astrophys. Letters $\underline{7}$, 221 (1971).

24. M. F. Barnothy and J. M. Barnothy, Bull. Am. Astron. Soc. $\underline{4}$, 239 (1972).

25. H. Arp, Astrophys. J. Supp. $\underline{14}$, 1 (1966).

26. B. A. Vorontsov-Velyaminov, Atlas and Catalogue of Interacting Galaxies, Vol. 1 (Sternberg Institute, Moscow State University, Moscow, 1959).

27. G. de Vaucouleurs and A. de Vaucouleurs, Reference Catalogue of Bright Galaxies (University of Texas Press, Austin, 1964).

28. C. D. Shane and C. A. Wirtanen, Publications of the Lich Observatory, Vol. XXII, Part 1 (1967).

29. E. M. Burbidge and G. R. Burbidge, Astrophys. J. $\underline{134}$, 244 (1961).

30. E. M. Burbidge and G. R. Burbidge, Astron. J. $\underline{66}$, 541 (1961).

31. G. A. Tammann, Astrophys. Letters 7, 111 (1970).

32. R. J. Allen, Astron. and Astrophys. 7, 330 (1970).

33. C. R. Lynds, in External Galaxies and Quasi-Stellar Sources, p. 376 (ed. by D. S. Evans, IAU Symposium No. 44, Springer-Verlag New York, Inc./New York, 1972).

34. H. C. Arp, in External Galaxies and Quasi-Stellar Sources, p. 380 (ed. by D. S. Evans, Springer-Verlag New York, Inc./New York, 1972).

35. S. van den Bergh, Astron. J. 66, 549 (1961).

36. F. Zwicky and C. T. Kowal, Catalogue of Galaxies and of Clusters of Galaxies, Vol. 6, p. 128 (California Institute of Technology, Pasadena, 1968).

37. H. Arp and J. Kormandy, Astrophys. J. 178, L101 (1972).

38. H. Shapley and A. Ames, Harvard Circular No. 294 (1926).

39. G. de Vaucouleurs, Astrophys. J. Supp. 6, 213 (1961).

40. G. de Vaucouleurs and A. de Vaucouleurs, Astron. J. 68, 96 (1963).

41. E. Holmberg, Astron. J. 66, 620 (1961).

42. C. T. Kowal, Proc. Astr. Soc. of Pacific 81, 608 (1969).

43. G. de Vaucouleurs, Nature 236, 112 (1972).

44. W. G. Tifft, Astrophys. J. 175, 613 (1972).

45. H. Arp, Nature 225, 1033 (1970).

46. B. M. Lewis, Nature Phys. Sci. 230, 13 (1971).

47. S. M. Simkim, Nature 239, 43 (1972).

48.   G. de Vaucouleurs and A. de Vaucouleurs, Nature
      236, 166 (1972).

49.   A. Sandage, The Hubble Atlas of Galaxies (Carnegie
      Institution of Washington, Washington, D. C., 1961).

50.   A. Sandage, IAU Highlights of Astronomy, p. 33
      (Reidel Publishing Co., Dordrecht, Holland, 1967).

51.   A. Sandage, Quart. J. Roy. Astr. Soc. 13, 282 (1971).

52.   M. Schmidt, The Observatory, 91, 209 (1971).

53.   M. Schmidt, Quart. J. Roy. Astr. Soc. 13, 297 (1972).

54.   J. N. Bahcall, Astron. J. 76, 283 (1971).

55.   J. Kristian, Astrophys. J. 179, L61 (1973).

56.   A. Sandage, Astrophys. J. 178, 25 (1972).

57.   A. Sandage, Astrophys. J. 180, 687 (1973).

58.   J. N. Bahcall, M. Schmidt, and J. E. Gunn, Astro-
      phys. J. 157, L77 (1969).

59.   N. A. Bahcall and J. N. Bahcall, Proc. Astr. Soc.
      of Pacific 82, 1276 (1970).

60.   J. N. Bahcall and N. A. Bahcall, Proc. Astr. Soc.
      of Pacific 82, 721 (1970).

61.   J. E. Gunn, Astrophys. J. 164, L113 (1971).

62.   L. B. Robinson and E. J. Wampler, Astrophys. J.
      171, L83 (1972).

63.   A. Oemler, Jr., J. E. Gunn, and J. B. Oke, Astro-
      phys. J. 176, L47 (1972).

64.   N. A. Bahcall, J. N. Bahcall, and M. Schmidt,
      Astrophys. J. 183 (August 1, 1973).

65.   J. N. Bahcall and R. E. Hills, Astrophys. J. 179,
      699 (1973).

66. R. E. Hills and J. N. Bahcall, Proceedings of the 6th Texas Symposium on Relativistic Astrophysics, New York, December 1972 (to be published).

67. G. K. Miley, Monthly Notices Roy. Astr. Soc. 152, 477 (1971).

68. G. M. Richter, Astrophys. Letters 13, 63 (1973).

69. R. G. Strom, Nature Phys. Sci. 239, 19 (1972).

70. H. Arp, Astrophys. J. 148, 321 (1967).

71. H. Arp, Astron. J. 75, 1 (1970).

72. H. van der Laan and F. N. Bash, Astrophys. J. 152, 621 (1968).

73. E. M. Burbidge, G. R. Burbidge, P. Solomon, and P. Strittmater, Astrophys. J. 170, 233 (1971).

74. J. N. Bahcall, C. F. McKee, and N. A. Bahcall, Astrophys. Letters 10, 147 (1972).

75. C. Hazard and N. Sanitt, Astrophys. Letters 11, 77 (1972).

RESPONSE TO DR. BAHCALL'S
PRESENTATION

BY

HALTON ARP

RESPONSE TO DR. BAHCALL'S
PRESENTATION

In replying to the points raised by Dr. Bahcall, I would like to make four major comments:

First, it may be significant that the one type of galaxy for which I find no evidence of discordant redshifts is the brightest giant E or cD galaxy in rich clusters. Since these are just the kind of galaxies on which the Hubble relation is conventionally based, I therefore have no evidence contradicting the usual interpretation of the expanding universe. But, by this same token, any arguments that are made about these kinds of objects proving a redshift-distance relation do not apply to the special galaxies that I am considering. I hate to discount the whole first part of Dr. Bahcall's talk, but I suppose that this is the purpose of such debates—to make certain we are all talking about the same thing.

Not all galaxies are giant E's, however, and it is just such objects as quasars, compact galaxies, peculiar galaxies, and some spirals that do not obey a low scatter, redshift-apparent magnitude relation. It is for such objects that I find a great deal of evidence for nonvelocity redshifts.

This brings me to the second point: Is there a pre-defined criterion which will unbiasedly select objects to test for discordant redshifts?

Approaching from a broad view, we should remark that if a phenomenon is to be discovered that violates some current underlying assumptions, then we can only learn about this new phenomenon inductively. In other words, no amount of deduction from accepted principles will illucidate the nature of a new effect. Now, the first systematic observational evidence indicating the existence of nonvelocity redshifts was published by Arp in a short note in *Science* in 1966. The conclusion of that article was that high-redshift quasi-stellar radio sources and small radio galaxies of relatively high redshift were associated with relatively low-redshift nearby galaxies.

From the subsequent evidence it was inductively concluded that whenever higher-redshift objects were associated with large, relatively low-redshift galaxies these companions tended to be fainter, of higher surface brightness, and dynamically or evolutionarily younger. This conclusion certainly encompassed the original evidence and all the subsequent evidence. In other words, a pre-specified criterion can be stated: "Smaller, younger (i.e., not giant E) galaxies are preferentially found to be physically associated with larger, lower-redshift galaxies." (The probability of a galaxy being a radio emitter is also taken to be correlated with young, active stages in small galaxies.) Looking back, we can see that every case of discordant redshift was just another instance of testing these same criteria with different sets of data.

The third remark to be made is that all these conclusions are consistent with other observational (but not

always other theoretical) results. For example, the conclusion that the quasars and small radio galaxies were e-
jected from the larger, central galaxies is consonant with
the conventional belief that blank field radio sources are
ejected in pairs, oppositely from central galaxies. Disturbed central galaxies can be taken as evidence for this
recent cataclysmic event. (It should be noted here that recent theories of spiral galaxies hold that spiral structure
is the result of recent ejection activity.) Consider also the
question of small compact or peculiar galaxies in the vicinity of larger galaxies. From where could they have originated? It is difficult enough under even primeval conditions to condense a galaxy out of an intergalactic medium.
But from where could these have come so recently? It is
certainly plausible to have these originate as recent ejecta
or fission products from the neighboring larger galaxies
since there is abundant observational evidence for ejection
of luminous material from galaxies. The one thing that is
new is that this recently-ejected matter seems to have some
component of intrinsic redshift. If these results were due
to accidentally associating background galaxies with nearby
galaxies, then we would expect the smaller associated galaxies to be mostly morphologically regular normal-surface-
brightness objects, which is, of course, exactly what they
are not. For Dr. Bahcall to show a photograph of a normal
galaxy with a star projected near the end of one arm misses
the point quite badly. The point is that the associations I
have shown, like NGC 7603, involve peculiar galaxies with
very peculiar filaments leading to unusual companions—
events that have negligible chance of occurring by accident

Fourth, I would comment that the results on dis-

cordant redshifts most definitely do predict phenomena that
are later observed. For example, the close associations
of quasars with bright galaxies discussed by Burbidge, Burbidge, Strittmatter, and Solomon predicted that some radio
sources with low initial accuracy of position would have been
mistakenly identified with galaxies instead of a nearby quasar. In fact, it later turned out that the radio source 3C 455
was really a high-redshift quasar only 23 arc sec from the
originally identified galaxy. There may be more of such
objects. The conclusion that satellite galaxies were ejected
from large nearby galaxies actually led Holmberg to investigate nearby spirals. He found companions actually associated preferentially along their minor axes. As I predicted,
the great majority of these are now turning out to be of considerably higher redshift. In the case of Stephan's Quintet,
it was unusual as it stood because the discordant redshift
was the low redshift compared to the four high redshifts in
comparable size galaxies. The prediction was that the high
redshifts should be associated with some nearby large galaxy
of low redshift. Now several lines of evidence indicate the
Quintet is indeed associated with the large low-redshift
NGC 7331. Moreover, these results predicted that similar
multiple interacting systems in fact do fall very close to
large galaxies, most of which are much lower redshift.
These results in turn predicted that interacting doubles of
higher redshift would be more numerous around bright
spiral galaxies. Preliminary results also indicate this to
be true. The small systematic redshift differences between
accepted companions like M32 and M33 and the dominant
M31 would never have been looked for if they had not been
predicted by the discordant redshift results which we are
discussing. The anisotropy between north and south

galactic hemispheres in the properties of quasars, which was discussed in the main text, would be predicted as a consequence of the contribution of some quasars from the local group of galaxies and some from the local supercluster of galaxies.

Some additional miscellaneous comments are that all sides agree that there is a continuity of observable characteristics from quasars through normal galaxies via the compact N and peculiar galaxies. But the question being debated is whether this is a continuity of increasing or decreasing luminosity. Another comment is that the brightest apparent-magnitude quasar in each redshift interval may appear to form a line of the correct slope to be commensurate with a redshift-distance relation. But work by Burbidge *et al.* shows this is not true if the second or third brightest quasar is selected, nor is it true if the very singular quasar 3C 273 is omitted.

I should like to note that Dr. Bahcall says he has estimated what the probabilities are that the associations I have discussed could be due to chance. Obviously there is always some finite chance that any one association could be accidental. If each association is considered separately, each could be dismissed on these grounds. But what is the chance that two or three or a half dozen could be accidental? Since these cases are independent, their improbabilities multiply, yielding in the end an extraordinarily low figure for the probability of chance occurrence. What value of probability would Dr. Bahcall accept as a demonstration establishing the case? Finally, I would like to ask him, seriously, if discordant redshifts do exist, what he would consider as an observation or a set of observations that he would accept as proof.

REPRINTS OF PAPERS
SELECTED BY
JOHN N. BAHCALL

INTRODUCTION TO REPRINTS

I have chosen 15 short articles to be reprinted in this volume. Four of the articles (by Hubble, Humason, Sandage, and Schmidt, respectively) were chosen for their great historical interest; they represent major landmarks in extragalactic astronomy. Most of the other articles were chosen because they contain important responses to questions in direct dispute in the Redshift Controversy and, in addition, describe techniques that can be applied to other problems in the controversy. Two of the papers (by Gunn and by Bahcall and Hills, respectively) represent the first stages of programs to extend to quasars the knowledge and techniques acquired in the study of galaxies.

John Bahcall

A RELATION BETWEEN DISTANCE AND RADIAL VELOCITY AMONG EXTRA-GALACTIC NEBULAE

By Edwin Hubble

Mount Wilson Observatory, Carnegie Institution of Washington

Communicated January 17, 1929

Determinations of the motion of the sun with respect to the extra-galactic nebulae have involved a K term of several hundred kilometers which appears to be variable. Explanations of this paradox have been sought in a correlation between apparent radial velocities and distances, but so far the results have not been convincing. The present paper is a re-examination of the question, based on only those nebular distances which are believed to be fairly reliable.

Distances of extra-galactic nebulae depend ultimately upon the application of absolute-luminosity criteria to involved stars whose types can be recognized. These include, among others, Cepheid variables, novae, and blue stars involved in emission nebulosity. Numerical values depend upon the zero point of the period-luminosity relation among Cepheids, the other criteria merely check the order of the distances. This method is restricted to the few nebulae which are well resolved by existing instruments. A study of these nebulae, together with those in which any stars at all can be recognized, indicates the probability of an approximately uniform upper limit to the absolute luminosity of stars, in the late-type spirals and irregular nebulae at least, of the order of M (photographic) = -6.3.[1] The apparent luminosities of the brightest stars in such nebulae are thus criteria which, although rough and to be applied with caution,

furnish reasonable estimates of the distances of all extra-galactic systems
in which even a few stars can be detected.

TABLE 1

NEBULAE WHOSE DISTANCES HAVE BEEN ESTIMATED FROM STARS INVOLVED OR FROM
MEAN LUMINOSITIES IN A CLUSTER

OBJECT	m_s	r	v	m_t	M_t
S. Mag.	..	0.032	+ 170	1.5	−16.0
L. Mag.	..	0.034	+ 290	0.5	17.2
N. G. C. 6822	..	0.214	− 130	9.0	12.7
598	..	0.263	− 70	7.0	15.1
221	..	0.275	− 185	8.8	13.4
224	..	0.275	− 220	5.0	17.2
5457	17.0	0.45	+ 200	9.9	13.3
4736	17.3	0.5	+ 290	8.4	15.1
5194	17.3	0.5	+ 270	7.4	16.1
4449	17.8	0.63	+ 200	9.5	14.5
4214	18.3	0.8	+ 300	11.3	13.2
3031	18.5	0.9	− 30	8.3	16.4
3627	18.5	0.9	+ 650	9.1	15.7
4826	18.5	0.9	+ 150	9.0	15.7
5236	18.5	0.9	+ 500	10.4	14.4
1068	18.7	1.0	+ 920	9.1	15.9
5055	19.0	1.1	+ 450	9.6	15.6
7331	19.0	1.1	+ 500	10.4	14.8
4258	19.5	1.4	+ 500	8.7	17.0
4151	20.0	1.7	+ 960	12.0	14.2
4382	..	2.0	+ 500	10.0	16.5
4472	..	2.0	+ 850	8.8	17.7
4486	..	2.0	+ 800	9.7	16.8
4649	..	2.0	+1090	9.5	17.0
Mean					−15.5

m_s = photographic magnitude of brightest stars involved.

r = distance in units of 10^6 parsecs. The first two are Shapley's values.

v = measured velocities in km./sec. N. G. C. 6822, 221, 224 and 5457 are recent
     determinations by Humason.

m_t = Holetschek's visual magnitude as corrected by Hopmann. The first three
     objects were not measured by Holetschek, and the values of m_t represent
     estimates by the author based upon such data as are available.

M_t = total visual absolute magnitude computed from m_t and r.

Finally, the nebulae themselves appear to be of a definite order of
absolute luminosity, exhibiting a range of four or five magnitudes about
an average value M (visual) = −15.2.[1] The application of this statistical
average to individual cases can rarely be used to advantage, but where
considerable numbers are involved, and especially in the various clusters
of nebulae, mean apparent luminosities of the nebulae themselves offer
reliable estimates of the mean distances.

Radial velocities of 46 extra-galactic nebulae are now available, but individual distances are estimated for only 24. For one other, N. G. C. 3521, an estimate could probably be made, but no photographs are available at Mount Wilson. The data are given in table 1. The first seven distances are the most reliable, depending, except for M 32 the companion of M 31, upon extensive investigations of many stars involved. The next thirteen distances, depending upon the criterion of a uniform upper limit of stellar luminosity, are subject to considerable probable errors but are believed to be the most reasonable values at present available. The last four objects appear to be in the Virgo Cluster. The distance assigned to the cluster, 2×10^6 parsecs, is derived from the distribution of nebular luminosities, together with luminosities of stars in some of the later-type spirals, and differs somewhat from the Harvard estimate of ten million light years.[2]

The data in the table indicate a linear correlation between distances and velocities, whether the latter are used directly or corrected for solar motion, according to the older solutions. This suggests a new solution for the solar motion in which the distances are introduced as coefficients of the K term, i. e., the velocities are assumed to vary directly with the distances, and hence K represents the velocity at unit distance due to this effect. The equations of condition then take the form

$$rK + X \cos \alpha \ \cos \delta + Y \sin \alpha \cos \delta + Z \sin \delta = v.$$

Two solutions have been made, one using the 24 nebulae individually, the other combining them into 9 groups according to proximity in direction and in distance. The results are

	24 OBJECTS	9 GROUPS
X	$-\ 65 \pm 50$	$+\ \ 3 \pm 70$
Y	$+226 \pm 95$	$+230 \pm 120$
Z	-195 ± 40	-133 ± 70
K	$+465 \pm 50$	$+513 \pm \ \ 60$ km./sec. per 10^6 parsecs.
A	$286°$	$269°$
D	$+\ 40°$	$+\ 33°$
V_0	306 km./sec.	247 km./sec.

For such scanty material, so poorly distributed, the results are fairly definite. Differences between the two solutions are due largely to the four Virgo nebulae, which, being the most distant objects and all sharing the peculiar motion of the cluster, unduly influence the value of K and hence of V_0. New data on more distant objects will be required to reduce the effect of such peculiar motion. Meanwhile round numbers, intermediate between the two solutions, will represent the probable order of the values. For instance, let $A = 277°$, $D = +36°$ (Gal. long. $= 32°$, lat. $= +18°$), $V_0 = 280$ km./sec., $K = +500$ km./sec. per million parsecs. Mr. Strömberg has very kindly checked the general order of these values by independent solutions for different groupings of the data.

A constant term, introduced into the equations, was found to be small and negative. This seems to dispose of the necessity for the old constant K term. Solutions of this sort have been published by Lundmark,[3] who replaced the old K by $k + lr + mr^2$. His favored solution gave $k = 513$, as against the former value of the order of 700, and hence offered little advantage.

TABLE 2

NEBULAE WHOSE DISTANCES ARE ESTIMATED FROM RADIAL VELOCITIES

OBJECT		v	v_s	r	m_t	M_t
N. G. C.	278	+ 650	−110	1.52	12.0	−13.9
	404	− 25	− 65	..	11.1	..
	584	+1800	+ 75	3.45	10.9	16.8
	936	+1300	+115	2.37	11.1	15.7
	1023	+ 300	− 10	0.62	10.2	13.8
	1700	+ 800	+220	1.16	12.5	12.8
	2681	+ 700	− 10	1.42	10.7	15.0
	2683	+ 400	+ 65	0.67	9.9	14.3
	2841	+ 600	− 20	1.24	9.4	16.1
	3034	+ 290	−105	0.79	9.0	15.5
	3115	+ 600	+105	1.00	9.5	15.5
	3368	+ 940	+ 70	1.74	10.0	16.2
	3379	+ 810	+ 65	1.49	9.4	16.4
	3489	+ 600	+ 50	1.10	11.2	14.0
	3521	+ 730	+ 95	1.27	10.1	15.4
	3623	+ 800	+ 35	1.53	9.9	16.0
	4111	+ 800	− 95	1.79	10.1	16.1
	4526	+ 580	− 20	1.20	11.1	14.3
	4565	+1100	− 75	2.35	11.0	15.9
	4594	+1140	+ 25	2.23	9.1	17.6
	5005	+ 900	−130	2.06	11.1	15.5
	5866	+ 650	−215	1.73	11.7	−14.5
Mean					10.5	−15.3

The residuals for the two solutions given above average 150 and 110 km./sec. and should represent the average peculiar motions of the individual nebulae and of the groups, respectively. In order to exhibit the results in a graphical form, the solar motion has been eliminated from the observed velocities and the remainders, the distance terms plus the residuals, have been plotted against the distances. The run of the residuals is about as smooth as can be expected, and in general the form of the solutions appears to be adequate.

The 22 nebulae for which distances are not available can be treated in two ways. First, the mean distance of the group derived from the mean apparent magnitudes can be compared with the mean of the velocities corrected for solar motion. The result, 745 km./sec. for a distance of 1.4×10^6 parsecs, falls between the two previous solutions and indicates a value for K of 530 as against the proposed value, 500 km./sec.

Secondly, the scatter ·of the individual nebulae can be examined by assuming the relation between distances and velocities as previously determined. Distances can then be calculated from the velocities corrected for solar motion, and absolute magnitudes can be derived from the apparent magnitudes. The results are given in table 2 and may be compared with the distribution of absolute magnitudes among the nebulae in table 1, whose distances are derived from other criteria. N. G. C. 404

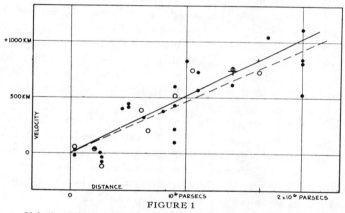

FIGURE 1

Velocity-Distance Relation among Extra-Galactic Nebulae.

Radial velocities, corrected for solar motion, are plotted against distances estimated from involved stars and mean luminosities of nebulae in a cluster. The black discs and full line represent the solution for solar motion using the nebulae individually; the circles and broken line represent the solution combining the nebulae into groups; the cross represents the mean velocity corresponding to the mean distance of 22 nebulae whose distances could not be estimated individually.

can be excluded, since the observed velocity is so small that the peculiar motion must be large in comparison with the distance effect. The object is not necessarily an exception, however, since a distance can be assigned for which the peculiar motion and the absolute magnitude are both within the range previously determined. The two mean magnitudes, -15.3 and -15.5, the ranges, 4.9 and 5.0 mag., and the frequency distributions are closely similar for these two entirely independent sets of data; and even the slight difference in mean magnitudes can be attributed to the selected, very bright, nebulae in the Virgo Cluster. This entirely unforced agreement supports the validity of the velocity-distance relation in a very evident matter.   Finally, it is worth recording that the frequency distribution of absolute magnitudes in the two tables combined is comparable with those found in the various clusters of nebulae.

The results establish a roughly linear relation between velocities and distances among nebulae for which velocities have been previously published, and the relation appears to dominate the distribution of velocities. In order to investigate the matter on a much larger scale, Mr. Humason at Mount Wilson has initiated a program of determining velocities of the most distant nebulae that can be observed with confidence. These, naturally, are the brightest nebulae in clusters of nebulae. The first definite result,[4] $v = + 3779$ km./sec. for N. G. C. 7619, is thoroughly consistent with the present conclusions.   Corrected for the solar motion, this velocity is $+3910$, which, with $K = 500$, corresponds to a distance of 7.8×10^6 parsecs.   Since the apparent magnitude is 11.8, the absolute magnitude at such a distance is -17.65, which is of the right order for the brightest nebulae in a cluster. A preliminary distance, derived independently from the cluster of which this nebula appears to be a member, is of the order of 7×10^6 parsecs.

New data to be expected in the near future may modify the significance of the present investigation or, if confirmatory, will lead to a solution having many times the weight.   For this reason it is thought premature to discuss in detail the obvious consequences of the present results.   For example, if the solar motion with respect to the clusters represents the rotation of the galactic system, this motion could be subtracted from the results for the nebulae and the remainder would represent the motion of the galactic system with respect to the extra-galactic nebulae.

The outstanding feature, however, is the possibility that the velocity-distance relation may represent the de Sitter effect, and hence that numerical data may be introduced into discussions of the general curvature of space.   In the de Sitter cosmology, displacements of the spectra arise from two sources, an apparent slowing down of atomic vibrations and a general tendency of material particles to scatter.   The latter involves an acceleration and hence introduces the element of time.   The relative importance of these two effects should determine the form of the relation between distances and observed velocities; and in this connection it may be emphasized that the linear relation found in the present discussion is a first approximation representing a restricted range in distance.

[1] *Mt. Wilson Contr.*, No. 324; *Astroph. J., Chicago, Ill.*, **64**, 1926 (321).
[2] *Harvard Coll. Obs. Circ.*, 294, 1926.
[3] *Mon. Not. R. Astr. Soc.*, **85**, 1925 (865–894).
[4] These Proceedings, **15**, 1929 (167).

Contributions from the Mount Wilson Observatory, No. 426

Reprinted from the *Astrophysical Journal*, Vol. LXXIV, pp. 35–42, 1931

Printed in the U.S.A.

APPARENT VELOCITY-SHIFTS IN THE SPECTRA OF FAINT NEBULAE

By MILTON L. HUMASON

ABSTRACT

Apparent velocity-shifts of the spectral lines of 46 extra-galactic nebulae have been observed at Mount Wilson, 9 of them by F. G. Pease. Most of these objects are fainter and more distant than any heretofore observed; approximately half of them are cluster nebulae. With one exception these observations confirm Hubble's velocity-distance correlation and provide numerical data which may be used in discussions of the significance of the red-shift of spectral lines.

For all faint nebulae the displacements are large and toward the red, the maximum of the continuous spectrum shifting also.

The largest apparent velocity-shift observed, +19,700 km/sec., is that of the brightest nebula in W. H. Christie's cluster in Leo. Its photographic magnitude is 16.8.

With the exception of N.G.C. 205, classified as F5, the spectral types of all the nebulae having absorption lines fall within the narrow limits G1–G5.

In 1929 Hubble found a relation connecting the velocities and distances of the extra-galactic nebulae for which spectra were then available.[1] The spectra were, in general, those of the nearer and brighter nebulae, and the relation was thus established out to the nearest of the great clusters of nebulae—the Virgo cluster at a distance of the order of two million parsecs. A program of investigation was immediately planned with a view to testing the validity of the relation over as great a range in distance as could be covered with the 100-inch reflector. Spectra of 46 of the fainter nebulae have now been observed. With one exception, possibly the velocity of an isolated object seen in projection on a remote cluster, the new data fully confirm the velocity-distance relation[2] previously formulated and extend the observational range to a distance of about thirty-two million parsecs. This phase of the investigation will be presented in a joint paper by Hubble and myself. The present discussion deals primarily with the spectra themselves.

[1] *Mt. Wilson Communications*, No. 105; *Proceedings of the National Academy of Sciences*, **15**, 168, 1929.

[2] It is not at all certain that the large red-shifts observed in the spectra are to be interpreted as a Doppler effect, but for convenience they are expressed in terms of velocity and referred to as apparent velocities.

MILTON L. HUMASON

THE SPECTROGRAPH

In general the spectra were photographed at the Cassegrain focus of the 100-inch reflector. Nebulae very rarely present uniform surfaces, and for the smaller condensations at least, the large aperture of the 100-inch is more efficient photographically than that of the 60-inch reflector. Moreover, by selecting the condensed elliptical nebulae with highly concentrated semistellar nuclei, it was possible to obtain with the 100-inch practically the full advantage which it offers in the case of stars.[1]

The first spectrograph used was provided with a 24-inch collimator, two prisms of light flint glass, and a 3-inch camera giving a dispersion of 170 A per millimeter at λ 4350. This combination proved too slow for efficient observation of faint nebulae. With prolonged exposures it was possible to record the region $\lambda\lambda$ 4200–5000; in general, however, the G band (λ 4303) was the only feature strong enough to be seen clearly, but, on account of the large displacements encountered, the identification was very uncertain.[2] For this reason one prism was removed, thus giving a dispersion of about 340 A per millimeter at λ 4350. The increased speed made it possible to register the spectra of faint nebulae to the violet of λ 4000 and hence to use the unmistakable H and K lines as a basis for identification of the other features.

The reduction in scale proved so effective that an even smaller dispersion seemed desirable. This was made possible by Dr. W. B. Rayton, of the Bausch and Lomb Optical Company, who designed a spectrograph objective having a ratio of F/0.6. The Rayton lens with two prisms gives a dispersion of about 418 A per millimeter at

[1] Hubble, *Mt. Wilson Contr.*, No. 398; *Astrophysical Journal*, 71, 231, 1930. The steepness of the luminosity gradient in elliptical nebulae is illustrated by Fig. 1a, which gives transparency-curves for photographic images of N.G.C. 3379. The semistellar nucleus is indicated by the sharp maxima shown by the two shortest exposures.

[2] The case of N.G.C. 4884 is an example. The first spectrum obtained with the 60-inch reflector was very weak. The stronger of the two lines measured was assumed to be λ 4383 *Fe*, and gave the velocity +1500 km/sec. announced in *Summary of the Year's Work at Mount Wilson* for 1928. This seemed the most probable identification, since the highest velocity then known was V. M. Slipher's value, +1800 km/sec. for N.G.C. 584. A later spectrogram, showing the H and K lines of calcium, proved the true red-shift to be +6700 km/sec., and the strong line originally measured to be the G band.

PLATE VII

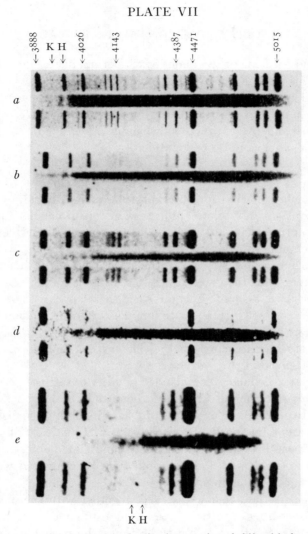

Nebular spectra showing increase in red-shift with decreasing apparent brightness corresponding to increasing distance. Arrows indicate positions of H and K in (*a*) and (*e*). For (*a*) to (*d*) the enlargement is 28 times the scale of the original negatives; for (*e*), photographed with about one-half the dispersion used for the others, 47 times.

a) Sky; normal position of H and K; wave-lengths of several of the helium comparison lines are indicated.
b) N.G.C. 221 (M 32); apparent velocity, −185 km/sec.
c) N.G.C. 385; apparent velocity, +4900 km/sec.
d) N.G.C. 4884; apparent velocity, +6700 km/sec.
e) Brightest nebula in Leo cluster; apparent velocity, +19,700 km/sec.

VELOCITY-SHIFTS OF FAINT NEBULAE

λ 4500, and with one prism, about 875 A per millimeter for the same region. This is much the fastest combination in actual use at Mount Wilson.[1] The definition is excellent, and four lines are generally recognizable in nebular spectra. These lines are H and K, $H\delta$ (λ 4101), and the G band (λ 4303.14). Occasionally $H\gamma$ (λ 4340), λ 4383 Fe, and $H\beta$ (λ 4861) can also be identified.

OBSERVING PROGRAM

The nebulae selected for observation are about equally divided between clusters and isolated objects. Clusters of nebulae offer the great advantage that fairly reliable distances can be derived from the mean luminosities of the many individual members, while the observations can be restricted to the several brightest members of each cluster. Thus the greatest possible distance for a given apparent luminosity is assured. Further, since there appears to be no correlation of red-shift with absolute luminosity among nebulae whose distances are known, the several brightest members of a cluster may safely be assumed to represent the cluster as a whole in respect to line displacement.

The isolated nebulae were included in order to test the possibility of a systematic difference between them and the cluster nebulae, and later to afford data for special problems involving the distances. Since apparent luminosity furnishes only a statistical criterion of distance, it was necessary to observe enough isolated objects to form several groups. Mean velocities could then be compared with the mean distances of the groups.

Table I lists the nebulae observed, together with the measured apparent velocities, spectral types, and estimated uncertainties. The uncertainties are possibly three times the probable errors as formally derived from the few lines measured and are believed to be a fair indication of the reliability. They depend on the scale, the exposure, and the number of lines that could be measured.

Spectra of 9 of the 46 nebulae were photographed by F. G. Pease, who has kindly placed the spectrograms at my disposal for

[1] For a further description of the extraordinarily efficient Rayton lens, together with some account of its performance, see Rayton, *Astrophysical Journal*, **72**, 59, 1930, and Humason, *Mt. Wilson Contr.*, No. 400; *Astrophysical Journal*, **71**, 351, 1930.

MILTON L. HUMASON

TABLE I

APPARENT VELOCITY-SHIFT AND SPECTRAL TYPE

N.G.C.	Apparent Velocity-Shift	Estimated Uncertainty	Spectral Type	Remarks
	km/sec.	km/sec.		
380.....	+ 4400	75	G5	Group in Pisces; not one of the large clusters but a group of about 25 nebulae
383.....	+ 4500	100	G3	
384.....	+ 4500	100	G5	
385.....	+ 4900	100	G5	
1270.....	+ 4800	100	G4	Cluster in Perseus
1273.....	+ 5800	75	G5	
1275.....	+ 5200	25	G+P (pec.)	
1277.....	+ 5200	75	G3	
2562.....	+ 5100	100	G4	Cluster in Cancer
2563.....	+ 4800	100	G4	
*.......	+ 19700	300†	G5	Christie's cluster in Leo
‡.......	+ 11700§	200	G5	Baade's cluster in Ursa Major
4192‖....	+ 1150§	100	G2	Cluster in Virgo
4374‖....	+ 1050	100	G4	
4853‖....	+ 7600§	100	G1	Cluster in Coma Berenices; N.G.C. 4865 may not be a member; velocity of N.G.C. 4884 previously announced as +1500 km/sec.
4860‖....	+ 7900§	75	G3	
4865.....	+ 5000§	75	G3	
4872.....	+ 6900	200†	G3	
4874.....	+ 7000	200†	G4	
4881.....	+ 6900	200†	G3	
4884.....	+ 6700§	75	G3	
4895.....	+ 8500	200†	G4	
I.C. 4045.....	+ 6600	200†	G1	
7611.....	+ 3400	75	G2	Cluster in Pegasus
7617.....	+ 3900	100	G1	
7619.....	+ 3800§	75	G3	
7623.....	+ 3800	125*	G2	
7626‖....	+ 3700	100	G3	
205.....	− 300§	50	F5	Distant companion of Andromeda
2859.....	+ 1500	100	G3	
2950.....	+ 1500	75	G4	
3193.....	+ 1300	100	G3	
3227.....	+ 1150§	30	G+Pd	
3610.....	+ 1850	75	G2	
4051.....	+ 650§	40	G+Pb	
5457.....	+ 300¶	25	G+Pd	

* The nebula observed is the brightest one in the cluster.
† The large uncertainty is due to the dispersion, 875 A per millimeter.
‡ The velocity is from the brightest nebula, Baade No. 24.
§ Velocity previously announced.   ‖ Spectrum obtained by F. G. Pease.
¶ The velocity previously published was +200 km/sec. Recently the nucleus and two emission patches have been observed, making the mean +300 km/sec.

W

PLATE VIII
N

Small region in the Leo cluster; R.A. 10^{h}24^{m}1, Dec. +10°47' (1930). The nebula observed, the brightest in the cluster, photographic magnitude about 16.8, is indicated by the arrow. The bright star in the upper right-hand corner is B.D. +11°2230. Scale, 1 mm = 7".

VELOCITY-SHIFTS OF FAINT NEBULAE

TABLE I—*Continued*

N.G.C.	Apparent Velocity-Shift	Estimated Uncertainty	Spectral Type	Remarks
	km/sec.	km/sec.		
6359.....	+ 3000§	75	G3	
6658.....	+ 4100	75	G3	
6661.....	+ 3900	100	G5	
6702‖....	+ 2250	75	G4	
6703‖....	+ 2000	75	G3	
6710.....	+ 5100	100	G3	
6822.....	− 150§	25	Pd	Emission patch in N.G.C. 6822
6824.....	+ 3200	75	G4	
7217‖....	+ 1050	100	G4	
7242‖....	+ 5000	200	G3	The large uncertainty is on account of a weak plate

measurement and discussion. Velocities of 13 of the nebulae have previously been published, 4 of them by Pease and myself jointly. The table thus includes all recent measures made at Mount Wilson. The velocities previously available, for the most part those made by V. M. Slipher at the Lowell Observatory, but including 7 measures from Mount Wilson and 3 from Lick Observatory, have been collected by G. Strömberg.[1] The two lists together include all velocities published to date.[2]

The present list gives velocities of 24 nebulae in 7 clusters, 4 in a group in Pisces, and 18 isolated objects. The nebulae range from irregular objects (N.G.C. 6822) and late-type spirals (M 101) to the early elliptical nebulae which predominate in the clusters. The largest displacement found is that for the brightest nebula in Christie's cluster in Leo,[3] which has a photographic magnitude of about 16.8. The observation was made with the smallest dispersion used, but the measured displacement of +19,700 km/sec. is believed correct within a few hundred kilometers. The single spectrogram available was exposed 13 hours and is one of the best so far obtained.

Where several velocities have been measured in a single cluster, the range is small compared with the mean in all except the Coma

[1] *Mt. Wilson Contr.*, No. 292, p. 2; *Astrophysical Journal*, **61**, 354, 1925.

[2] Slipher reports spectra for three additional nebulae in the Virgo cluster, but has published no velocities as yet.

[3] This cluster was found by W. H. Christie, on plates taken at Mount Wilson with the 60-inch reflector. The 1930 position is: R.A. 10ʰ24ᵐ1, Dec. +10°47′.

MILTON L. HUMASON

cluster. There, around a mean of +7300 km/sec., 8 nebulae show a range of 1900 km/sec., which is the maximum among all the clusters. A ninth object (N.G.C. 4865) gives +5000 km/sec. On direct photographs this object is in no way distinguishable from the other bright cluster nebulae, but the velocity derived from two spectrograms is the one conspicuously discordant result in the table.

MEASUREMENTS

On the average, the velocities depend on measures of about three lines. These are generally H and K and the G band ($\lambda\,4303$), with occasionally one or more of the lines $H\delta$ ($\lambda\,4101$), $H\gamma$ ($\lambda\,4340$), $\lambda\,4384$ Fe, and $H\beta$ ($\lambda\,4861$), according to the density of the spectrograms in the region of the lines. As an example of the results obtained from spectra having a dispersion of 875 A per millimeter, the individual measurements are listed for the brightest nebula in

	E.M.	M.H
3933 (K).........	+19,890 km/sec.	+19,925 km/sec.
3968 (H).........	19,571	19,708
4101 ($H\delta$)........	19,609	19,615
4303 (G).........	19,778	19,276
4340 ($H\gamma$)........	+19,815	+19,579
Mean........	+19,733	+19,621

Christie's cluster. In Table I the probable uncertainty is given as 300 km/sec., and the apparent velocity entered to the nearest 100 km/sec.

The wave-length of the blend which forms the G band ($\lambda\,4303.14$) was derived from spectra of standard velocity stars having as nearly as possible the same type as the nebulae. Before this wave-length was obtained, the value $\lambda\,4307.91$ was used in reducing measures of N.G.C. 4853, 4860, and 4865, but their velocities have since been corrected by an average of +233 km/sec.

Each spectrogram has been measured twice, once by Miss Elizabeth MacCormack and once by the writer. These duplicate measures are in good agreement. Velocities less than 2000 km/sec. are given to the nearest 50 km/sec., those larger to the nearest 100 km/sec.

Valuable confirmation that the velocity displacements, $\delta\lambda/\lambda$, do

VELOCITY-SHIFTS OF FAINT NEBULAE

not vary appreciably with the wave-length is afforded by the emission spectra from the nucleus of N.G.C. 1275 in the Perseus cluster of nebulae. The measured lines range from λ 3727 to λ 4861, with no systematic difference in the velocities.

DETERMINATION OF SPECTRAL TYPES

Except for the bright-line nebulae, classifications have been based on a comparison with spectra of N.G.C. 221 (M 32). This nebula was chosen as the standard type because plates of high dispersion were available, from which the spectrum was classified as dG3 by Adams, Joy, and the writer. Additional spectrograms were obtained with the different dispersions in order that comparisons might always be made between spectra having the same scale.

The criteria used in assigning types to the bright-line nebulae are those adopted by the International Astronomical Union and given in the *International Critical Tables*.

Nebulae having absorption lines show only a small dispersion in type, all of them, except N.G.C. 205, falling within the narrow limits G1–G5. N.G.C. 205 has been classified as F5.

Bright-line spectra are of two different types, according as the continuous spectrum is strong or weak in comparison with the emission. The relative intensity of the continuous spectrum in the first type appears the same as in spectra having absorption lines. Examples are N.G.C. 3227, 4051, and 1275—all nebulae having bright stellar nuclei. The emission lines in the spectrum of N.G.C. 1275 are shifted to the red by the same amount as the absorption lines in the spectra of other members of the cluster to which it belongs, namely, $+5200$ km/sec.

Examples of the second type, in which the bright lines predominate, are obtained from emission patches in the outer regions of large nebulae, for instance, N.G.C. 5457 (M 101) and N.G.C. 6822.

Wide, shallow absorption lines have been observed in high-dispersion spectra of M 31 and M 32, and have been found approximately twice as wide in M 32 and almost four times as wide in M 31 as the lines in the spectrum of skylight. A widening of the lines seems noticeable in spectra of the fainter nebulae, but no definite statement in regard to this can be made, on account of the small scale.

MILTON L. HUMASON

The maximum of the continuous spectrum is shifted by an amount equal to the displacement due to the velocity. This suggests a color excess, which has actually been found from extra-focal photographic and photovisual magnitudes and which in some cases exceeds the amount expected. Nebulae in the Perseus cluster have an excess color-index of half a magnitude. Indications are that color-index appears to depend upon galactic latitude rather than distance.

FUTURE INVESTIGATIONS

An attempt will be made to extend the observed range in distance by measures of fainter clusters of nebulae. Some extension seems quite possible, but the limit with the 100-inch reflector will be reached at about photographic magnitude 17.5. Exposures necessary for the fainter nebulae are not so long as the magnitudes would indicate because the red-shift is so large that the H and K lines are brought into the region to which the photographic plate is highly sensitive. Further, lower dispersion can be used, for, since the red-shift is larger, a larger probable error can be tolerated. The main difficulty arises from the fact that at photographic magnitude $17.5\pm$ the nebulae become so faint visually at the Cassegrain focus of the 100-inch reflector that they cannot be seen on the slit of the spectrograph.

In order to test thoroughly the agreement of the velocities for individual members of a cluster, a larger number of velocities will be observed in the Virgo cluster.

High-dispersion spectra of some of the brightest nebulae will be obtained in order to investigate further the widening of the absorption lines which appears in such objects as M 31 and M 32.

I wish to express my thanks to Mr. T. A. Nelson, night assistant at the 100-inch telescope, and to Mr. Glenn Moore, relief night assistant, for their aid in obtaining the spectrograms.

CARNEGIE INSTITUTION OF WASHINGTON
    MOUNT WILSON OBSERVATORY
        March 1931

COSMOLOGY:
A SEARCH FOR TWO NUMBERS

Precision measurements of the rate of expansion
and the deceleration of the universe may soon provide
a major test of cosmological models

ALLAN R. SANDAGE

As recently as the 1950's about all that observational cosmology had succeeded in establishing was that galaxies exist and the universe expands. But beginning in the 1960's a flood of new discoveries has enriched our picture of the universe and has begun to provide a basis on which to distinguish between competing cosmological models. There has been a 30-year effort, now drawing to a close, to get precise measurements of two parameters that will provide a crucial test for cosmological models. The two key numbers are the rate of expansion (the Hubble constant H_0) and the deceleration in the expansion (q_0). The hope is that current research, by determining the extragalactic distance scale for nearby galaxies and searching for exceedingly distant clusters where the redshift is large, will measure both of these numbers to a precision of 15%.

New discoveries of the 1960's, spurred by the sophistication of new instruments and ideas, include:
- Black-body radiation predicted by George Gamow, Ralph A. Alpher and Robert Herman and left over from the big-bang "creation" event
- Isotropic extragalactic x-ray and γ-ray background flux
- Quasars with redshifts greater than 2, which imply recession velocities greater than 80% of the speed of light
- Absorption lines in quasar optical spectra resulting perhaps from an intergalactic medium or from the passage of radiation through clusters of galaxies
- Evidence that the helium abundance is about 30% by mass in all primeval matter.

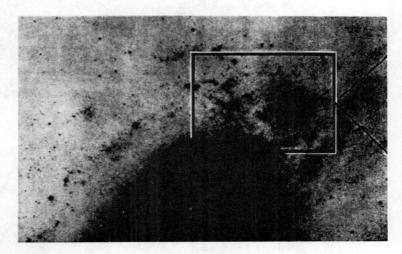

THREE CEPHEIDS in NGC2403 from blue plates taken 359 days apart with the 200-inch telescope. Periods are 46.460, 20.260 and 34.354 days for variables 5, 6 and 8. Magnitudes are 22.07 and 22.32 for variables 5 and 6 in the upper right and 22.02 for variable 8 in the lower right panel.   —FIG. 1

Cosmological models

These discoveries relate directly to the two major classes of cosmological models. In the big-bang models of A. Friedmann, Sir Arthur S. Eddington, Georges Lemaître, and Gamow, the expansion began from a singularity in space and time, emerging from that state a finite time ago amidst conditions of extreme density and pressure. On the other hand, the steady-state universe of Hermann Bondi, Thomas Gold, and Fred Hoyle had no beginning and no end, but rather continuously remakes itself according

7 November 1950

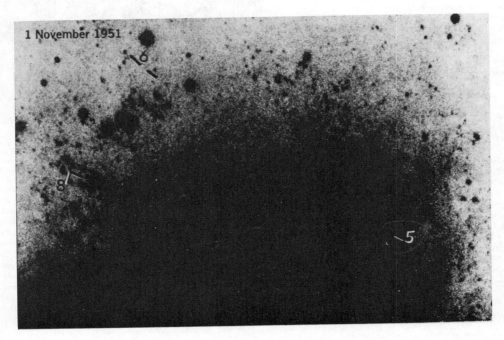

1 November 1951

to a fixed and immutable pattern.

If the residual 3 K background flux discovered by Arno A. Penzias and Robert W. Wilson,[1] and studied with vigor by the Princeton group[2] is indeed degraded fireball radiation, then a big-bang origin would seem possible. Gamow, Rodger Taylor, Hoyle, and James Peebles emphasized that a similar conclusion follows if the pristine helium abundance is indeed 30%, because so much helium apparently can only be made in the primeval physical conditions immediately after a Friedmann-type singular state. Must we then take the big bang seriously?

Perhaps, but especially if we could establish that the time scale when all this might have happened agrees with other related events such as the age of the chemical elements and the age of the oldest stars. Unfortunately there is still incomplete knowledge of all parameters required to date accurately the oldest stars as well as lack of precise values for H_0 and q_0, but current research is working in this direction.

Hubble's expansion law

From 1929 to the present, extensive observations of redshifts and *relative* distances of galaxies have established the form of expansion law. Distances may be indicated by the Cepheid variable stars shown in figure 1. Redshifts increase linearly with distance when we can neglect the effects of looking back in time as we look out in space. For such "local" distances the expansion law of Edwin P. Hubble is

$$c \ \Delta\lambda/\lambda_0 = H_0 D_0 \qquad (1)$$

where c is the velocity of light, $\Delta\lambda/\lambda_0$

Allan Sandage has been a staff member at the Mount Wilson and Palomar Observatories since he received his PhD from Cal Tech in 1953. His main interests are stellar evolution, observational cosmology, form of the redshift laws, quasars and distance scales. In 1960 Sandage and Thomas Matthews were the first to isolate the quasars.

is the observed fractional wavelength shift of spectral lines in galaxies relative to their laboratory values λ_0 ($\Delta\lambda$ is positive, hence redshifted, for galaxies that recede), and D_0 is the present distance. The Hubble "constant" H_0 is written with a subscript to denote its instantaneous, present-epoch value. In all models (except the steady state) H is a function of cosmic time because the redshift generally decreases with time due to the self gravity of the universe, and distances increase due to the expansion.

Equation 1 is deceptively simple. It is but the first term in a rigorous derivation from the theory surrounding each of the isotropic-world models we have mentioned. The more complete equation, valid for arbitrarily large distances, contains q_0 in a term that increases with redshifts. Added complication comes, of course, because in such remote galaxies we observe the expansion rate, and hence the deceleration, as it was a very long time ago.

Once we find H_0 from "local" galaxies using equation 1, and q_0 from observations of very distant systems, the time elapsed since the big-bang singularity follows immediately[3] from a straightforward calculation. Figure 2 shows how this birthday can be obtained for the simplest world models previously mentioned. Although the observer's problem, to find H_0 and q_0, is easy to state, it has defied solution for 40 years.

Measurement of H_0

The first complications in measuring H_0 concern the observed redshifts. Nature has added spurious velocity vectors to the systematic expansion motion. Galaxies possess random motions of the order of 200 km/sec caused by local gravitational perturbations on a scale of about two-million light years, or the order of the cluster sizes of small galaxy groups. This random effect can be averaged out, but only if the sample of calibrating galaxies is large.

Secondly, the possible larger-scale anisotropy pointed out by Gerard H. de Vaucouleurs[4] distorts the velocity field in some directions for redshifts smaller than about 4000 km/sec. The anisotropy presumably arises from an abnormal concentration of groups of galaxies such as the Virgo cluster complex on a local scale of around 30-million light years. Until recently observations of galaxies in the Southern

Hemisphere, crucial to map adequately this anisotropy, have been entirely lacking.

A final complication is the rotational motion of the sun about our galactic center, amounting to approximately 300 km/sec in the direction of Cygnus. That part of the rotation in the direction of any given nearby field galaxy is often an appreciable fraction of the observed redshift for nearby galaxies and appears as an added anisotropy in the observed velocity field. Its precise mapping and accurate subtraction also require currently unavailable data for nearby galaxies in the Southern Hemisphere.

Determination of distances

Distance calibration is a stepwise procedure, with the errors proliferating with each step. First one measures the apparent brightness of certain well defined objects, the distance indicators, in the nearby resolved galaxies. If the absolute brightness of these indicators is known from a reliable previous calibration, the distance follows from the inverse-square intensity fall off. Because a unique relation exists between the period and absolute luminosities of Cepheid variable stars these stars are excellent distance indicators.

The crucial distance range within which H_0 can be determined is quite narrow. It extends between 10^7 light years, which is remote enough so that expansion velocities begin to dominate the spurious velocity effects, and 6×10^7 light years, which is the upper limit for the indicators to be resolved in nearby galaxies with the 200-inch Hale telescope. In this range, indicator objects include the brightest resolved red and blue supergiants, the angular size of HII regions, normal novae, and perhaps, after much new calibration, supernovae.

Each of these classes must first be calibrated in even nearer galaxies, less than 10^7 light years away, where the more precise distance indicators of Cepheid variables can be measured. The important nearby galaxies in this range include members of our Local Group, galaxies in the M81 group, and members of the South Polar group.

1936 value of H_0

From 1936 to 1952, available calibrations determined the value of H_0 to be 165 km/sec per million light years. For the simplest model of an empty, unaccelerated universe, this value

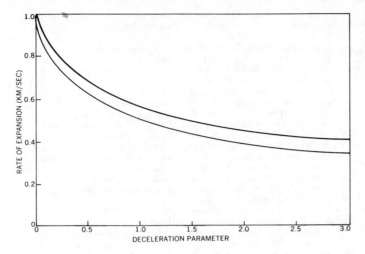

AGE OF BIG-BANG UNIVERSE T₀
can be determined, once the Hubble constant H₀ and the deceleration parameter
are known, from plots of the rate of expansion, H₀T₀. Curves illustrate a radiation-field space (color) and a matter-filled universe (black).   —FIG. 2

gives a big-bang birthday, H_0^{-1}, of only 1.8×10^9 years.

Even in the late 1930's this abnormally short time was known to violate the age of the earth's crustal rocks, and the lower limit of 7×10^9 years for the age of the earth's radioactive elements.

Clearly, either the simplest big-bang (Friedmann) models were incorrect or the value of H_0 was wrong. Many astronomers preferred the former choice. Lemaitre and Eddington devised new big-bang models with a repulsive "cosmological force" increasing with distance as $F_c = +1/3\Lambda D$, where the "cosmological constant" Λ determines the field strength of the new force. Such models still possessed the singularity in space and time present in the Friedmann models, but they had arbitrarily long creation times relative to H_0^{-1}. The time-scale discrepancy also gave the *modus vivendi* of the new but highly revolutionary steady-state model, which has no datable birthday.

Part of the justification for these new models collapsed in 1952 when Walter Baade[5] showed that the Cepheid period–luminosity (P–L) relation was in error by about 1.5 magnitudes. Baade's revision increased the distance to Local Group members in which Hubble had originally discovered Cepheids, and accordingly increased the Hubble time by about a factor of two. The exact value was uncertain because Baade's method was accurate enough only to establish a discrepancy and not to recalibrate the P–L relation precisely.

The impasse was broken in 1955 when John B. Irwin, during an expedition to South Africa, rediscovered and emphasized Peter Doig's[6] earlier result that Cepheids in our galaxy occur in certain star clusters. Distances to such star clusters can be obtained with high precision by photometric methods developed in the 1950's by Olin J. Eggen, Harold L. Johnson, and others. Soon after Irwin pointed out that the Cepheid U Sagittarii was a member of the cluster M25, and that S Norae occurred in NGC6087, many astronomers began a feverish search for other Cepheid-cluster associations, uncovering a total of 13 such calibrators. The Cepheid relation was recalibrated[7,8] from the cluster distances and also from a new understanding of the intrinsic scatter in the P–L relation[9] discovered earlier by Harlow Shapley.

Stars can pulsate only when their surface temperatures and absolute luminosities occur in a narrow, but finite, range.[10] This pulsational instability strip is denoted by the colored lines on the conventional Hertzsprung-Russell (H–R) diagram shown in figure 3a. In this diagram the visual luminosity M_V of a star is plotted against its color B–V which is a measure of its temperature. Here

$$B–V = 2.5 \log (l_V/l_B) + \text{const.}$$

and l_V and l_B are the luminosities integrated over the visual and the blue range of the spectrum, respectively. All stars within the strip pulsate with periods that increase on the average with increasing luminosity. But be-

cause the gray lines of constant period *slope* through the strip, Cepheids of the same period can differ in absolute luminosity by about one magnitude, the bluer Cepheids being brighter. Observations[8] indeed confirm that the scatter in the P–L relation illustrated in figure 3b is correlated with color and can be calibrated out once Cepheid colors have been measured.

With the new absolute calibration of the P–L color relation, distances to the Local Group were determined from the tedious and exacting photometric studies of M31,[11] IC1613,[12] NGC6822[13] and the Large and Small Magellanic Clouds.

Additional corrections to H_0

Distances within the Local Group do not in themselves determine H_0 but are used in calibrating brighter indicators at greater distances until the crucial distance range defined earlier is reached.

After Baade discovered an error in the distance to M31, corrections to the larger distances were found.[9] These involved the detection of errors[14] in the apparent-magnitude scale from magnitudes of 16 to magnitudes of 21, the problem of separating stars from HII regions in nearby resolved spirals and the calibration of absolute luminosities of the brightest resolved stars that are used at distances where the Cepheids fade below plate limit.

Beyond the Local Group new observations are clearly required. The next more distant group contains some 10 highly resolved systems centered on the giant spiral M81. When the 200 inch went into routine operation in 1949 it began a search for measurement of Cepheids in NGC2403, a member of the group, which is shown in figure 1. By 1968 the distance of the M81 group was found to be[15] $(1.1 \pm 0.06) \times 10^7$ light years, which is a factor 4.5 more distant than the value adopted in 1950.

But even the M81 group is too near to use equation 1 because spurious velocities mask the systematic expansion field. The importance of the group lies rather in giving increased precision to the calibration of still brighter distance indicators to be used further out.

The 200-inch telescope has recently begun to study this distance interval with some precision. The first results, from direct photometry of the brightest stars and measurements of the angular sizes of HII regions, show that we can expect improved distances to galaxies at least as remote as the Virgo Cluster (redshift 1136 km/sec) in the coming few years.

Although final results are not yet available, we can extrapolate the partial data for galaxies now being studied. The first plates of the new program suggest that H_0 may be as low as 15 km/sec per million light years. These data, together with previous work by Erik B. Holmberg, Jose L. Sersic, Sidney van den Bergh, de Vaucouleurs and our Mount Wilson-Palomar group, place H_0 tentatively in the interval $15 \leq H_0 < 40$ km/sec per 10^6 light years.

The range excludes Hubble's 1936 value $H_0 = 163$ km/sec per 10^6 light years, representing a distance scale increased by almost a factor of 10. The corresponding Hubble time increases in direct ratio, changing from H_0^{-1} (1936) $= 1.8 \times 10^9$ years to between 7.5×10^9 and 19.5×10^9 years. The current program at the Mount Wilson-Palomar Observatory is expected to reduce the uncertainty in H_0 to about 15%. But what about q_0?

Measurement of q_0

The predictions concerning the expansion properties of all world models can be expressed as functions of the two observables of redshift and apparent luminosities for galaxies of fixed absolute luminosity. Wolfgang Mattig was the first to show that all Friedmann models ($\Lambda = 0$) obey a rigorous relation between apparent bolometric magnitude and redshift of the form

$$m_{bol} = 5 \log q_0^{-2} \left[q_0 z + (q_0 - 1) \right.$$
$$\left. \left[(1 + 2q_0 z)^{1/2} - 1 \right] \right] + const. \quad (2)$$

where $z = \Delta\lambda/\lambda_0$, where q_0 contains the second time derivative of the spatial expansion factor and where the constant depends on H_0 and the mean absolute luminosity of galaxies in the sample. (Look-back time has been taken into account.)

A similar but more complicated expression exists[16] for the more general

models where $\Lambda \neq 0$, but the equation contains one more parameter related to Λ. In principle, if the observations are sufficiently precise and extensive in z, both q_0 and Λ can be determined directly from the data. This, of course, is the observer's ultimate goal, but many more (m, z) number pairs are needed if Λ is also to be determined.

From 1929 to 1960, the principal goal of observers was to test the as-

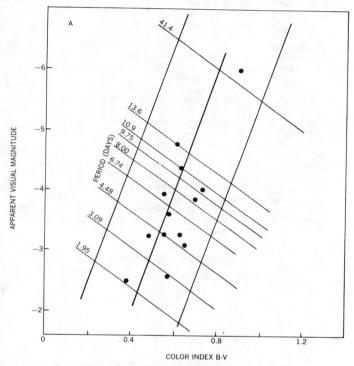

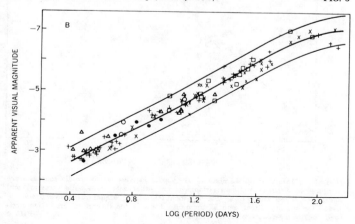

SCATTER IN THE PERIOD-LUMINOSITY RELATION (below) is caused by the color variation of the Cepheid stars (above) among lines of constant period (sloping grey lines). Pulsating stars occur only within a region of instability (colored lines). In the lower diagram, the calibrating Cepheids are the galactic cluster Cepheids (solid circles) and the h and Perseus association (open circles). Other Cepheids belong to galaxies of the Local Group (other symbols). —FIG. 3

ymptotic form of equation 2 for small z. Expansion to first order in z gives

$$m_{bol} = 5 \log cz + 1.086(1 - q_0)z$$
$$\ldots + \text{const.} \quad (3)$$

In the limit as z approaches zero, this equation reduces to equation 1 if one notes that apparent magnitude m and distance D are related by $m = 5 \log D$ + const. from the definition of astronomical magnitudes.

All observations since 1936 have confirmed the leading term on the righthand side of equation 3. Figure 4 shows the most recent observational data for 42 clusters of galaxies with the first-ranked cluster member as the distance indicator. The small horizontal dispersion about the line proves the near constancy of absolute luminosity for galaxies chosen in this way. The line, drawn with the theoretical slope of 5, fits the data well.

Unfortunately this first approximation does not allow us to make a choice of cosmological models. Only in the next approximation, which involves q_0,

do the models differ. For big-bang universes of the Friedmann type ($\Lambda = 0$), q_0 must be positive; the sign implies deceleration from self gravity of the universe. For the steady-state model, $q_0 = 1$. In models with Λ greater than zero but still with a Friedmann-type singularity (the Eddington-Lemaître cases), q_0 can range from -1 to any positive value, depending on the position of the present epoch of observation in the expansion history of the universe.

The curvature of the three-space submanifold of the four-dimensional space–time is determined by q_0, providing that the time coordinate is suitably defined. Separation of space–like and time–like coordinates in the way proposed by Howard P. Robertson and A. G. Walker[17] determines the Riemannian spatial curvature c/R as $c^2/R^2 = H_0^2 \; (2q_0 - 1)$. For q_0 greater than $1/2$, c/R is positive and the three-dimensional space is non-Euclidean and closed (finite volume). For q_0 less than $1/2$ the universe is

open (infinite volume), with negative curvature. Only for $q_0 = 1/2$ is the three-dimensional space flat.

Can q_0 be found from the data now available in figure 4? Only marginally. The data extend only to redshifts of $z = 0.46$, whereas equations 2 and 3 show that the intensity difference at this redshift between models which differ by one in the value of q_0 is only 0.5 magnitudes, comparable to the scatter of galaxies about the mean line. Furthermore, uncertainties in the various corrections to observed magnitudes are themselves of the order of 0.2 magnitude. Finally, to interpret figure 4 properly, we require knowledge of the change of galaxian absolute luminosities in the look-back time due to evolution of their stellar content.

Two major corrections must be applied to the measured intensities. The first correction results from the fact that galaxies do not have well defined boundaries and hence the intensity varies with the measuring aperture. Thus all measurements must be corrected to some adopted "standard galaxy diameter." The second correction is for the redshift, which causes the photoelectric photometers, with fixed-frequency band passes, to measure different regions of the spectrum for redshifted galaxies than for galaxies at rest. Proper corrections (called the K dimming) can be calculated once the energy distributions $I(\lambda_0)$ of the galaxies are known. Because all giant elliptical galaxies, which are the "test" objects in figure 4, have similar distribution curves, a universal K correction may be adopted for all first-ranked cluster members. A new determination of K for such galaxies[18] indicates that this correction is now known accurately enough to reduce the errors in q_0 to 15%.

The evolutionary correction

Some workers have believed that the most serious uncertainty is the change of galaxian absolute luminosity due to evolution of stellar content during the look-back time. Fifteen years ago, before the observational aspects of stellar evolution and galaxian content were sufficiently understood, the correction was ignored. But recent progress has illumined and perhaps has even solved the problem.

All galaxies in figure 4 are giant ellipticals (E galaxies) because virtually no spirals occur in rich clusters of galaxies. To discover the stellar content

PAST PROPER TIME (FRACTION OF HUBBLE TIME)

HUBBLE DIAGRAM for 42 first-ranked galaxies in clusters. Nonradio sources were measured by William A. Baum (triangles) and other data, both radio (crosses) and nonradio (closed circles), by the 200-inch telescope. Magnitudes have been corrected for aperture effects and K dimming. Data fit the theoretical prediction (colored line) of a Friedmann model with $q_0 = 1$ and $\Lambda = 0$.   —FIG. 4

of E galaxies, Hyron Spinrad, David E. Wood, Robert McClure, van den Bergh and I followed the classical work of Baade in 1944. Our studies pointed to an old, primeval stellar population whose H–R diagram, illustrated by the dark-colored curve in figure 5, resembles that of the old open cluster NGC188 in our galaxy. No new star formation has occurred for the past $T_G/2$ years, where T_G is the galaxy age, which appears to be nearly the same for all ellipticals.

As stars evolve the temperature decreases and the spectrum moves away from the blue region. At age zero the stellar population would have been arrayed along the black line in figure 5 and at the present time might be represented by the light-colored curve. Because we know how such a configuration evolves in time, we can estimate the luminosity change of an entire elliptical galaxy over the relevant look-back time. This time, called τ, is primarily a function of redshift, with a weak dependence on q_0.

Simple calculation[19] shows that $\tau(z,q_0)$ varies with redshift for various world models according to figure 6. For example, galaxies with $z = 0.4$ in a $q_0 = +1$ model are observed as they were 0.25 Hubble times ago. If $H_0^{-1} = 13 \times 10^9$ years, and if all galaxies are 10^{10} years old, systems at this redshift have a fractional look-back time of $\tau/T_G = 0.33$. It can be shown that in this interval, the main sequence termination point illustrated in figure 5 has burned to its present position from a starting luminosity only 0.54 magnitudes brighter.

We now reach an apparent paradox. Although the more distant galaxy, seen τ proper years earlier, has a brighter termination point in its H–R diagram (dark-colored curve in figure 5), the integrated luminosity of the entire galaxy need not be brighter at the earlier epoch for two reasons. First, much of the integrated light comes from unevolving dwarf stars fainter than a magnitude of 4 in which no luminosity changes occur in individual stars in time τ. The 0.54 magnitude change is thus diluted by the fraction of the unevolved light. Secondly, the number of stars along the younger (upper) sequence is smaller than along the older (lower) sequence because of the increase in the luminosity function $\varphi(M)$, or the number of stars per magnitude interval along the main sequence that feeds the evolving portions of the sequences. Because $\varphi(M)$ is such a steeply rising function toward fainter M, the higher intensity of the younger sequence can be overcome by the increased number of stars along the older sequence.

Calculation of these two effects from Spinrad's[20] synthesis of the stellar content of E galaxies shows that the evolutionary rate of change in absolute magnitudes is

$$\frac{dM}{d\tau} = -0.044 \pm 0.02 \text{mag}/10^9 \text{ yrs}$$

$$(4)$$

in the sense that younger galaxies are fainter in integrated luminosity.

The correction is relatively small and it could be zero. The effect on q_0 is the same order of uncertainty as that from errors in the aperture and K corrections. The conclusion at the moment is that the evolutionary correction is not very serious. It can be calculated with some confidence, and should be known even better when current work on stellar content by Spinrad and others is more secure in the next few years.

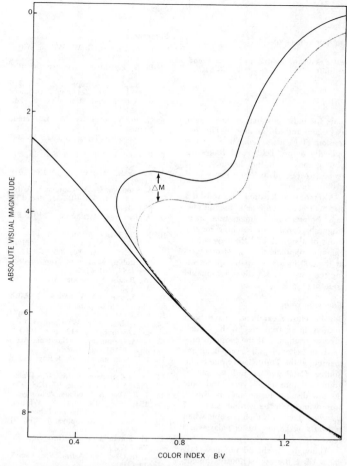

HERTZSPRUNG-RUSSELL DIAGRAM for the stellar content of ellipsoidal galaxies at age zero (black), at the present epoch (light color) and τ years ago (dark color) as seen in distant galaxies.   —FIG. 5

To find more distant clusters

To find q_0 by extending figure 4 to larger redshifts requires the discovery of more distant clusters. The problem

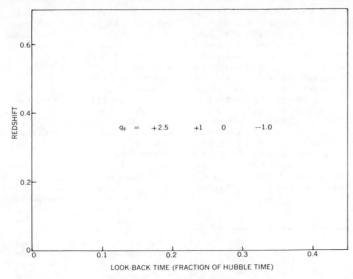

LOOK-BACK TIMES calculated for the simplest Friedmann models with $\Lambda = 0$ and with various values of q_0 are functions of the redshift.   —FIG. 6

more, combining this value of q_0 and a value of H_0^{-1} between 7×10^9 and 19.5×10^9 years with figure 1 gives a big-bang birthday, for models with $\Lambda = 0$, which ranges from 4.2 to 11.7×10^9 years. This range is within the time scale of the oldest stars (about 10^{10} years) and the chemical elements (about 10^{10} years), and suggests that some logical connection exists between the clocks, as required in big-bang models.

But the present discussion is only a prelude to the coming decade. If work now in progress is successful, better values for both H_0 and q_0 (and perhaps even Λ) should be found, and the 30-year dream of choosing between world models on the basis of kinematics alone might possibly be realized.

References

1. A. A. Penzias, R. W. Wilson, Astrophys. J. **142**, 419 (1965).
2. R. B. Partridge, American Scientist **57**, 37 (1969).
3. A. Sandage, Astrophys. J. **133**, 355 (1961); Ya. B. Zeldovich, Advances in Astron. and Astrophys. **3**, 242 (1965).
4. G. de Vaucouleurs, W. L. Peters, Nature **220**, 868 (1968).
5. W. Baade, Trans. I.A.U. **8**, 397 (1952).
6. P. Doig, J. Brit. Astron. Soc. **35**, 201 (1925).
7. R. Kraft, Astrophys. J. **134**, 616 (1961).
8. A. Sandage, G. A. Tammann, Astrophys. J. **151**, 531 (1968); *ibid* **157**, 683 (1969).
9. A. Sandage, Astrophys. J. **127**, 513 (1958).
10. R. F. Christy, Astrophys. J. **144**, 108 (1966); D. S. King, J. P. Cox, Pub. Astron. Soc. Pac. **80**, 365 (1968); R. S. Stobie, Mon. Not. Roy. Astron. Soc. **144**, 461 (1969).
11. W. Baade, H. Swope, Astron. J. **68**, 435 (1963).
12. W. Baade (ed. A. Sandage), Astrophys. J. (to be published).
13. S. Kayser, Astron. J. **72**, 134 (1967).
14. J. Stebbins, A. E. Whitford, H. Johnson, Astrophys. J. **112**, 469 (1950).
15. G. A. Tammann, A. Sandage, Astrophys. J. **151**, 825 (1968).
16. J. E. Solheim, Mon. Not. Roy. Astron. Soc. **133**, 321 (1966).
17. H. P. Robertson, A. G. Walker, Pub. Astron. Soc. Pac. **67**, 82 (1955).
18. J. B. Oke, A. Sandage, Astrophys. J. **154**, 21 (1968); A. E. Whitford, Astrophys. J. (to be published).
19. A. Sandage, Astrophys. J. **134**, 916 (1961).
20. H. Spinrad, Pub. Astron. Soc. Pac. **78**, 367 (1966).
21. J. B. Oke, Pub. Astron. Soc. Pac. **81**, 11 (1969).
22. A. Sandage, Yearbk. Carnegie Inst. Wash. **65**, 163 (1966); J. V. Peach, Astrophys. J. (to be published). □

is not trivial because the field of view of the 200-inch telescope is only 0.05 square degrees, and the probability of chance discovery by random surveys is exceedingly small. However, two promising new approaches are now being tried.

A new restricted-sky survey is under way with the Palomar 48-inch Schmidt telescope and the recently developed low-noise Eastman III-aJ photographic plate. The exceedingly fine plate grain increases signal-to-noise ratio sufficiently to permit detection of images fully 1.5 magnitudes fainter than conventional emulsions. Clusters with redshifts of about 0.6 are seen as smudges at the plate limit of the wide-angle Schmidt, and the regions are then photographed with the smaller-field 200 inch which has a three-magnitude advantage over the Schmidt.

An even more promising method is to photograph, with the 200 inch, sky positions that contain unidentified radio sources. Because 30% of radio galaxies are in clusters, many distant clusters should be discovered in this way. But precise radio-source positions must be known for the method to work. Using long base-line radio interferometers, Cambell M. Wade at the National Radio Astronomy Observatory has recently achieved the unprecedented accuracy of ±0.2 arc sec-

onds for radio positions in both right ascension and declination. The extension of Wade's procedure to the optically empty fields, and follow-up photography with the 200-inch telescope is expected to locate many new distant clusters.

After optical location, the redshifts and corrected apparent magnitudes must be measured. Image-tube spectrographs capable of reaching magnitudes of about 22 and the new multi-channel photoelectric spectrometer[21] at the 200 inch are capable of detecting the necessary signals in reasonable counting times.

The value of q_0

Several cluster candidates have been located in a trial run with Wade's precise positions. If the program proceeds on schedule, all empty fields remaining in the Third Cambridge Radio Source Catalogue should be photographed within several years, and some 30 new distant clusters may be found. Improvement in the current value[22] of $q_0 = +1.2 \pm 0.4$ is then possible when redshifts of 0.8 for cluster galaxies are reached.

Although we should not now take $q_0 = 1.2 \pm 0.4$ very seriously, the current value does differ from the steady state prediction of $q_0 = -1$ by five times the probable error. Further-

THE ASTROPHYSICAL JOURNAL, 176: L5-L8, 1972 August 15
© 1972. The American Astronomical Society. All rights reserved. Printed in U.S.A.

ON THE EVIDENCE FOR A PHYSICAL CONNECTION BETWEEN MARKARIAN 205 AND NGC 4319

ROGER LYNDS
Kitt Peak National Observatory,* Tucson, Arizona

AND

A. G. MILLIKAN†
Emulsion Research Division, Eastman Kodak Company, Rochester, New York
Received 1972 May 19

ABSTRACT

New direct photographs and spectroscopic material are discussed in relation to the finding by Arp that there is a filament connecting Markarian 205 and NGC 4319, objects of greatly differing redshift. The new material appears to raise doubts concerning the nature of the feature reported by Arp and tends, to a small degree, to support a more conventional interpretation of Markarian 205 as a Seyfert galaxy located at its indicated Hubble distance.

The point at issue and the motivation for the observations reported here is the possible occurrence of non-Doppler or non-Hubble redshifts in extragalactic objects. In addition to various statistical arguments suggesting a physical association between large-redshift quasars and relatively nearby galaxies, there have been several cases of an apparent morphological connection between extragalactic objects having markedly different redshifts. One of the more outstanding of these cases is the subject of this paper.

Arp (1971) has presented evidence for a possible morphological connection between NGC 4319 and Markarian 205. NGC 4319 is a relatively nearby spiral galaxy, for which Arp reports an apparent radial velocity of 1700 km s^{-1} (corrected for solar motion), whereas Markarian 205 is a compact object (classified by Arp as a quasi-stellar object) for which Weedman (1970) finds a Seyfert-like spectrum with a redshift of 0.07 or an apparent radial velocity of 20,250 km s^{-1}. Markarian 205 is located approximately 40″ nearly due south of the nuclear region of the galaxy and inside the faint outer spiral structure. The new evidence presented by Arp consists of direct photographs showing a luminous filament connecting Markarian 205 with NGC 4319. This feature was reported to be visible on several plates and is especially easy to see on a reproduction of a plate taken through a narrow interference filter that included the wavelength expected for Hα at the redshift of NGC 4319. Arp suggested that the feature is radiating in Hα at a redshift near that of the galaxy. The connecting feature as shown on the published illustration of a IIIaJ photograph was somewhat less convincing because of the possibility that the effect could arise either from the overlap of the brightness distributions of NGC 4319 and Markarian 205 or from the possible presence of a faint background galaxy between the two objects.

On 1970 July 6, prior to the publication of Arp's paper, a limit spectrogram of the region of the filament was obtained with an image-tube spectrograph on the Kitt Peak National Observatory 2-meter telescope. The spectral resolution was approximately 8 Å, and the slit of the spectrograph was placed exactly through Markarian 205 at a position angle of 0°. Neither Hα nor any other emission lines were found to be present

* Operated by the Association of Universities for Research in Astronomy, Inc., under contract with the National Science Foundation.

† On Study Research Leave at Kitt Peak National Observatory and the University of Arizona.

in the space between Markarian 205 and the galaxy. When they appeared in print, the photographs by Arp showed that the filament in question should have fallen, at least partially, within the spectrograph slit. Therefore, it is surprising that Hα was not recorded, if indeed the filament is emitting in Hα as Arp's photograph indicates. However, even though the spectrograph slit was 2″ wide, the filament was probably not precisely centered because it appears to lie along a nearly north-south line approximately 1″ east of the center of Markarian 205. Consequently, a new spectrogram was obtained on 1972 March 12 with the slit at a position angle of 3° and passing 0″.5 to the east of the center of Markarian 205. The spectral resolution on this spectrogram is 3.5 Å, and the information content is adequate to show Hα in the telluric airglow spectrum. Nevertheless, this spectrogram shows no convincing evidence for the presence of Hα in the region of Arp's filament. (There may be some weak emission in the nucleus of the galaxy.) Simple calculations indicate that, for an emission feature less than 3.5 Å wide, the signal-to-noise ratio on the spectrogram should be at least equal to that on Arp's Hα photograph. However, there are uncertainties in making such a comparison; it would be an overstatement to say that there is absolutely no emission present in the space between the two objects.

Prior to obtaining the second spectrogram, an attempt was made to secure a new Hα photograph of the filament so that measurements could be made to facilitate placing the spectrograph slit exactly on the feature. An image-tube photograph was obtained with the 2-meter telescope and an interference filter centered at λ6583 and having a half-transmission width of 50 Å. This photograph failed to record the filament, even though features are shown that are as faint as or fainter than those shown on the comparable photograph by Arp. Because of this failure to record the filament, further photographic material was obtained. The remainder of this paper will be devoted to a discussion of this material in relation to the interpretation of the photographs published by Arp.

The only explanations that have occurred to us for the failure to record the filament on our Hα photograph are (a) the wavelength of the Hα emission has a critical value that makes important any differences in the spectral transmission characteristics of the filter used for our photograph and that used for Arp's photograph, (b) the filament is emitting primarily continuum radiation that could have been transmitted by Arp's filter if it were not well blocked outside the main passband, and (c) the filament is largely spurious. Concerning the first possibility, we present in table 1 the transmission characteristic of the filter used for our Hα photograph. The filter is well blocked at all wavelengths outside the main passband to which the image tube is appreciably sensitive. The characteristics of the main passband are seen to be generally similar to Arp's description of his filter, except that the passband is somewhat narrower and there is not as much alteration in the transmission characteristics with the F/7.5 beam of the 2-meter telescope as there is with the faster focal ratio of the Hale telescope. The expected wavelength of Hα at the redshift of the galaxy (with respect to the Sun) is 6596 Å, within the filter passband. However, without more detailed information on the characteristics of Arp's filter, explanation (a) cannot be definitely excluded—although it does appear unlikely in view of the absence of any convincing evidence for discrete Hα emission on the two spectrograms. Such information is also necessary for an evaluation of explanation (b).

TABLE 1

SPECTRAL TRANSMISSION OF λ6583 FILTER

Wavelength (Å)	6530	6545	6560	6575	6590	6605	6620
Transmission	0.06	0.21	0.64	0.67	0.70	0.33	0.08

A question now arises as to the reality of the connecting feature claimed by Arp to be present on his IIIaJ photograph. Although an inspection of the reproduction of the photograph leaves one with the unmistakable impression that there is a luminous feature of some kind between Markarian 205 and NGC 4319, there is some question concerning the validity of such an impression. One of the most obvious alternative interpretations is that the overlap of the outer regions of the two objects produces the appearance of a connection. Such an effect has been produced in laboratory experiments in which photographs have been made of a series of pairs of test objects of differing relative sizes and separations. The details of these experiments will be given in a forthcoming publication.

Although the laboratory experiment demonstrated that a spurious impression of a connection between two objects can be produced, the conditions of the experiment were not exactly analogous to those of an actual observation. Therefore, a new IIIaJ photograph (using no filter) was obtained with the 2-meter telescope. This photograph, together with an isodensitracing of the relevant portion, is reproduced in figure 1 (plate L1). The contour interval on the isodensitracing is 0.06 in photographic density. The scale of the original plate ($12\rlap{.}{''}7$ mm^{-1}) is essentially the same as for the Hale telescope, and, except for somewhat poorer seeing, the new photograph is roughly comparable with the IIIaJ photograph published by Arp. One can imagine the presence of a connecting feature on the new photograph, and we find that it is possible to illustrate the feature to almost any desired degree of prominence by using sufficiently high contrast and by carefully choosing the exposure level in the photographic reproduction process. However, a more reliable evaluation of the evidence for the connecting feature may be obtained by analyzing the isodensitracing.

Let us hypothesize that there is no connecting feature and assume that the brightness distribution of the galaxy is smooth in the vicinity of Markarian 205 and that the brightness distribution of Markarian 205 is symmetrical about its center. Then we find, from the photometric calibration of the photograph, that the galaxy contour passing through the center of the apparent connecting feature has a surface brightness of 0.19 in units of sky brightness. The contour of Markarian 205 passing through this same point has a surface brightness of 0.06. The sum of these, 0.25, falls only a little short of the calculated surface brightness, 0.27, of the apparent connecting feature. It therefore appears that, to within the accuracy of the measurements, it is not unreasonable to explain the entire impression of a connecting feature as arising from the overlap of the brightness distributions of the two objects. Furthermore, there is no indication of a systematic disturbance of the inner isophotes of NGC 4319 of a type that would support Arp's suggestion that a connecting feature extends in to the nucleus of the galaxy.

On the other hand, the measurements also admit the possibility of a small contribution from an additional object. As a matter of fact, when viewed from a distance, figure 1 gives as much the impression of a discrete feature between the two objects as of a connecting filament. Furthermore, figure 1 and the original plate show that the field is liberally populated with faint extended objects—presumably distant galaxies—and that the probability of the chance location of such an object in the region between Markarian 205 and NGC 4319 is not entirely negligible. The actual presence of such an object is supported by other 2-meter telescope photographs made at 2.5 times larger plate scale. Two of the best of these were obtained with an image tube on IIIaJ emulsion: one was obtained through a filter consisting of an aqueous solution of CuSO$_4$, and the other was unfiltered. The spectral transmission characteristic of the copper sulfate filter and the approximate spectral response of the image tube are given in table 2.

These two photographs show essentially the same features, and a superposition of the two is reproduced in figure 2 (plate L2). (The irregular feature just above the center of the right-hand edge of the illustration is a flaw on one of the photographs.) The

L8   ROGER LYNDS AND A. G. MILLIKAN

TABLE 2

SPECTRAL TRANSMISSION OF CuSO₄ FILTER AND IMAGE-TUBE SPECTRAL SENSITIVITY

	WAVELENGTH (Å)						
	3500	4000	4500	5000	5500	6000	6500
Transmission...	0.91	0.93	0.93	0.91	0.77	0.40	0.06
Sensitivity*....	7.2	24.0	40.5	40.0	35.0	30.5	25.0

* Milliamperes per watt.

limiting sensitivity of the composite photograph appears to be somewhat better than that of the IIIaJ photograph in figure 1 and at least as good as that of the published reproduction of Arp's IIIaJ photograph. The composite gives the impression that the space between Markarian 205 and NGC 4319 is occupied not by a connecting filament but by a roughly circular, diffuse feature. If this feature is real, then it represents, in conjunction with the effect resulting from the overlapping distributions of the objects, a reasonable explanation for the appearance of a connecting feature shown by some of the photographs obtained at a smaller plate scale. However, it must be stressed that the feature is very subtle and that its true nature cannot be established by the present material. In particular, the possibility cannot be ruled out that the feature represents a morphological characteristic of either Markarian 205 or NGC 4319.

A somewhat more definite feature shown by the composite photograph is a low-surface-brightness envelope surrounding Markarian 205. This distribution is shown on each of the constituent photographs of the composite, as well as on several other photographs and, to a small extent, on the photograph reproduced in figure 1. The elliptical feature is concentric with but relatively distinct from the higher-surface-brightness, circularly symmetric halo of Markarian 205. The major axis of the distribution is about 19″ at a position angle of approximately −30°. For a Hubble parameter of 100 km s⁻¹ Mpc⁻¹, the major-axis dimension is 19 kpc if the feature is located at the indicated Hubble distance of Markarian 205. Thus, the observations are not inconsistent with an interpretation of Markarian 205 as a luminous Seyfert galaxy having a low-surface-brightness disk structure and having a distance corresponding to its Hubble distance.

In conclusion, we find it possible to interpret the feature shown on Arp's IIIaJ photograph as something other than an actual physical connection. However, we do not understand why we have not been able to confirm, either spectroscopically or photographically, Arp's detection of an Hα connecting feature. These observations have been presented in the hope that, in conjunction with the photographs of Arp, they might lead, through further observational effort, to a secure understanding of this remarkable juxtaposition of objects.

We wish to thank Mr. Richard Kron for making several isodensitracings and to acknowledge conversations with Dr. Thomas Adams who has been working on the same problem.

REFERENCES

Arp, H. 1971, *Ap. Letters*, **9**, 1.
Weedman, D. W. 1970, *Ap. J.* (*Letters*), **161**, L113.

Astrophysical Letters, 1972, Vol. 12, pp. 139–141

Observations of NGC 4319 and Markarian 205

HOLLAND C. FORD and HARLAND W. EPPS   *University of California, Los Angeles, California, USA*

A narrow-band, sky-limited, Hα photograph taken at the prime focus of the Lick 120-in telescope during excellent seeing does not show a connection between NGC 4319, whose redshift is 0.006, and Markarian 205, whose redshift is 0.070. A wide-band, sky-limited, green-light photograph taken during good seeing does not show a connection at a photographic density that can be considered significant. We suggest that Markarian 205 is an N-type galaxy at cosmological distance.

Astrophysical Letters, 1972, Vol. 12, pp. 143–146

Limits on Hα Emission in NGC 4319 Associated with Markarian 205

T. F. ADAMS   *Yerkes Observatory, Williams Bay, Wisconsin, USA*

R. J. WEYMANN   *Steward Observatory, University of Arizona, Tucson, Arizona, USA*

Photographic plates of Markarian 205 taken through a narrow-band filter centered on Hα at the redshift of the nearby galaxy NGC 4319 fail to reveal the 'connection' reported by Arp. The observational limits on the emission measure are used to rule out a model in which Markarian 205 was ejected in the plane of NGC 4319. Other models in which Markarian 205 is not in the plane of NGC 4319, but still at the same distance, cannot be rejected without a substantial improvement in the observations.

Astrophysical Letters, 1970, Vol. 7, pp. 111–113   © Gordon and Breach, Science Publishers Ltd.

Stephan's Quintet or Quartet?

G. A. TAMMANN   *Hale Observatories, Carnegie Institution of Washington, California Institute of Technology, Pasadena, California, U.S.A.*

Arguments are given that NGC 7320 is not a physical member of Stephan's quintet but is rather a companion galaxy to the foreground galaxy NGC 7331.

The galaxies NGC 7317–7320 are generally known as Stephan's quintet, although Stephan (1877) distinguished only four galaxies, counting the interacting galaxies NGC 7318a and NGC 7318b as one object. The question of the physical association of the five galaxies is of greatest interest since NGC 7320 has a corrected redshift of 1073 km sec^{-1} (Burbidge and Burbidge 1961a) as compared to the mean corrected redshift for the remaining four galaxies of 6735 ± 400 (standard deviation) km sec^{-1} (Humason *et al.* 1956). The membership of NGC 7320 was discussed by Burbidge and Burbidge (1961a, 1961b) and these authors seemed to favour the membership in spite of the discordant redshift. It is the aim of this note to show that NGC 7320 is most likely a foreground object. The arguments are the following.

(1) Visual inspection of Arp's (1966) excellent 200-inch photograph of the quintet shows clearly, even on the reproduction, that NGC 7320 is resolved. Besides a number of individual stars or starlike objects near the plate limit, the galaxy shows a coarse texture that is typical for fairly nearby galaxies. On the other hand the texture of NGC 7319 is very smooth, suggesting a much greater distance. The case of NGC 7318a/b, which shows a number of exceedingly bright knots in its spiral arms and somewhat finer substructure within these arms than NGC 7319, is less clear; however, if it must be decided whether this pair lies at the distance of NGC 7319 or at that of NGC 7320, the overall appearance and the small angular diameter of the central regions favour the former alternative.

(2) The redshift of NGC 7320 is almost identical with that of NGC 7331, a bright giant Sb galaxy, which lies, as pointed out by S. van den Bergh (Burbidge and Burbidge 1961b), less than half a degree away. From the optical rotation curve the corrected redshift of NGC 7331 is 1128 km sec^{-1} (Rubin *et al.* 1965), in good agreement with the corresponding radio values of 1130 km sec^{-1} (Roberts 1969) and 1093 km sec^{-1} (Gouguenheim

1969). The most natural conclusion from this is that NGC 7331 and NGC 7320 lie at almost identical distances.

(3) De Vaucouleurs (1969) finds that about 20 or possibly only 10 per cent of all galaxies are isolated in space. It is therefore somewhat surprising that no companion of NGC 7331 is yet known. A very faint anonymous galaxy at R.A. $\approx$ 22 hr 34.9 min, $\delta \approx 34°\ 22'$ (1950.0), separated by two diameters of NGC 7331 from that galaxy, seems to be a likely candidate for physical companionship, but this object is an extremely dwarfish Ir system, comparable to the tiny irregular companion NGC 5477 of M 101 (see e.g. Beale and Davies 1969), and it would be unusual if this were the only companion to NGC 7331. Holmberg (1969) concluded from statistical arguments that NGC 7331 has four physical companions with diameters $\geqslant 1$ kpc within a rather narrow radius of about 15 arc min. It is therefore tempting to identify NGC 7320 as an additional companion that lies slightly farther out, but is still separated by less than 4.5 diameters of NGC 7331 from this galaxy. The redshift of 1492 km sec^{-1} (corrected) of NGC 7343 is too uncertain (Humason *et al.* 1956, de Vaucouleurs and de Vaucouleurs 1964) for us to decide whether this galaxy also might be a member of the supposed group around NGC 7331.

(4) The galaxies NGC 7318a, NGC 7318b, and NGC 7319 are strongly interacting. The appearance of the distorted spiral arms corresponds well with the results of Toomre (private communication), who investigated the gravitational effects on galaxies in close encounter. If NGC 7320 were at the same distance, it would be strange, as remarked by A. Poveda (Burbidge and Burbidge 1961b), if this galaxy did not show a detectable distortion. This is suggestive, although the lack of distortion could be explained either by assuming that the true distance between this galaxy and the remaining quartet is somewhat larger than the true distances between the interacting galaxies (an argument

G. A. TAMMANN

which would not hold for the past of this galaxy if ejection is assumed), or/and that the direction of the relative motion, which influences the effects of gravitational interaction (Toomre 1970), is such as to prevent a distortion of NGC 7320.

(5) Burbidge and Burbidge (1961a) determined the chance for a galaxy such as NGC 7320 to fall within 2 arc min of a given quartet of galaxies as only 1 : 10^3. However, they did not consider the nearness of the bright giant NGC 7331. If a typical giant galaxy has on the average five companions within 100 kpc (corresponding to ~ 36 arc min at the distance of NGC 7331 as discussed below), the chance that one companion would fall within a given area of 4π (arc min)^2 is $\sim 1 : 65$. It is therefore not very surprising to find a foreground object in the immediate neighbourhood of the quartet.

The quartet itself lies in an extended region that includes an unusually high number of moderately bright galaxies. This region was identified by Zwicky and Kowal (1968) as a cluster of galaxies (2231.2 + 3732), which they describe as 'open' and 'near' and in which they counted 782 galaxies. The quartet could well be a member of this cluster. This view is somewhat supported by the fact that the only additional published redshift within the cluster area is that of NGC 7335, whose value $v_0 = 6576$ km sec^{-1} (Humason et al. 1956) agrees very well with the mean velocity of the quartet.

The combined weight of these arguments seems fully consistent with the assumption that NGC 7320 is a companion galaxy of NGC 7331 and that these galaxies happen to be seen projected on a more distant cluster of which NGC 7317–7319 may very well be members.

In this case, the question remains as to the true distance of NGC 7331 and NGC 7320. From their redshifts and $H = 75$ km sec^{-1} Mpc^{-1} (Sandage 1968, Tammann 1969), a distance of 14 Mpc seems indicated, but this result is unreliable since the velocities are small. NGC 7331 has the same luminosity class, Sb I–II, as NGC 224 and NGC 3031 (van den Bergh 1960b). Using Holmberg's (1958) photographic magnitudes of these galaxies, we find that NGC 7331 is 5.98 and 2.06 mag fainter than NGC 224 and NGC 3031, respectively. The galactic front absorption for NGC 224 was taken from Baade and Swope (1963), that for NGC 3031 from Tammann and Sandage (1968), and the tentative value $A_B \approx 0.60$ mag NGC 7331 was taken from de Vaucouleurs and Malik (1969). With distance moduli of $(m - M)^\circ = 24.20$ mag for NGC 224 (Baade and Swope 1963) and $(m - M)^\circ = 27.55$

mag for NGC 3031 (Tammann and Sandage 1968), for NGC 7331 there results a mean of $(m - M)^\circ = 29.9 \pm 0.4$ mag, corresponding to a distance of ~ 10 Mpc. The projected linear distance between NGC 7331 and NGC 7320 then becomes 80 kpc, which seems to be a most reasonable value. The phenomenological appearance of NGC 7320 is affected by the front absorption; it seems not dissimilar to NGC 2403 (cf. Sandage 1961), but it is probably somewhat fainter than the latter. Its short spiral arms of low surface brightness would indicate that the galaxy belongs in van den Bergh's (1960a) luminosity class Sc III–IV. It would therefore be expected to be $\sim 1.9 \pm 0.7$ mag fainter than an Sb I–II galaxy (van den Bergh 1960b). Zwicky and Kowal (1968) estimated the actual magnitude difference between NGC 7331 and NGC 7320 to be 3.4 mag. This discrepancy of two standard deviations is not taken to be serious since the attribution of the luminosity class for NGC 7320 is hampered by the considerable galactic absorption. This absorption may also explain why Burbidge and Burbidge (1961a, b) found a low mass, but a surprisingly high mass density within the observable bright part of this galaxy.

New redshift observations in the field of NGC 7331 would be important for identifying more members of this nearby group and for substantiating the suspicion that the quartet is a member of the cluster 2231.2 + 3732. In addition, absorption determinations in this field would be of great interest; the above-mentioned absorption was determined for optically selected galaxies; the average absorption at this low latitude is, however, 2.5 times higher according to de Vaucouleurs and Malik (1969).

It is quite possible that similar cases exist where projection effects lead to a conspicuous grouping of galaxies with strongly deviating redshifts.

Note. After this paper was submitted to the Editor, the author received a paper by R. J. Allen, 'Stephan's Quartet?' (*Astron. Astrophys.*, 7, 330, 1970). On the basis of radio observations Allen (1) confirmed the low redshift of NGC 7320, and (2) concluded that NGC 7320 lies most probably at a distance of ~ 10 Mpc. The latter conclusion is absolutely independent from the arguments presented here. He also received preprints by H. C. Arp and C. R. Lynds of papers presented at the *IAU* Symposium No. 44 '*External Galaxies and Quasi-Stellar Sources*' in Uppsala. Lynds determined new redshifts for 13 galaxies in the vicinity of NGC 7331; their mean redshift (including the published value of NGC 7335) is $\bar{v}_0 \approx 7350 \pm 1010$ (standard deviation) km sec^{-1}. These redshifts give strong support to the reality of at least the south-following part of the cluster 2231.2 + 3732, and Lynds concluded that Stephan's quartet is physically associated with these galaxies, for which he adopted a distance of ∼90 Mpc. His conclusion of NGC 7320 being a foreground galaxy and a companion to NGC 7331 agrees very well with the present results. On the other hand, Arp proposed that all or most galaxies surrounding NGC 7331, including Stephan's quintet, are ejected from this galaxy and are therefore at virtually identical distances. It should be noted that this unconventional interpretation is contradicted by the present arguments (1), (2), and (4), which indicate that the distance of NGC 7331 and NGC 7320 differs significantly from that of the remaining quartet. In addition, Dr. S. van den Bergh (private communication) has tentatively classified NGC 7337, the galaxy second nearest to NGC 7331, on a IIIa-J plate taken with the 48-inch Schmidt telescope and he finds at least for this SBb galaxy that its high luminosity class is much more consistent with a distance of ∼90 Mpc than with the small distance of NGC 7331.

The author is indebted to Drs. H. C. Arp and C. R. Lynds for making their preprints available to him, and to Dr. S. van den Bergh for the information concerning NGC 7337. He acknowledges that this work was supported by the National Science Foundation under grant GP-14801.

REFERENCES

Arp, H., 1966, *Atlas of Peculiar Galaxies* (California Institute of Technology, Pasadena) No. 319, p. 54.
Baade, W., and Swope, H. H., 1963, *Astron. J.*, **68**, 435.
Beale, S., and Davies, D., 1969, *Nature*, **221**, 531.
Bergh, S. van den, 1960a, *Astrophys. J.*, **131**, 215.
Bergh, S. van den, 1960b, *Publ. David Dunlap Obs.*, **2**, 159.
Burbidge, E. M., and Burbidge, G. R., 1961a, *Astrophys. J.*, **134**, 244.
Burbidge, E. M., and Burbidge, G. R., 1961b, *Astron. J.*, **66**, 541.
Gouguenheim, L., 1969, *Astron. Astrophys.*, **3**, 281.
Holmberg, E., 1958, *Medd. Lunds Astron. Obs.*, Ser. II, Nr. 136.
Holmberg, E., 1969, *Ark. Astron.*, **5**, 305.
Humason, M. L., Mayall. N. U., and Sandage A. R., 1956, *Astron. J.*, **61**, 97.
Roberts, M. S., 1969, in *Galaxies and the Universe*, ed. A. and M. Sandage (Univ. Chicago Press), in press.
Rubin, V. C., Burbidge, E. M., Burbidge, G. R., and Crampin, D. J., 1965, *Astrophys. J.*, **141**, 759.
Sandage, A., 1961, *The Hubble Atlas of Galaxies* (Carnegie Institution, Washington), p. 36.
Sandage, A., 1968, *Astrophys. J.*, **152**, L149.
Stephan, E., 1877, *C R Acad. Sci.* **84**, 641; *Monthly Not. Roy. Astron. Soc.*, **37**, 334.
Tammann, G. A., 1969, *Mitt. Astron. Ges.*, No. 27, p. 55.
Tammann, G. A., and Sandage, A., 1968, *Astrophys. J.*, **151**, 825.
Toomre, A., 1970, The Spiral Structure of Our Galaxy (*IAU Symposium No. 38*), ed. W. Becker and G. Contopoulos (Reidel Publishing Co., Dordrecht, and Springer-Verlag, New York), p. 334.
Vaucouleurs, G. de, 1969, *Galaxies and the Universe*, ed. A. and M. Sandage (Univ. Chicago Press), in press.
Vaucouleurs, G. de, and Malik, G. M., 1969, *Monthly Not. Roy. Astron. Soc.*, **142**, 387.
Vaucouleurs, G. de, and Vaucouleurs, A. de., 1964, *Reference Catalogue of Bright Galaxies* (University of Texas Press, Austin), p. 215.
Zwicky, F., and Kowal, C. T., 1968, *Catalogue of Galaxies and of Clusters of Galaxies* (California Institute of Technology, Pasadena), Vol. 6, p. 128 ff.

Received 11 September 1970

THE RADIAL VELOCITIES OF GALAXIES NEAR NGC 7331

C. R. LYNDS

Kitt Peak National Observatory, Tucson, Ariz., U.S.A.

Abstract. Radial velocity determinations for galaxies in the vicinity of NGC 7331, including the members of 'Stephan's Quintet', show clearly a superposition of at least two distinct physical systems. One includes NGC 7331, NGC 7320 (a Quintet member), and NGC 7343, while the other – seven or eight times more distant – contains the four remaining 'Quintet' constituents and six of the seven optical 'companions' of NGC 7331.

I report here the results of a partially completed program of radial velocity determinations for galaxies in the vicinity of NGC 7331 – a region that includes the group known as Stephan's Quintet. The motivation for the investigation and the context of this discussion have to do with the reality of the physical association of galaxies, where suggestions of such associations are based on apparent geometrical grouping. The discussion is specifically directed at two apparent groups: (a) Stephan's Quintet; and (b) NGC 7331 together with six or seven apparent companions.

Stephan's Quintet is located approximately $0°.5$ south-preceding NGC 7331 and consists of the galaxies NGC 7317, 7318a, 7318b, and 7319 having radial velocities in the neighborhood of 6500 km s^{-1} (de Vaucouleurs and de Vaucouleurs, 1964) and NGC 7320 having a velocity of approximately 800 km s^{-1}. The highly discordant velocity of NGC 7320 was discovered by E. M. and G. R. Burbidge (1961) who also discussed the possibility of the galaxy's membership in the group and the consequent very large internal energy of the system. There appear to be two alternatives: (a) NGC 7320 is a physical member and the system is highly unstable; or (b) NGC 7320 is a foreground galaxy.

NGC 7331 is an Sb spiral galaxy of moderately large angular extent (approximately $10'$ along the major axis) having no companions of comparable apparent size in its vicinity; the nearest comparable galaxies are more than $6°$ distant. On the other hand, there are several galaxies of substantially smaller angular dimensions nearby, and the nearer of these together with NGC 7331 constitutes one of the 174 systems discussed by Holmberg (1969) as possible physical systems. Holmberg notes seven companions to NGC 7331 within a radius of approximately $16'$ and concludes that, on a statistical basis, four among these are likely to be physically associated with the larger galaxy. However, one of the apparent companions, NGC 7335, is known to have a radial velocity near 6300 km s^{-1}, as compared with the much smaller value of approximately 800 km s^{-1} for NGC 7331 (de Vaucouleurs and de Vaucouleurs, 1964). The only other galaxy in the field to have a published radial velocity is NGC 7343; it has a velocity of about 1200 km s^{-1} but is outside Holmberg's survey area.

The similarity between the radial velocities of NGC 7331, NGC 7343, and the discordant member of Stephan's Quintet, on the one hand, and between the velocities of the other members of the Quintet and that of one of the optical companions of

D. S. Evans (ed.), External Galaxies and Quasi Stellar Objects, 376–379.

NGC 7331, on the other, led me to investigate further the possibility that the distribution of galaxies in the field results from the superposition of a nearby and a distant group of galaxies. To this end, radial velocities have been determined for several of the numerous other galaxies in the field. These observations, together with the few existing determinations already mentioned, are indicated in Figure 1. This illustration shows a 1°.7 × 1°.7 field centered near the large spiral galaxy NGC 7331; the radial velocities are given to two significant figures and are uncorrected for the rotation of our galaxy.

A careful inspection of the field reveals what appears to be a large association of

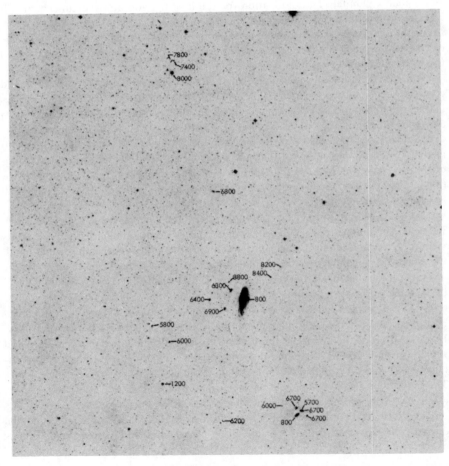

Fig. 1. A 1°.7 × 1°.7 field centered near the large spiral galaxy NGC 7331. North is at the top; East is toward the left. The radial velocities that are known for galaxies in the field are indicated to two significant figures.

galaxies having apparent dimensions of the order of those of the members of Stephan's Quintet and the optical companions of NGC 7331. It will also be noted that the radial velocities for those galaxies that have been observed show a marked segregation into two groups: one at about 1000 km s^{-1} which includes NGC 7331, NGC 7343, and NGC 7320; and one at about 7000 km s^{-1} which includes the remaining 18 galaxies.

The simplest and most straightforward interpretation of the observations is that there are two distinct physical groups of galaxies: (a) NGC 7331, together with two confirmed physical companions, NGC 7320 and NGC 7343, at a distance of 11 Mpc (based on the distance modulus of 30.2 adopted by Holmberg for NGC 7331); and (b) a background cloud at approximately 90 Mpc (based on a Hubble parameter of 80 km s^{-1} Mpc^{-1}, the value adopted by Holmberg) involving a large fraction of the galaxies in the field as members. Six of the seven optical companions to NGC 7331 noted by Holmberg appear to be members of the background cloud. (There is no velocity determination for the seventh optical companion, which is evidently the low surface-brightness galaxy approximately 12′ north of NGC 7331.) The same may be concluded for the four high-velocity members of Stephan's Quintet. There is some indication of a division in the velocity distribution for the distant cluster, but this cannot be regarded as well-founded at present. In any case, the velocity dispersion among the background galaxies, taken together, is not unexpectedly large for a cluster. A more complete census and further velocity determinations in the field can be expected to delineate the composition of the two associations more clearly.

The interpretation of the observations given here appears to me to be in complete agreement with known facts and does not seem to require any appreciable modification of accepted and well-founded concepts concerning the phenomenon of clustering or the expansion of the Universe. Calculations of the probability of the chance superposition of a foreground or background galaxy on an apparent group of galaxies, such as that made by the Burbidges for NGC 7320 in the case of Stephan's Quintet, are interesting but appear to me to be of little assistance in establishing membership in an isolated instance. A study of a large sample of apparent groups of galaxies might establish whether or not there is a statistically meaningful tendency for more than the expected number of groups to have a member with a discordant velocity, but an affirmative conclusion, although far-reaching and significant, would only be statistical in nature and not necessarily applicable in drawing conclusions concerning any particular group in the sample. A similar statement may be made concerning the very exhaustive study of groups of galaxies made by Holmberg. There is nothing in the paper by Holmberg or that by the Burbidges that indicates that these authors significantly disagree with this point of view.

In partial reconciliation of the present observations with the statistical prediction of Holmberg, it is worth noting that the background cluster of galaxies discussed here does not appear to extend as far as the comparison areas used in Holmberg's survey. And, in connection with what I now prefer to call 'Stephan's Quartet', a certain amount of 'sub-clustering' may well be a characteristic of the background cluster, for it will be noticed that the three northernmost galaxies for which velocities were determined appear to be members of a compact apparent septet that may later be established as a physical group. However, even if all members of this group have similar radial velocities, there is no assurance that some of the apparent members are not simply members of the larger cloud projected on the group.

The essence of the method applied in this investigation has been simply the addition, through the Hubble law, of radial distance to the two tangential coordinates as a metric dimension in which the observed distribution of galaxies may be used as a clustering criterion.

Acknowledgements

This paper was written while the author was visiting the Astronomy Centre, University of Sussex. The author acknowledges helpful discussions with H. C. Arp and T. D. Kinman during the course of this investigation.

References

Burbidge, E. M. and Burbidge, G. R.: 1961, *Astrophys. J.* **134**, 244.
Holmberg, E.: 1969, *Arkiv Astron.* **5**, No. 20.
de Vaucouleurs, G. de Vaucouleurs, A.: 1964, *Reference Catalog of Bright Galaxies*, University of Texas Press, Austin.

Discussion

Mrs Rubin: Is the scale of the galaxies in Stephan's Quintet sufficient to study the stellar population, or other distance discriminants, to establish the distances of the galaxies, independently of their velocities?

Mrs Burbidge: It occurred to Geoff and me that it should be possible to look at the diameters of the H II regions in NGC 7331 and NGC 7320; NGC 7331 is of similar redshift to the Virgo cluster galaxies. There might be other discriminants that could be used in the classifications by van den Bergh and by de Vaucouleurs.

Allen: Using the large radio telescope at Nançay, we have succeeded in observing neutral hydrogen in NGC 7320. I have compared the hydrogen mass and total mass derived from the observations with the values obtained on other Sbc galaxies. The result is that NGC 7320 is most probably at a distance of about 10 Mpc, corresponding to its redshift, and not at 67 Mpc which corresponds to the redshift of the other four members of Stephan's Quintet. (This research will be published in *Astron. Astrophys.*).

(Reprinted from Nature, Vol. 239, No. 5366, pp. 43–44,
September 1, 1972)

Small "Non-velocity" Redshifts in Galaxies

THE question of whether or not the small excess redshifts observed in some companion galaxies[1,2] have a non-velocity origin has recently been raised. The de Vaucouleurs[3] have pointed out that small, systematic errors in the redshifts of E and SO galaxies might arise because the lines in the low dispersion spectra on which they are measured do not have well-determined effective wavelengths. The data reported here support this possibility. In the three cases I present, systematic differences arise between velocities derived from measurements of CA II, H(λ3968.5) and K(λ3933.7) carried out "classically" (with a visually positioned, two coordinate measuring engine) and velocities derived from either "classical" measurements of other spectral lines in the same object or measurements of the same H and K lines using numerical techniques.

The strongest evidence for such systematic velocity differences comes from a measurement technique which finds the rotation curve for a galaxy by cross-correlating segments of its spectrum with that of a star of similar spectral type. For this method to work the star and galaxy spectra must be taken with the same instrument under similar conditions, measured with a digital microdensitometer, converted to relative intensities and

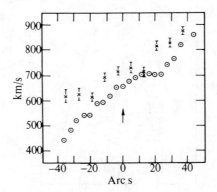

Fig. 1 Central rotation curve for NGC 5866. Points with error bars ($\times$) are from "classical" measurements, those with open circles ($\circ$) from cross-correlation with the spectrum of η Draconis. Size of bars and circles is 2σ. The arrow indicates the nucleus.

the stellar spectrum numerically "broadened"[4]. The results of such measurements for the linear part of the rotation curve of NGC 5866 (SO spectral type K) are plotted with open circles in Fig. 1. The comparison star used for these was η Draconis (G8 III). The measurements are all for a single galaxy plate, a 130 Å/mm, image tube spectrum taken at Kitt Peak, whose zero point was determined by measuring night sky lines.

Fig. 2 shows the cross-correlation calculated for three limited spectral regions of the galaxy's nucleus and suitably normalized. There is good agreement between the redshift determined from the region of H and K, where the continuum slopes steeply, and that found using the region 4000–4200 Å, which is relatively flat and void of prominent lines. The spectral regions 3800–4000 Å, 3800–4200 Å, and 4000–4200 Å, give nuclear redshifts of 642, 664, and 649 km/s, respectively.

The points plotted with error bars in Fig. 1 were obtained by measuring H and K in the "classical" way: by setting on the absorption lines visually with a two-coordinate measuring engine. A smooth curve through these points gives a nuclear velocity of 725 ± 30 km/s.

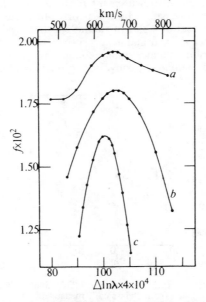

Fig. 2   Cross-correlation between spectra of η Draconis and of the nucleus of NGC 5866. Zero point for the ordinate is arbitrary.   a, 4000–4200 Å;   b, 3800–4200 Å;   c, 3800–4000 Å.

Because the cross-correlation measurements for the H and K lines (3800–4000 Å) agree with those for the "flat" region (4000–4200 Å), the difference between the "classical" measurements and the "impersonal" computer calculation cannot be intrinsic to the CA II lines themselves. Moreover, the only possible source of systematic error between the two curves is the velocity of η Draconis relative to its comparison spectrum, which provides the zero point for the cross-correlation measurements. This velocity, however, is well determined (with a standard error of 13 km/s) and any error it introduces should be well within the observed difference of 73 km/s. Thus, the systematic shift between the "classically" measured points and those measured with the cross-correlation technique must arise from the different method of measurement. In short, measurements of Ca H and K absorption lines carried out "classically" require a different effective wavelength than those made using a template constructed from a stellar spectrum which closely mimics that of the galaxy.

The second line of evidence for a systematic shift comes from determining the centroids of the galaxy H and K lines numerically. This procedure gives a recessional velocity of 635 km/s when the slope of the continuum is left in, and 690 km/s when it is removed numerically. The latter value is in good agreement with a similar measurement by de Vaucouleurs[5] of 692 km/s and is clearly an upper limit to the redshift. Both of these calculations lead to curves whose shapes are identical to that defined by the circles in Fig. 1. In this case, since both the "classical" measurements and the profile measurements were done on the same plate, there is no possibility of a zero point difference.

Finally, earlier "classical" measurements on NGC 2903 (my unpublished data) and NGC 5194[6] have shown a systematic difference between velocities from hydrogen lines (Hβ emission and Hθ absorption) and those from Ca K absorption lines. These shifts, again, show Ca K velocities to be 40–60 km/s larger than those for Hθ absorption or Hβ emission. Because all three of these lines are measured on the same plate simultaneously, there is no way in which a systematic zero point error could arise.

All these data suggest that velocities obtained from the H and K absorption lines of Ca II by "classical" methods give redshifts that are systematically too high by 40–70 km/s. This shift may arise from a tendency of the measurer to set the cross hairs on the denser, or redward edge of the sloping continuum in the region of these lines. Because redshifts for spiral galaxies are usually measured from emission lines while those for E and SO galaxies come principally from the absorption lines of CA H and K and the G band[7] one might well expect to find a systematic velocity difference between these two classes of objects. The possibility that the "classically" measured redshifts for E and SO galaxies are systematically too high could most easily be investigated by means of a device, such as that described by Lowrance et al.[8] which records galaxy spectra directly in intensity, thus facilitating the type of analyses described here. An investigation of this type on NGC 224, 3031, 5128 and their companions would either support or disprove the small redshift discrepancies within these groups noted by Arp[1]. But until such observations are conducted the evidence presented here would seem to suggest that any small "non-velocity" redshifts between spirals and SOs should be attributed to nothing more exotic than errors of measurement.

The microdensitometry and computing for this work were done at the Goddard Institute for Space Studies, NYC. Partial support was provided by a NSF grant.

SUSAN M. SIMKIN

Department of Astronomy,
Columbia University,
New York, NY,
and Kitt Peak National Observatory

Received June 22, 1972.

[1] Arp, H., *Nature*, **225**, 1033 (1970).
[2] Jaakkola, T., *Nature*, **234**, 534 (1971).
[3] de Vaucouleurs, G., and de Vaucouleurs, A., *Nature*, **236**, 166 (1972).
[4] Minkowski, R., *Problems of Extragalactic Research* (edit. by McVittie, G.), 112 (Macmillan, New York, 1962).
[5] de Vaucouleurs, G., *IAU Synposium No. 30* (edit. by Batten, A., and Heard, J.), 16 (Academic Press, New York, 1967).
[6] Simkin, S., *Bull. Amer. Astron. Soc.*, **1**, 362 (1969).
[7] Humason, M., Mayall, N., and Sandage, A., *Astron. J.*, **61**, 97 (1956).
[8] Lowrance, J., Morton, D., Zucchino, P., Oke, J., and Schmidt, M., *Astrophys. J.*, **171**, 233 (1972).

Reprinted from *The Observatory*, Vol. **91**, No. 985, pp. 209–214
December 1971

QUASARS AND COSMOLOGY

By Maarten Schmidt

The Halley Lecture for 1971, *delivered in Oxford on May* 6

MR. VICE-CHANCELLOR, LADIES AND GENTLEMEN,—

Modern observational cosmology was started by Hubble in 1929 when he showed that the velocities of recession of galaxies as measured by their redshifts were roughly proportional to their distances. The largest radial velocity in Hubble's material was 1090 kilometres per second. Humason set himself the task of detecting larger redshifts by observing fainter galaxies. This required much telescope time with the slow photographic plates then available. Humason's spectra of the twelfth magnitude galaxy NGC 7619 required exposure times of around 40 hours in 1929. The radial velocity was measured at 3779 kilometres per second, corresponding to a redshift $\Delta\lambda/\lambda_0$ of the spectral lines of about 0·0126. Eventually he obtained redshifts up to 0·13 with the Mount Wilson 100-inch telescope in 1936. Soon after the completion of the Palomar 200-inch telescope Humason obtained a redshift of 0·20 for galaxies in the Hydra cluster. These redshifts were based on shifts of absorption lines in the spectra of the galaxies. The night sky emits a complex spectrum that in the observations interferes with the galaxy spectrum. For redshifts larger than 20 per cent the galaxies are so faint that their absorption lines cannot be identified with certainty against the night-sky spectrum.

Larger redshifts are usually only detectable in spectra with emission lines. Radio astronomy became involved at this stage since a large fraction of the galaxies identified as strong radio sources have an emission-line spectrum. Minkowski managed to obtain a redshift of 0·46 for the radio galaxy 3C 295 in 1960, still the largest galaxy redshift to date.

Accurate radio positions determined at Cambridge and at Caltech's Owens Valley Radio Observatory in the early sixties pointed to the identification of some sources with "stars" rather than galaxies. The prototype of the *quasars*, as they are now called, is 3C 273. This star-like object of thirteenth magnitude shows a spectrum containing six emission lines. These were identified as hydrogen Balmer lines and some others, at a redshift of 0·16. Such a redshift is normally seen only in galaxies of magnitude 17 or fainter. Much larger redshifts of quasars, up to around 2, were detected in 1965. At these large redshifts strong emission lines of hydrogen (Lyman-α at 1216 Ångströms) and triply ionized carbon (at 1550 Ångströms) become observable in the photographic part of the spectrum. The emission lines are usually some 40 Ångströms wide, corresponding to a velocity range of about 3000 kilometres per second. The largest redshift to date is 2·877 for the quasar 4C 05·34, observed by Lynds and Wills, or more than six times Minkowski's record redshift for a galaxy.

We define a quasar as an object with a dominant star-like component (some have associated faint nebulosity), with an emission-line spectrum showing a large redshift. The property described as "star-like" in practice means that the angular diameter is smaller than one second of arc. The light of most, and perhaps all, quasars is variable over time intervals between one week and several years. This important property indicates that much of the light originates in a volume less than a light week or a few light years across, depending on the rapidity of the variations.

All quasars that have been identified through their radio radiation show an ultra-violet excess relative to ordinary faint stars. This property has been used to search optically for quasars. Surveys by Luyten and Sandage show that there are around 0·4 quasars per square degree brighter than eighteenth magnitude. The number rises steeply toward fainter magnitudes and the total number in the sky is probably of the order of one or ten million. If there exist quasars with no ultra-violet excess, their detection will be very difficult.

Quasars with detectable radio emission (quasi-stellar radio sources) are relatively rare. Only one in 300 quasars brighter than eighteenth magnitude is a radio source strong enough to be included in the 3CR catalogue. The reason that most of the quasars studied so far are quasi-stellar radio sources is that the radio emission serves so effectively in singling out quasars, as it did in the discovery of these objects.

The radio properties of quasars are similar to those of the radio galaxies. In particular, most quasars show the double structure so characteristic of radio galaxies. At frequencies above 1000 megahertz some quasi-stellar sources are variable with time scales of months to years, especially those that have a flat or inverted radio spectrum. Interferometry with large base lines of many thousands of kilometres has shown the existence of very small components; in some cases measured angular diameters are less than 0·001 seconds of arc.

Most of the energy radiated by quasars appears in the infra-red. The prototype quasar 3C 273 radiates 10 to 100 times more energy in the infra-red than in the visual and ultra-violet regions of the spectrum. This quasar has also been observed as an X-ray source.

The origin of the large redshifts observed in quasars has been a central and somewhat controversial issue ever since their discovery. Most astronomers accept the cosmological hypothesis, which states that the redshifts are due

December 1971   *Quasars and Cosmology*   211

to the general expansion of the Universe. Alternative explanations are often called the "local" hypothesis, because they place the quasars at much smaller distances. Different versions of the local hypothesis assume that the redshift is due to an explosion centred on our Galaxy, that it is gravitational in nature, or leave the redshift unexplained.

The emission-line spectrum exhibited by quasars allows the derivation of some important properties of quasars depending on the interpretation of their redshift. The crucial observation is that most of the quasars show the forbidden $\lambda 3727$ line of ionized oxygen, when the line is in an accessible wavelength region. This line is observable with a strength comparable to that of the hydrogen lines only if the electron density in the emitting gas is 10^4 per cm^3 or lower. The hydrogen emission lines are a consequence of the recombination of protons and electrons in the tenuous gas. At an electron density of 10^4 per cm^3 and a reasonable electron temperature of 20,000° K the intensity of the hydrogen $H\beta$ emission line will be 6×10^{-18} ergs per cm^3. If the volume of the gas cloud is V (cm^3) and the distance of the quasar is r (cm) then we will measure a strength of the $H\beta$ line of $5 \times 10^{-19} \ V/r^2$ ergs/sec through an area of 1 cm^2. The observed flux in the $H\beta$ emission line for the typical quasar 3C 48 is 1×10^{-13} ergs/sec per cm^2. Hence, for this quasar

$$V/r^2 = 200,000 \qquad\qquad (1)$$

independently of the nature of the redshift.

Cosmological Hypothesis. If the redshift of 3C 48, measured to be 0·37, is cosmological, then its distance r is around 1100 megaparsecs, or 3.5×10^{27} cm. Hence $V = 2 \times 10^{60}$ cm^3, corresponding to a radius of 25 parsecs or 80 light years if the volume has a spherical shape. The mass of the gas cloud equals 20 million solar masses. The source of all the energy needed to ionize the gas cloud and to emit the intense continuum in optical and infra-red wavelengths is presumably inside the gas cloud. Its mass is uncertain. If the line-emitting gas cloud is stable the mass may be estimated from the size and the velocity dispersion of the gas cloud. The resulting mass of almost 10^{10} solar masses is similar to that of galaxies.

Gravitational Hypothesis. The gravitational redshift, which corresponds to the loss of energy suffered by the photon in leaving the gravitational field of an object, is small in "ordinary" astronomical objects. Since it is to first order proportional to the ratio of mass/radius the gravitational redshift can become large in objects of large mass or small radius. If 3C 48 had a mass of one solar mass, its radius would have to be less than 4 kilometres to allow the gas to exhibit a redshift of 0·37. At the corresponding volume of 270 km^3 or 2.7×10^{17} cm^3, formula (1) requires a distance of about 10^6 cm or 10 kilometres!

Larger distances can only be obtained by adopting a larger mass for the quasar. For a mass of 10^5 solar masses 3C 48 would still be inside the solar system. For a galactic-type mass of 10^{11} solar masses the distance according to formula (1) would be 10 kiloparsecs and 3C 48 would rival the gravitational field of our own Galaxy. Larger masses will place the quasars outside our Galaxy and their contribution to the mass density of the Universe becomes of importance. The mass density due to quasars only becomes smaller than the critical value 2×10^{-29} grams per cm^3 when the average quasar mass is at least 2×10^{13} solar masses, or ten times larger than the largest known mass of a galaxy. At this minimum mass the nearest quasar would be at a distance of at least 4 megaparsecs (about six times the distance to the Andromeda galaxy) and the 10^6 quasars that must exist would have to be distributed out to a distance of 400 megaparsecs. At this distance the cosmological redshift already amounts to 0·13.

Local Doppler Hypothesis. The hypothesis proposed by Terrell is that the quasars were ejected from the centre of our Galaxy about five million years ago. Distances are reduced by a factor of about 2000 compared to those based on the cosmological hypothesis. The volume of the gas cloud of 3C 48 (see formula (1)) is reduced to 5×10^{53} cm³ and the mass of the gas to 5 solar masses. If the fast-moving gas is stable, then the total mass of the quasar is around 4×10^7 solar masses. If the gas envelope is escaping then mass loss occurs that may be shown to amount to at least 10^4 solar masses since the explosion took place.

The kinetic energy carried by 10^6 quasars of at least 10^4 solar masses each at velocities close to the velocity of light is around 10^{64} ergs. The generation of this energy requires a mass of around 10^{10} solar masses at 100 per cent efficiency, or 10^{11} solar masses at a still optimistic efficiency of 10 per cent. The latter mass represents essentially all the mass in our Galaxy. This excessive minimum requirement essentially rules out the local-Doppler hypothesis for the quasars.

The above considerations based on the emission-line spectra all depend on the electron density for which we used the maximum value of 10^4 per cm³ derived from the observed $\lambda 3727$ emission. For smaller values of electron density the individual quasar masses derived for the local hypothesis become even larger. Also, for the gravitational hypothesis we have assumed that the entire volume inside the radius discussed has almost the same gravitational redshift. This would be the case in the relativistic cluster models proposed by Fowler and Hoyle. If this is not the case, then masses and distances derived for the gravitational hypothesis have to be increased.

There is little direct evidence that quasars are at cosmological distances. A redshift-magnitude diagram shows only a weak relationship with much scatter, in the sense that fainter quasars show on the average larger redshifts. This is no argument against the cosmological distances since the scatter may well be due to differences in the absolute magnitude of quasars. The correlation between the radio angular diameter of quasars and their redshifts found by Miley does not directly yield information on distances as it may be due to a redshift-dependent property of the quasars. Other evidence may be obtained from associations with galaxies or clusters of galaxies, or from similarities between quasars and other astronomical objects.

Associations of quasars with other kinds of objects would be informative. No quasars have been found associated with very rich clusters of galaxies yet. Some cases of quasars associated with small groups of galaxies have been reported. The best case is that of the quasar PKS 2251+11, redshift 0·32, reported by Gunn. The quasar is seen in the direction of a small group of galaxies, the brightest of which Gunn found to have essentially the same redshift as the quasar. This result constitutes the strongest support of the cosmological hypothesis.

Arp, on the other hand, has claimed that in several cases a luminous link exists between a quasar and a galaxy of much lower redshift. A statistical evaluation of such associations and of others that have been suggested for galaxies of different redshift is very difficult, but obviously of great importance.

December 1971   *Quasars and Cosmology*   213

Similarities have been noticed between the nuclei of Seyfert galaxies and quasars, in properties such as the ultra-violet excess of the continuum, the wide emission lines, the strong infra-red radiation, and in some cases the rapid variability. G. R. Burbidge feels that this requires that Seyfert nuclei and quasars have essentially the same absolute magnitude, leading to typical quasar distances of around 100 Megaparsecs. This would leave the redshifts unexplained, of course. It seems quite arbitrary to require the same absolute magnitude: galaxies showing similar properties are known to have luminosities that may differ by a factor of 100.

On the basis of cosmological redshifts quasi-stellar sources and radio galaxies have many similarities. In particular, the quasi-stellar sources have radio luminosities up to around 10^{45} ergs per second and double components separated by up to around 500 kiloparsecs, essentially the same as observed for radio galaxies. Very small, often variable components are occasionally found in quasi-stellar sources and in radio galaxies, though more frequently in the former. The similarities are so strong that radio astronomers cannot with certainty distinguish quasi-stellar sources from radio galaxies in individual cases. It is attractive on the cosmological hypothesis to consider a possible evolutionary relationship between these two major kinds of radio sources.

Recent observations by U.S. radio astronomers with long-base-line interferometers show that small components of the radio sources 3C 279 and 3C 273 are separating from each other with very high speeds. For 3C 279 the expansion rate, computed on the basis of a cosmological redshift distance, equals six times the velocity of light. At a smaller distance the expansion rate is reduced, of course, and one may argue that this observation speaks in favour of the local hypothesis. However, the observations are *not* in conflict with special relativity. As shown by Rees, apparent super-relativistic transverse velocities can be observed if the components are ejected with relativistic speeds at small angles relative to the line of sight.

Absorption Spectra. The observation of absorption lines in the spectra of quasars has only added to their enigma. First, the *absence* of absorption due to intergalactic neutral atomic hydrogen is to be noticed. This absorption would have appeared below the Lyman-α emission line in quasars of sufficient cosmological redshift. The absence requires either that the density of intergalactic neutral atomic hydrogen is less than 10^{-11} atoms per cm^3 at redshift 2, or that the redshifts are not cosmological.

Narrow absorption lines are observed in the spectra of a number of quasars, especially in those of large redshift. Many of the lines can be identified as due to hydrogen, carbon, silicon, etc., in various stages of ionization. Their redshifts are usually a little smaller than the emission line redshift and one would tend to interpret them as caused by material expelled by the quasar. In a few quasars, however, some 50 to 100 narrow absorption lines are observed, and in these cases there is evidence for many different redshift systems. A typical case is the 16·6 magnitude quasar PHL 957 which has an emission-line redshift of 2·69 and in which some 80 absorption lines provide evidence for the existence of eight redshift systems, with redshifts between 2·0 and 2·7.

If the absorbing material has been ejected by the quasar then ejection velocities of some one or two tenths of the velocity of light, and in some cases even more, are required to explain the difference in redshifts. The velocity dispersion in each redshift system cannot be more than 25 kilometres per second so as to account for the small widths of the absorption lines in PHL 957. It is difficult to understand how material ejected at some 50,000 kilometres per second can have a velocity dispersion of less than 25 kilometres per second. An alternative hypothesis is that the different absorptions are due to galaxies or concentrations of matter at a smaller cosmological redshift than the quasars. The observed number of absorption systems is larger than one would expect on this hypothesis.

Distribution of Quasars. The distribution of quasars over the sky appears to be isotropic. This follows from the observed isotropy of faint radio sources around the sky, 20 to 30 per cent of which are quasi-stellar sources. This isotropy is to be expected in the cosmological hypothesis. In the gravitational hypothesis it would require that we are situated close to the middle of the postulated quasar cloud of 400 megaparsecs' radius.

The distribution of quasars *in depth* appears to be far from uniform. Counts of quasars detected optically, from their ultraviolet excess, show that around magnitude 18 their number increases by a factor of about 6 per magnitude. If quasars were uniformly distributed through space, then at the cosmological distances involved their number should increase by a factor of only 2 per magnitude. The observations thus suggest that at large distances there are many more quasars than nearby. Or, since large distances observed correspond to long light travel times, that at early cosmic times the number of quasars was much larger, by a factor of perhaps 1000, than at present.

The derivation of the space distribution of *radio* quasars is complicated by the fact that optical and radio selection effects operate simultaneously, since the redshift of a radio quasar can only be determined at optical wavelengths, and hence requires optical identification. An accurate method can be used in which the position of each quasar in the volume available to it within the sample limits is considered. Application of the method to quasars in the Revised 3C catalogue shows that these radio-intense quasars have a space distribution similar to quasars detected optically.

As it seems likely that quasars have cosmically short life times, of perhaps 10^6 to 10^8 years, the numbers seen at a given epoch represent the birth rate of quasars. There are indications that the increase in density may not continue beyond a redshift of 2·5. This would suggest that the greatest activity in quasar births took place some 1 or 1·5 thousand million years after the expansion of the Universe started. If galaxies were forming at that time, then one may speculate that quasars represented a stage in the formation of the galaxies or in that of their nuclei.

THE ASTROPHYSICAL JOURNAL, **164**:L113–L118, 1971 March 15

ON THE DISTANCES OF THE QUASI-STELLAR OBJECTS

JAMES E. GUNN
Hale Observatories, California Institute of Technology,
Carnegie Institution of Washington
Received 1971 January 16

ABSTRACT

It is shown that Ton 256 ($z = 0.131$) and PKS 2251+11 ($z = 0.323$) are associated with galaxies of essentially the same redshift, thus implying that these objects are at cosmological distances. The nature of Ton 256 is questionable, and it is argued that it represents a transition between Seyfert galaxies and QSOs. There is no doubt that PKS 2251+11 is a bona fide QSO; it is bright ($M_v = -24.7$), blue ($B - V = 0.20$, $U - B = -0.84$), has a starlike image under high resolution, and is a strong radio source. A peculiar emission structure near PKS 2251+11 is also discussed.

The question of distances to the QSOs has been open since their nature was first discussed in the light of their large redshifts (Greenstein and Schmidt 1964). The hypothesis that they are "local," with distances of the order of tens of megaparsecs (Hoyle and Burbidge 1966; Hoyle and Fowler 1967; Terrell 1964; Arp 1966), has been supported mainly on the grounds that their variations are very difficult to explain if they are at the cosmological distances implied by their redshifts. The discovery of luminous, optically variable N-type galaxies and Seyfert nuclei which are certainly distant has somewhat weakened these theoretical arguments, and it is now widely—though certainly not universally—accepted that they are at cosmological distances, but as yet no conclusive positive evidence has been offered.

It has been shown (Bahcall, Schmidt, and Gunn 1969) that the object B264 is in a cluster of galaxies with the same redshift. The object, however, is significantly less luminous than bright galaxies, and Arp (1970) has shown it to be nonstellar; it is probably an N-type galaxy. The positive association of a cluster with a bona fide QSO would provide clear, direct confirmation of the cosmological nature of these objects. It is the purpose of this Letter to report two such cases of association: one, as before, somewhat clouded by the uncertain nature of the object (Ton 256), the other seemingly conclusive (PKS 2251+11).

That the cluster Zw 1612.7+2624 was superposed on Ton 256 was first noted by Bahcall and Bahcall (1970), who give estimated magnitudes for the members. The object Ton 256 was included in the original list of radio-quiet QSOs by Sandage (1964), but was recently found by Arp (1970) to be nonstellar, and, in fact, clearly to possess an envelope of some 15–20 kpc extent. Surface photometry is impossible because of the bright nucleus, but the photograph suggests that the nonthermal source is in the nucleus of a giant elliptical galaxy. The case for this object is quite different than for B264, however, since even if the contribution from Hβ and the nebular lines is omitted, the absolute rest-frame V-magnitude of the source is -22.7 if one takes $z = 0.131$, $V = 15.4$ (Iriarte 1959), $H = 75$, and 0.3 mag correction for the emission lines. Unpublished scans by J. B. Oke corroborate this value. The brightest galaxies have a mean visual absolute magnitude of -22.0 with a standard deviation of about 0.3 mag (Sandage 1968); thus the nucleus of Ton 256 is probably at least as bright as the brightest galaxies. If it is shown that the really bright QSOs are at cosmological distances, Ton 256 suggests itself as an excellent example of an intermediate case between the Seyfert nuclei and the QSOs. It should be noted that III Zw 2 is another such overluminous, presumably nonthermal, source with a companion with measured redshift (Sargent 1970).

All of this, of course, assumes that the redshift of Ton 256 itself is cosmological in origin, and since any stellar spectral features from the underlying galaxy are completely obliterated by the nonthermal radiation from the nucleus, this is not by any means obviously the case. Spectra at 190 Å mm^{-1} have been obtained with the 200-inch Cassegrain image-tube spectrograph for galaxies 1, 2, and 11 of Bahcall and Bahcall, and for a galaxy (B) 8″ south-preceding Ton 256, visible clearly on Arp's published photograph.

The results for objects 1, 2, and B are summarized in Table 1; the spectrogram of galaxy 11 was of very poor quality, and no redshift was obtained; it seems certain, however, that it is very different from 0.131. For all three of the others, the H- and K-lines and the G-band were easily measurable, and the agreement with the emission redshift of Ton 256 is excellent. Galaxy B, in particular, which is only 20 kpc distant in projection, agrees exactly to within the errors, about 300 km sec^{-1}. *If* the real distance is the projected distance, and *if* the motion is in the line of sight, this implies a maximum mass for Ton 256 of $10^{12} \mathfrak{M}_\odot$; there is, of course, no reason to assume the above conditions true, but it seems unlikely that the mass is enormously greater than ordinary galactic masses.

There seem to be no grounds for reservations about the nature of PKS 2251+11. A photograph of PKS 2251+11 obtained with an ITT 4708 image converter at the prime focus of the 200-inch telescope with a filter combination passing the 6000–7000 Å band is shown in Figure 1 (Plate L2). The image is starlike within the resolution, except for a jet-like protuberance to the south, the reality of which is doubtful. Arp (1970) has concluded on the basis of a 200-inch plate of comparable resolution that the object is starlike and that, if one insists on this quality for "real" QSOs, it is a bona fide QSO.

A scan obtained with the multichannel spectrometer (Oke 1969) on the 200-inch is presented in Figure 2 (a continuum scan has been published earlier by Oke and Neugebauer 1970, and the emission spectrum has been studied by Wampler 1968). If its redshift of 0.323 is cosmological, its continuum flux alone gives an absolute visual magnitude of -24.7 in its rest frame (for $q_0 = \frac{1}{2}$, $H = 75$), safely in the category of very powerful sources. The spectrum is quite flat, the continuum spectral index through the optical and near-ultraviolet being about -0.5. This index extrapolates the spectrum neatly to the 2000-MHz radio point, but the infrared (2 μ) point is a bit high. The corresponding colors are quite blue ($B - V = 0.20$, $U - B = -0.84$), in the middle of the range of "normal" QSO colors (Burbidge and Burbidge 1967).

It is evident from Figure 1 that PKS 2251+11 is superposed on a small, compact cluster of galaxies, largely contained in the small, field of the image-tube plate. Most of these galaxies are visible on Arp's plate of this object. The brightest of these is galaxy 1, 28″ distant from the QSO. A scan of this object is presented in Figure 3. The wavelength resolution is 80 Å between λλ5400 and 6000, 160 Å to the red and blue of those points. The sharp drop characteristic of galaxies at $\lambda \sim 4000$ Å, due mainly to the H- and K-lines of Ca II, occurs clearly at about λ5320, a fit to the "standard" spectrum of Oke and Sandage (1968) yielding a redshift of 0.33 ± 0.01.

This "standard" elliptical galaxy, synthesized from the nuclei of M31 and NGC 3379,

TABLE 1

TONANTZINTLA 256 CLUSTER-GALAXY REDSHIFTS

Object	Ca II H	Ca II K	G-band	Mg I "b"	Average
B&B* No. 1	0.118	0.127	0.110	0.126	0.1205 ± 0.005
B&B No. 2	0.1308	0.1313	0.1313	0.134	0.1318 ± 0.001
Ton 256B	0.1313	0.1319	0.1320	...	0.1317 ± 0.001

* Bahcall and Bahcall (1970).

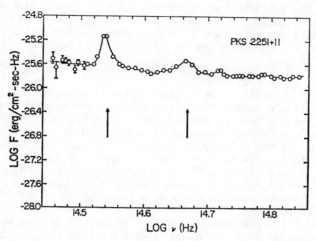

FIG. 2.—Multichannel scan of PKS 2251+11. *Arrows,* Hα and Hβ. Diaphragm size was 10″

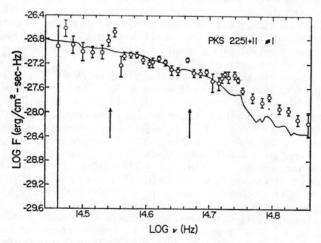

FIG. 3.—Multichannel scan of 2251+11 galaxy 1. *Light solid line,* a "standard" elliptical spectrum redshifted to $z = 0.323$. (See text.) Diaphragm size was 7″.

is shown by comparison, redshifted to $z = 0.323$. The drop at rest wavelength ∼5200, due mainly to the Mg ɪ *b*-lines, is also seen, as well as the structure to the blue of H and K. The only discrepancy seems to be the weakness of the G-band and the uniformly higher ultraviolet. It is clear from the photograph, however, that galaxy 1 is not an elliptical, but is highly elongated, with a pronounced centered nucleus. An Sb spiral with its younger stellar population would have exactly the required spectral properties, and it is concluded that the system is probably a middle-to-late spiral. The redshift is clinched by the appearance of Hα and Hβ in emission; the arrows indicate the predicted locations of the lines for $z = 0.323$.

We conclude finally that the redshift of galaxy 1 is the same as that of the QSO, to within an estimated total uncertainty of 3 percent, 0.01 in z, or about 3000 km sec⁻¹.

L116   JAMES E. GUNN   Vol. 164

The appearance of the Balmer lines in emission at the observed level is puzzling. The nucleus of the galaxy is bright but extended, and it seems unlikely that it is a Seyfert. The Hβ flux is 5×10^{-15} ergs cm^{-2} sec^{-1}, corresponding to a luminosity of 1×10^{42} ergs sec^{-1} in the frame of the galaxy. Taking the Hβ emissivity to be (Aller and Liller 1968)

$$j_\beta \simeq 10^{-25} N_e{}^2 (10^4/T) \text{ ergs sec}^{-1}$$

and taking the gas in the galaxy to be in a disk 10 kpc in radius and 200 pc thick (the radius corresponding to the measured image, the thickness a blind guess that our Galaxy is representative), we find that the desired value of $\langle N_e{}^2 \rangle$ is about 5 for temperatures around 10^4 ° K. Thus, *most* of the hydrogen must be ionized if the galaxy is typical of our own. What provides the flux of ionizing radiation? The obvious candidate is the QSO. For the observed Hβ flux, the number of recombinations leading to the emission of Balmer photons is of order 10^{54} sec^{-1}. The extrapolated luminosity for the QSO is 1.6×10^{30} ergs sec^{-1} Hz^{-1} at the Lyman limit, corresponding to about 5×10^{53} ionizing photons per second incident on the galaxy if its distance from the QSO is about equal to its projected distance of 100 kpc.

The galaxy is easily calculated to be very thick in the Lyman lines, so Menzel and Baker's case B applies, for which one Balmer photon is created per recombination, and within a very uncertain factor of 2 the ionizations expected from the present QSO flux balance the observed recombinations. The primary uncertainties are the solid angle, projection effects, and the obvious fact that the flux seen by the galaxy refers to an epoch perhaps 3×10^5 years earlier than the measured QSO flux. The weakness of the forbidden lines, particularly $\lambda 3727$ and the nebular lines, might be explained either by low temperatures or very high ionization resulting from the flat ultraviolet spectrum of the QSO. This problem will be pursued in a later paper.

The absolute proper visual magnitude of the galaxy for $H = 75$, $q_0 = \frac{1}{2}$ is -21.2 from the scan; the apparent visual magnitude is about 20.3. The other galaxies in the group are fainter yet, probably prohibitively faint for either spectra or scans; the scan of galaxy 1 required nearly 4 hours of telescope time, and comparable accuracy on even galaxy 2 would require a whole night. Thus, while a check on other members of the group would be desirable, it seems an impossible task with present instrumentation.

One can, however, ask the following question: What is the probability of finding a galaxy of absolute magnitude -21 or brighter within 30'' of a given random position and within 3000 km sec^{-1} of a given redshift near 0.3? The answer is a realistic assessment of the probability of a chance coincidence, if possible physical connections such as the ionization in the galaxy are ignored. The calculation of this probability would be trivial if the luminosity function for galaxies were well known, since the absolute magnitude and the volume (about 1.3 Mpc^3 transformed to the present epoch) is known. There is evidence that the luminosity function in clusters differs from that in the field, and neither is well determined. If we use Kiang's (1961) luminosity function for field galaxies as probably the best compromise, the a priori probability is about 1.0×10^{-3}. This should, of course, be multiplied by the number of trials, and this is difficult to assess realistically. Image-tube plates of comparable quality have been obtained for six objects, these being selected on the basis of right ascension and redshift ($z \leq 0.35$) alone. Of these, four (PKS 2251+11, PHL 1093, Ton 256, PKS 2135−14) have "suspicious" associated galaxies. For the objects reported on here, the galaxies have concordant redshifts. Of the others, only PKS 2135−14 has been investigated spectroscopically, but no plates of sufficiently good quality to measure redshifts were obtained. The reader may come to his own conclusions; the author feels that a realistic estimate of the probability is no larger than about 2×10^{-3}.

There is one further piece of supporting evidence. The nebulous object labeled W (Wisp) on the plate was scanned. The result (Fig. 4) has a rather poor signal-to-noise

No. 3, 1971   DISTANCES OF QUASI-STELLAR OBJECTS   L117

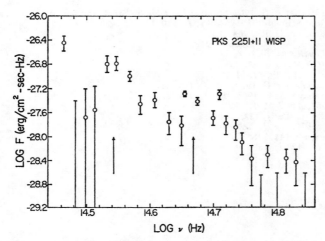

Fig. 4.—Multichannel scan of 2251+11 Wisp. *Arrows, redshifted* locations of Hα and Hβ. Diaphragm size was 7″.

ratio, and no comments on the fascinating possibilities of the object's nature and origin will be made here, except to note that it seems to be, or to be connected with, an extended H II region, presumably also excited by the QSO, and showing peaks at the redshifted wavelengths of Hα, Hβ, and possibly Hγ. The probable errors are large, and by itself the data would not be convincing, but they add some weight to the case. If better data corroborate the general features of Figure 4, gravitational-redshift models can be ruled out even without the results on galaxy 1.

We conclude that the object PKS 2251+11 satisfies the most stringent requirements for inclusion in the class of quasi-stellar objects and that it is almost certainly at the same distance as a galaxy of essentially the same redshift. The case for such objects as Ton 256 and III Zw 2 being intermediate between QSOs and Seyfert nuclei is strengthened. It should be noted, finally, that a galaxy even somewhat brighter than galaxy 1 could quite easily be superposed on PKS 2251+11 and be completely undetectable. The clear association of QSO-like activity with galactic nuclei at all luminosity levels up to and including the luminosity of the brightest galaxies, together with the evidence that at least one "real" QSO has a cosmological redshift, makes the picture of QSOs as events *in* galaxies compelling, and indeed at this point the most conservative of the various possibilities offered to date.

The author would like to thank Professors Maarten Schmidt and John Bahcall for useful discussions on this problem, and Gary Tuton for expert observing assistance. Thanks go especially to Professor J. B. Oke for instruction on the use of the spectrometer, for assistance and discussions on the tedious task of data reduction, and interesting discussions about the problem itself.

REFERENCES

Aller, L. H., and Liller, W. 1968, in *Nebulae and Interstellar Matter*, ed. B. M. Middlehurst and L. H. Aller (Chicago: University of Chicago Press).
Arp, H. C. 1966, *Science*, **151**, 1214.
———. 1970, *Ap. J.*, **162**, 811.
Bahcall, J. N., and Bahcall, N. 1970, *Pub. A.S.P.*, **82**, 721.
Bahcall, J. N., Schmidt, M., and Gunn, J. 1969, *Ap. J. (Letters)*, **157**, L77.

L118   JAMES E. GUNN

Burbidge, E., and Burbidge, G. 1967, *Quasi-stellar Objects* (San Francisco: W. H. Freeman & Co.).
Greenstein, J. L., and Schmidt, M. 1964, *Ap. J.*, **140**, 1.
Hoyle, F., and Burbidge, G. R. 1966, *Ap. J.*, **144**, 534.
Hoyle, F., and Fowler, W. 1967, *Nature*, **213**, 373.
Iriarte, B. 1959, *Lowell Obs. Bull.*, **4**, 130.
Kiang, T. 1961, *M.N.R.A.S.*, **122**, 263.
Oke, J. B. 1969, *Pub. A.S.P.*, **81**, 11.
Oke, J. B., and Neugebauer, G. 1970, *Ap. J.*, **159**, 341.
Oke, J. B., and Sandage, A. R. 1968, *Ap. J.*, **154**, 21.
Sandage, A. R. 1964, *Ap. J.*, **141**, 1560.
————. 1968, *Ap. J.* (*Letters*), **152**, L149.
Sargent, W. L. W. 1970, *Ap. J.*, **160**, 405.
Terrell, J. 1964, *Science*, **145**, 918.
Wampler, E. J. 1968, *Ap. J.*, **153**, 19.

PLATE L2

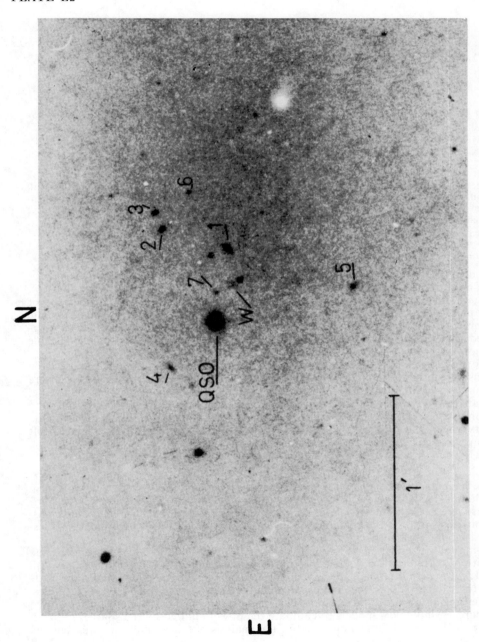

Fig. 1.—PKS 2251+11 and associated objects. The exposure was 20 minutes with an ITT 4708 image tube on baked IIIa-J emulsion behind a filter made up of 3 mm BG38 + 4 mm OG2. The photograph was obtained at the prime focus of the 200-inch telescope with the use of the Ross f/3.67 corrector.

Gunn (see page L114)

THE ASTROPHYSICAL JOURNAL, **179**:699–703, 1973 February 1

THE HUBBLE DIAGRAM FOR THE BRIGHTEST QUASARS

JOHN N. BAHCALL*

Institute for Advanced Study, Princeton, New Jersey 08540

AND

RICHARD E. HILLS†

Radio Astronomy Laboratory, University of California, Berkeley, California 94720

Received 1972 June 12

ABSTRACT

The slope of the magnitude-redshift relation for the optically most luminous quasars with redshifts ranging from 0.2 to more than 2 is consistent with the value of 5 expected from the expansion of the universe if luminosities are evaluated assuming quasars are at the cosmological distances implied by their redshifts.

Subject headings: cosmology — quasi-stellar sources or objects — redshifts — relativity

I. INTRODUCTION

The apparent-magnitude-versus-redshift diagrams of all quasars in various samples have been discussed by, among others, Sandage (1965), Hoyle and Burbidge (1966), Longair and Scheuer (1967), and Schmidt (1968); most authors agree that there is not a strong correlation between apparent magnitude and redshift for all quasars. The reason could be that there is a large spread in luminosities. McCrea (1972) has suggested that the brightest quasar at each redshift might have a significant Hubble relation.

We have grouped quasars according to their redshifts z, and have chosen the intrinsically brightest member of each group as a standard candle. Luminosities were calculated assuming that the redshifts are caused by the expansion of the universe. Our analysis resembles the more usual construction of a Hubble diagram by choosing the brightest galaxies in clusters (see, e.g., Sandage 1967) and enables us to estimate observational selection effects quantitatively. We find a tight relation between redshift and apparent magnitude for the brightest quasars (see fig. 1).

II. METHOD OF ANALYSIS

We have analyzed the catalog of quasars recently published by DeVeny, Osborn, and James (1971), excluding all objects with $|b^{II}| < 30°$, and those for which the UBV colors were not determined photoelectrically. The visual-apparent-magnitude-versus-redshift diagram for the remaining 156 objects resembles the scatter diagrams described by previous authors.

Using the prescriptions of Schmidt (1968, 1970), we computed first the optical flux density f_{2500} (referred to a standard emitted wavelength of 2500 Å and a bandwidth of 1 Hz) for each quasar from the redshift and UBV magnitudes [units of f are watts m^{-2} Hz^{-1}]. This takes account of what are called K-corrections in galaxy studies. Assuming quasars are at the cosmological distances implied by their redshifts, we used standard cosmological models to compute the power emitted by the source, F_{2500} [in W Hz^{-1}]. In order to make quantitative the analysis of selection effects, we excluded

* Supported in part by the National Science Foundation, grant GP-16147 A#1.

† Supported in part by the National Science Foundation, grant GP-3042X.

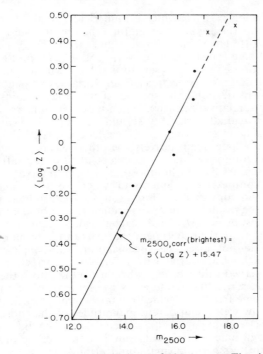

FIG. 1.—The Hubble diagram for the brightest of 105 quasars. The circles represent the intrinsically brightest quasars (for $q_0 = +1$) in successive redshift bins containing 15 objects each; the brightest objects are, respectively, 3C 273, 3C 232, PKS 1354+19, PKS 1252+11, PKS 1127−14, 3C 298, and TON 1530. The two crosses represent the large-redshift quasars PHL 957 ($z = 2.69$) and PHL 938 ($z = 2.88$); they were not included in the analysis. The parameters used are $m_v(\mathrm{max})$ = 19.5 mag, bin size = 15, and $\log F_{\min} = 22.4$ clusters. The plotted points include K- and B-corrections. The straight line in figure 1 corresponds to objects ~ 5 mag brighter than the brightest cluster galaxies (i.e., $\log F = 24.2$).

from further consideration any object with apparent magnitude fainter than some limit $m_v(\mathrm{max})$ (typically 19.5 mag) or with a power output F_{2500} smaller than some value $F(\mathrm{min})$ [typically $\log F(\mathrm{min}) = 22.4$, which corresponds to an absolute luminosity one magnitude brighter than the brightest cluster galaxies; see Bahcall (1971) for a definition of quasars in terms of absolute luminosity]. The remaining objects were ordered according to z and grouped into bins each containing b objects, so that the first bin contained the objects having the b smallest redshifts, and so on. *It is important to use bins containing the same number of objects in order to have a sampling of the luminosity function which is unbiased with respect to redshift.* This is *not* equivalent to taking the upper envelope of the (m_v, $\log z$)-curve since in an upper-envelope procedure one would presumably choose the brightest objects in redshift intervals irrespective of whether many or few objects were represented in the observational sample. For each bin the value of z_a, where

$$\log z_a = b^{-1} \sum_{i=1}^{b} \log z_i \, ,$$

and the luminosity of the brightest quasar, F_{2500} (brightest), was found. This latter was converted (for $q_0 = +1$) to an apparent magnitude m_{2500}(brightest) $= -2.5$

$\log F_{2500}$(brightest) $+ 5 \log z_a + 76$, which is numerically similar to a visual apparent magnitude, but represents the apparent magnitude the object would have if seen at z_a instead of its own redshift. The value of m_{2500} differs from the visual apparent magnitude by less than 0.4 mag for all the objects, except 3C 273, plotted in figure 1.

There are three selection effects or observational facts which complicate the interpretation of the Hubble diagram: some quasars are variables, the known quasars have a broad range of luminosities, and the spectra of intrinsically faint quasars can be obtained only if they are relatively nearby. When many quasars are included in each group, the effects of individual variability should be reduced; in any case, Penston and Cannon (1970) estimate that only 20 percent of quasars are optically violent variables. In order to evaluate quantitatively the other selection effects, we assumed that the quasar optical luminosity function is independent of redshift. Schmidt (1970) gives the numbers of quasars having luminosities lying in intervals of $\log F_{2500}$. We performed a series of computer experiments in which we drew randomly b objects from this distribution and converted the luminosity of the brightest into an absolute luminosity. Averaging the results of many experiments, we obtained a mean absolute magnitude of the brightest object, $\overline{M}_b(z_a)$. At each redshift z_a, the selection effects were introduced by excluding from the experiment that part of the luminosity function for which the objects would not meet the specifications $m_v < m_v(\text{max})$ and $F_{2500} > F(\text{min})$. Examination of how $\overline{M}_b$ varies with z_a yielded a correction $B(z_a)$ which we subtracted from the observed values of m_{2500}(brightest) to remove the selection effect. The correction is in the sense that it increases the luminosity of the nearby bright quasars because they come from bins which contain many intrinsically faint objects.

Next we examined how well the apparent magnitudes of the chosen objects fit the relation m_{2500}(brightest) $= S \log z_a + B(z_a) + $ constant. The best value of S, the slope of the redshift-magnitude relation, was found by the method of least squares for several sets of the parameters b, q_0, $m_v(\text{max})$, and $F(\text{min})$.

In order to obtain a reliable estimate of $\overline{M}_b$ from the numerical experiments, many independent samplings were used for each redshift. We thus obtained the variance of $\overline{M}_b$ and estimated the uncertainties, σ_b, in m_{2500}(brightest) that arise solely from the finite bin size. These uncertainties were compared to the actual deviations Δm of the data points from the best linear fit by evaluating

$$\chi^2 = \sum_{j=1}^{N} [\Delta m(z_{aj})/\sigma_b(z_{aj})]^2 ,$$

where N is the number of bins and it has been assumed that all the dispersion is in magnitude. Tables of the distribution of χ^2 were then used to calculate the probability of finding a value of χ^2 as large as the one obtained if the assumed dependence of m_{2500} on z and our error estimates were exactly correct.

III. RESULTS

We chose as our standard case $b = 15$, $q_0 = +1$, $m_v(\text{max}) = 19.5$ mag, and $\log F(\text{min}) = 22.4$. With no correction for selection effects (i.e., $B \equiv 0$), the values of m_{2500} versus $\log z_a$ were fitted to a straight line of slope 5 which had a standard deviation of 0.44 mag, assuming all the dispersion was in magnitude. The slope which gave a slightly better fit was $S = 4.0 \pm 0.6$. The numerical experiments gave the following values for the correction in the standard case: $B(z_a) = +1.5$, $+1.0$, $+0.5$, and $+0.0$ mag for the respective values $0.1 \leq z_a \leq 0.8, 0.8 < z_a \leq 1.25, 1.25 < z_a \leq 2.0$, and $2.0 < z_a$. The numerical estimates of the expected error $\sigma_b(z_a)$ fell from 0.5 mag to 0.25 mag over the same range of z_a. The corrected magnitudes for the standard case are shown in figure 1 together with the best-fitting line of slope 5, which has a standard

702   JOHN N. BAHCALL AND RICHARD E. HILLS   Vol. 179

TABLE 1

SOME RESULTS FOR THE HUBBLE DIAGRAM OF THE BRIGHTEST QUASARS

m_v (max) (1)	b (2)	q_0 (3)	log F (min) (4)	χ^2 probabilities for slope 5 and $B \neq 0$ (%)	χ^2 probabilities for slope 5 and $B \equiv 0$ (%)
19.5	15	+1	22.4	80	71
19.5	20	+1	22.4	63	63
18.0	15	+1	22.4	18	62
19.5	15	0	22.4	54	< 1
19.5	15	+2	22.4	60	87
19.5	15	+1	22.0	88	90
19.5	15	0	22.0	93	14

NOTE.—See text for an explanation of the symbols.

deviation of only 0.3 mag. The two crosses in figure 1 correspond to the individual objects PHL 957 ($z = 2.69$, Lowrance *et al.* 1972) and 4C 05.34 ($z = 2.88$, Lynds and Wills 1970); they were not included in the analysis because there are not enough quasars known with $z > 2.2$ to form a complete bin of 15 members. For the case shown in figure 1, $\chi^2 = 3.0$, corresponding to a probability of 80 percent. The assumed errors may have been overestimated because of the discreteness of the luminosity function.

Some results of parameter variations are indicated in table 1 by means of the χ^2 probabilities. The examples for $q_0 \neq +1$ are only illustrative because we used for our calculations of B and σ_b a luminosity function which was derived for $q_0 = +1$. Comparison of the relative χ^2 probabilities in columns (5) and (6) suggests that *the inclusion of the corrections B is not important for obtaining a good fit with slope 5 and the present data.* [The computed values for the corrections B are rather insensitive to the precise value of the bin size but depend somewhat more on log F_{min} and m_v(max). Corrections for all cases are available from the authors upon request.]

IV. DISCUSSION

Several conclusions and remarks are appropriate in connection with the results described above. First, our findings are consistent with the combined hypotheses of a cosmological origin for the redshifts and the absence of strong luminosity evolution. The Hubble diagram for the brightest quasars, which links small and large redshift objects, may provide the strongest evidence that the very large redshifts of quasars ($z \gtrsim 1$) are cosmological in origin. Our results imply limits on the amount of uniform brightening (or dimming) of the quasar luminosity function. For example, a luminosity evolution $\Delta m = c \log (1 + z)$ for all quasars is ruled out at the 99 percent confidence level for $|c| \geq 4$ mag. (Note that $c = -5$ corresponds to $q_0 = -1$, the steady-state theory, and no luminosity evolution.) Second, if quasars were "local" objects, then our procedure would not be expected to yield the slope of ~ 5 found in figure 1 (see also table 1). If quasar luminosities and distances were independent of redshift, then our analysis would yield an m_{2500} that depends only weekly on redshift [$\propto 5 \log (1 + z)$ without corrections], which is ruled out at more than the 99 percent confidence level. Third, it is important to measure more quasar redshifts and photoelectric magnitudes in order to test the Hubble relation reported here, preferably by obtaining a large sample of objects and their luminosity function with a uniform discovery technique. Fourth, systematic studies of the dependence of optical variability upon optical luminosity are required for further improvements. Fifth, our methods are applicable to any set of objects with a broad luminosity function and known redshifts.

This work was performed while one of us (J. N. B.) was a visiting professor in the astronomy department of the University of California at Berkeley. It is a pleasure to thank the students and faculty for their hospitality and stimulation. We are indebted to Adrian Webster for critical readings of the manuscript.

REFERENCES

Bahcall, J. N. 1971, *A.J.*, **76**, 282.
DeVeny, J. B., Osborn, W. H., and James, K. 1971, *Pub. A.S.P.*, **83**, 611. [The value of m_v (3C 280.1) is incorrectly given as 13.44 instead of 19.44 in this reference because of misprints in the original literature.]
Hoyle, F., and Burbidge, G. R. 1966, *Nature*, **210**, 1346.
Longair, M. S., and Scheuer, P. A. G. 1967, *Nature*, **215**, 919.
Lowrance, J. L., Morton, D. C., Zucchino, P., Oke, J. B., and Schmidt, M. 1972, *Ap. J.*, **171**, 233.
Lynda, R., and Wills, D. 1970, *Nature*, **226**, 532.
McCrea, W. H. 1972, *External Galaxies and Quasi-Stellar Sources*, ed. D. S. Evans (Copyright IAU), p. 283.
Penston, M. V., and Cannon, R. D. 1970, *R.O.B.*, **159**, 1.
Sandage, A. R. 1965, *Ap. J.*, **141**, 1560.
———. 1967, *Ap. J. (Letters)*, **150**, L9.
———. 1968, *ibid.*, **152**, L149.
Schmidt, M. 1968, *Ap. J.*, **151**, 393.
———. 1970, *ibid.*, **162**, 371.

THE ASTROPHYSICAL JOURNAL, Vol. 152, May 1968

RADIO SOURCES AND ARP'S PECULIAR GALAXIES

H. VAN DER LAAN* AND F. N. BASH†
National Radio Astronomy Observatory,‡ Green Bank, West Virginia
Received July 26, 1967

ABSTRACT

The claim that there is an excess of radio sources in the neighborhood of Arp's peculiar galaxies, arranged as aligned pairs centered on those galaxies, has been extensively tested. The distributions of radio sources and of radio-source pairs near positions in the Arp *Atlas* and near equally clustered but arbitrarily positioned control catalogue points were compared. Considering circles ranging in radius from 1° to 10° by the addition of successive annuli of 1° width, radio sources, source pairs, and aligned source pairs were identified and counted. In all cases the number of aligned pairs constitutes that fraction of all pairs expected for random radio-source distributions. There is no preferred alignment at all.

Among the control catalogues the Arp catalogue is statistically undistinguished for purposes of radio-source and source-pair counts in all circles from 1° to 3° and from 4° to 10°. In the 4°–3° annulus there is an excess of sources, 2 standard deviations above the population mean, an excess which has a statistical significance of 4 per cent. There is no excess of aligned pairs inside 4° circles. The fact that the excess count occurs in annuli whose angular diameter is 8° emphasizes the fluctuation's chance character. The number of radio-source pairs increases rapidly with increasing source number density, and we suggest that this steep dependence conspired with the chance fluctuation to give rise to Arp's hypothesis.

I. INTRODUCTION

There are several reasons for the lively response, both informal and in the literature, evoked by Arp's announcement (1966a) that certain peculiar galaxies are genetically related to non-thermal radio sources. First, this relation, if it were established with a significant measure of confidence, would have extraordinary repercussions in extra-galactic astronomy. Second, it would change, if not clarify, the current impasse in the area of quasi-stellar radio-source problematics. Third, it involves all the hazards inherent in statistical arguments which draw physical conclusions from geometric premises.

The preliminary discussion of this alleged relation (Arp 1966a) was followed by at least three independent, complementary publications which assessed several features of Arp's claim and their statistical significance (Lynden-Bell, Cannon, Penston, and Rothman 1966; Holden 1966; Wagoner 1967). None of these authors concur with Arp's far-reaching conclusions, although they do not entirely dismiss them. Residual effects of marginal significance appear to persist.

In a second article Arp discusses in detail individual peculiar galaxies, accompanied by a more or less aligned pair of radio sources or more complex configurations (Arp 1967). A close reading of that paper shows that § VI, entitled "The Probability of Radio Source Pairings," is pivotal in the entire argument. In that section, which we shall discuss later, Arp considers it highly probable that, within 6° circles centered on peculiar galaxies, at least 12 out of 16 radio-source pairs spanning peculiar galaxies numbered 100–150 of his peculiar galaxy catalogue (Arp 1966b) form physical associations rather than chance configurations. Normally, configurations in projection are, in astronomy, accepted as possible physical groupings only if consistent with kinematic evidence. But Arp asserts that the geometric evidence is so strong as to override kinematic objections and attributes redshifts which contradict his hypothesis to unspecified non-kinematic effects.

Sections I–V of Arp (1967) anticipate the conclusion of his § VI, and it is our conten-

* Present address: Sterrewacht, Leiden, Netherlands.
† Now at the Department of Astronomy, University of Texas.
‡ Operated by Associated Universities, Inc., under contract with the National Science Foundation.

tion that the entire case stands or falls with it. The extensive description that Arp provides, while not without astronomical interest insofar as it illumines peculiarities of disturbed galaxies, is circumstantial evidence which loses its relevance when the hypothesis does not have the foundation of a statistically significant excess of aligned radio-source pairs. This paper presents conclusive results which show that the hypothesis has no such basis.

II. METHOD AND SCOPE OF THE TEST

The 3C revised catalogue of radio sources (Bennett 1962) provides us with a homogeneous coverage of the northern hemisphere, where most galaxies in Arp's *Atlas* are found. Positions are taken from the list of Pauliny-Toth, Wade, and Heeschen (1966). It must be noted that Arp's case is built on radio-source pairs from the 3CR catalogue, while sources and groups of sources from catalogues compiled by Australian observers (the MSH and Parkes lists) are referred to only if they strengthen the case already made. The lack of consistency in Arp's appeals to these lists and the fact that only a small fraction of the peculiar-galaxies area overlaps that of the Parkes survey preclude any meaningful or reliable statistical test there. This study therefore deals with 3CR sources only.

The positions of the radio-catalogue and Arp-catalogue objects, precessed to the same epoch, were put in computer storage. For each peculiar galaxy with $|b^{II}| > 15°$ and $\delta > +05°$ all 3CR sources within a circle of specified radius were found, and for each the great-circle distance and position angle with respect to that galaxy were calculated. This provides a list of peculiar galaxies away from the galactic plane, well within the declination limit of the radio survey and, with each galaxy, a list of nearby radio sources, in order of increasing distance. For each galaxy, the $\frac{1}{2}n\,(n-1)$ possible pairs of the n sources in the circle centered on the galaxy are listed, with their computed angles of alignment, the sums and the differences of their distances to the galaxy, and an "improbability coefficient" for each configuration. (We return to this in § IV.)

For the entire set of circles centered on the catalogue of points provided by the peculiar-galaxy positions, the following data are obtained: (1) the number of radio sources per square degree in successive 1° annuli around the Arp galaxies, averaged over the whole catalogue; (2) the distribution of alignment angles, in 10° steps from 0° to 180°; (3) the number of aligned pairs per 10 successive catalogue positions. (A pair is considered aligned if the position angles of the pair's members differ by $180° \pm \Delta\phi$, where $\Delta\phi \leq 30°$; this range, necessarily arbitrary, is the one Arp [1967] adopts.)

Aligned pairs are conspicuously marked in the computer output for ease of individual checks and confirmation.

In order to obtain comparative results, we did not generate arbitrary lists of positions for at least two reasons: (1) to avoid introducing anything but repeatable results and (2) because the objects in the Arp catalogue are extremely clustered, a feature difficult to reproduce without error in a randomly generated catalogue. For these reasons, all comparison catalogues used in this study are obtained from the Arp catalogue by a rotation of the original one. We chose to rotate in galactic longitude, thus saving the catalogue points from moving into the zone of avoidance. Rotation below the declination limit of the radio survey does, of course, cause the number of points and the actual catalogue numbers retained to vary from set to set. Points near the north galactic pole are not moved very much by this rotation, but, by taking large enough steps (15° and 30°), peculiar-galaxy positions in that region were moved enough for any aligned pair not to be repeated in the next set.

Table 1 gives details of the original and the comparison catalogues. When, upon testing for various effects using just over 1000 points, a possible marginal effect was detected only in going from 3° to 4° circles, more comparison catalogues were produced, to include over 4000 points, allowing a statistically more reliable investigation.

III. RESULTS

a) 10° Circles

Source counts in 6°–7°, 7°–8°, 8°–9°, and 9°–10° annuli are identical for Catalogue A (the original unrotated Arp catalogue) and Catalogues B–F. All vary randomly and are within 2 sample standard deviations of the mean of about 0.0163 sources per square degree. The number of aligned pairs is within a few per cent of one-sixth of all counted pairs for each of the six catalogues, as expected for randomly distributed sources and our alignment criterion. The distribution of 3CR sources with respect to Arp's peculiar galaxies, at distances of 6°–10° from those galaxies, is therefore indistinguishable from the distribution with respect to any similarly clustered group of points arbitrarily positioned on the sky.

b) 6° Circles

It is within 6° circles that Arp claims to find a statistically significant excess of radio-source pairs aligned with peculiar galaxies. The ratio of aligned pairs to all possible pairs

TABLE 1

CATALOGUE SPECIFICATIONS

Catalogue	Rotation in l^{II} (deg)	Net No. Catalogue Positions in Test Area	Tests Involving These Catalogues
A..............	0	236	Peculiar galaxies from Arp's catalogue. Involved in all tests
B..............	30	218	Source counts in successive 1° annuli;
C..............	60	185	pairs, aligned pairs, and unique
D..............	90	149	aligned pairs counted in 3°, 4°, 5°, 6°,
E..............	120	129	and 10° circles
F..............	150	136	
Subtotal.......	....	1053	
G..............	180	139	Source counts in 4° circles. Aligned pairs
H..............	210	159	and unique aligned pairs counted in
I..............	240	170	3° and 4° circles
J..............	270	190	
K..............	300	220	
L..............	330	245	
Accumulated subtotal.....	....	2176	
Aa..............	15	231	Source counts in 4° circles. Aligned
Bb..............	45	201	pairs and unique aligned pairs count-
Cc..............	75	164	ed in 4° circles
Dd..............	105	139	
Ee..............	135	129	
Ff..............	165	134	
Gg..............	195	151	
Hh..............	225	167	
Ii..............	255	178	
Jj..............	285	203	
Kk..............	315	235	
Ll..............	345	246	
Subtotal.......	....	2178	
Total.........	....	4354	

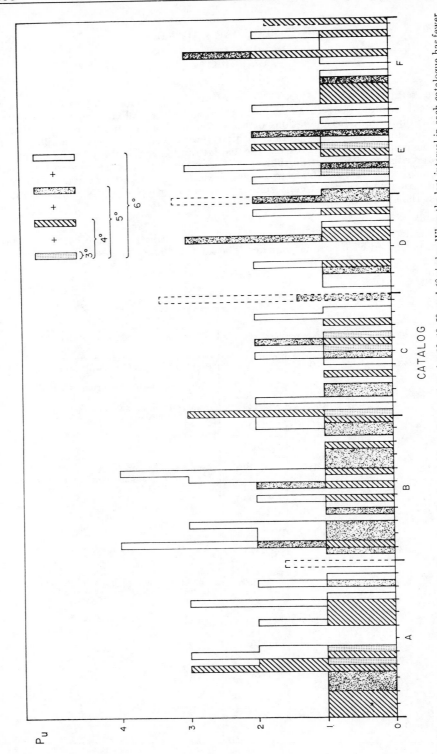

Fig. 1.—Cumulative histogram of unique aligned pairs per 10 catalogue points, for 3°, 4°, 5°, and 6° circles. When the last interval in each catalogue has fewer than the requisite number of points, the count, appropriately scaled, is shown in dashes. This is also the case in Figs. 3 and 4.

ARP'S PECULIAR GALAXIES   625

varies from about 0.14 to 0.19; for Catalogue A it is 0.165. Hence, there is no evidence whatever for any preferred angle of pair members with respect to the peculiar galaxies. The source counts in the 4°–5° and 5°–6° annuli are no different for Catalogue A, as compared to Catalogues C–F. Catalogue B has an excess source count of $2\frac{1}{2}$ sample standard deviations in the 5°–6° annulus.

In Figure 1 the number of unique aligned pairs per 10 successive catalogue points is given for Sets A–F. The histogram is cumulative, with 3°-circle pairs first, to which are added successively the additional pairs found when 4°, 5°, and 6° circles are searched. Several clarifying remarks are in order: (1) The Arp catalogue is a morphological one, so that the galaxy number has no relation to its position; a given number in several

TABLE 2

ARP NUMBER OF THE LAST CATALOGUE MEMBER IN
EVERY 10-POINT INTERVAL FOR CATALOGUES A–F

A	B	C	D	E	F
17	18	23	27	33	34
31	32	36	47	60	57
46	49	62	73	81	79
63	68	80	95	102	101
76	80	95	111	120	120
87	94	107	136	152	152
99	105	122	159	175	176
112	117	149	177	197	197
125	136	168	197	239	240
141	152	181	218	266	260
158	169	197	242	289	275
170	182	211	272	313	301
184	195	235	294	335	324
196	207	262	319	(9 only)	338
207	220	274	334		(5 only)
224	241	294	(9 only)		
239	265	311			
260	277	328			
272	293	337			
287	311	(6 only)			
300	326				
316	337				
330	(7 only)				
337					
(6 only*)					

* Number of points in the last interval.

catalogues refers to the same point in the catalogue, differing in position only by the rotation as given in Table 1; thus point $E(n)$ is point $A(n)$ rotated in galactic longitude by 120°. Since upon rotation various points disappear below the declination limit of the radio survey, the numerical limits of the 10-point intervals are not identical from one catalogue to the next (see Table 2). (2) A unique pair is one allowable in Arp's sense; thus, 2 catalogue positions may both have a pair of radio sources aligned with these positions as centers. But, if these pairs have one common radio source, they cannot both be counted for purposes of testing Arp's hypothesis. The pairs in the histogram of Figure 1 are unique. All interfering pairs have been found, and the best-aligned one with the most nearly central catalogue point has been retained in each case. (3) This histogram is presented only for the sake of comparison with Arp's results. As remarked above, the number of "aligned pairs" is found to constitute one-sixth of all pairs for each catalogue, including A.

The 6° histogram has three conspicuous peaks—one in Catalogue A, two in Catalogue B. Over intervals of 40 catalogue points these peaks contain 10, 11, and 10 pairs, respectively. Determining their statistical significance using Student's t-test and the sample standard deviation, we find a significance level for each peak of about 15 per cent. Since the distribution shown ranges over approximately 24 intervals of 40 catalogue points, our empirical results and the computed significance are consistent.

Part A of the histogram is to be compared directly with Figure 31 in Arp (1967). Arp's histogram presents all 42 aligned pairs rather than just the 29 unique aligned pairs allowed by the criteria of his own hypothesis (non-integer values of the histogram are not explained in Arp's text). That histogram Arp compares with the "empirically expected numbers of accidental associations" indicated by a dashed line, at a level of about 1.3 pairs per 10 catalogue points. The number of chance pairs that we obtain when using a sample of points more than 7 times as large as Arp's is 1.7 for aligned pairs and 1.24 for unique aligned pairs. *It appears therefore* that Arp has compared the histogram of

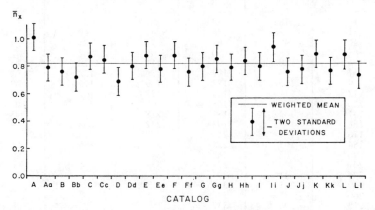

Fig. 2.—The average radio-source count per catalogue position inside 4° circles for the Arp catalogue and the twenty-three controls.

all aligned pairs with a *smoothed* expectation of *unique aligned pairs*, and, without any discussion of the significance of the peak from A101 to A158 as an ordinary statistical fluctuation, concluded that pairs in that peak in excess of his estimated number of chance pairs are "real identifications."

Further examination of Figure 1 shows that, when the pairs contributed by sources in the 5°–6° annulus are removed, the peak from A101 to A125 is the most prominent one remaining. This continues to be the case when pairs which have at least one source in the 4°–5° annulus are removed. Inside 3° circles very few pairs remain, either in Catalogue A or in any of the other catalogues. Therefore, if we wish to make Catalogue A as prominent as we can among the array of catalogues studied, we must look at the 4° circles. At any other separation the radio-source distribution near Arp's peculiar galaxies is indistinguishable from that distribution near comparison catalogue positions.

c) 4° Circles

In order to determine reliably the mean and the standard deviation of source and pair populations, an additional eighteen comparison catalogues were generated, as indicated in Table 1. In Figure 2 the average radio-source count per catalogue point inside 4° circles is shown for each of the twenty-four catalogues. It can be seen here that Catalogue

No. 2, 1968   ARP'S PECULIAR GALAXIES   627

A stands out inasmuch as it has the largest deviation from the mean, about 2 sample standard deviations.

It turns out that for Catalogue A the ratio of aligned pairs to all pairs is 0.169, so that also in the case of 4° circles there is no preference whatever for *aligned* pairs. In Figure 3 the histograms of unique aligned pairs inside 4° circles for twelve catalogues are presented.

There is an excess of pairs, and therefore of unique aligned pairs, in Catalogue A. To check whether this excess is simply a reflection of the excess source counts (as was the case in the correspondence of extra pairs and extra counts in the 6° circles for Catalogue B), the expected number of aligned pairs was computed. Assuming a Poisson distribution of radio sources, and using the empirically determined source number density for each catalogue, the expected number of aligned pairs for a group of N points with a source count of m sources per point is

$$P_c = \sum_{x=0}^{k>>m} \frac{n_x}{12} x(x-1) \cdot$$

where

$$n_x = N \frac{e^{-m} m^x}{x!} \cdot$$

Table 3 shows the results. They demonstrate that the only difference between Catalogue A and an arbitrary set of equally clustered points is that seen in Figure 2.

Substituting the Arp catalogue for the radio catalogue, we have found all Arp galaxies in the immediate neighborhood of each Arp galaxy. This reveals the extreme clustering of the objects in Arp's catalogue. It is this clustering which is responsible for the interference of aligned pairs. The last column in Table 3, giving the ratio of aligned to unique aligned pairs, is a measure of that interference. For a random distribution of radio sources, that ratio is expected to be comparable for equally clustered assemblies of points with comparable source counts

It remains to assess the significance of the one difference between Catalogue A and the other catalogues, shown in Figure 2. To determine this, we have found the standard deviation of the number counts of radio sources inside 4° circles, per 40 catalogue points, from a 60-interval sample shown in Figure 4.

Student's *t*-test gives 4 per cent level of significance to the 20 per cent excess in the radio-source counts for Catalogue A.

Similarly the population of unique aligned pairs per 10 points (shown in Fig. 3) can be used to assess the significance of the greater-than-average number of such pairs in Catalogue A. Using Student's *t*-test and the information in Figure 3 gives a 6 per cent significance, in agreement with the source-count test.

The peak in the unique-aligned-pair distribution centered approximately on A100–A150—which Arp gives so much emphasis—is, at the 40 per cent level, a quite ordinary fluctuation in a distribution such as that of Catalogue A. Figure 3 has several comparable fluctuations; for example, the one in Catalogue K is relatively more prominent than that in Catalogue A.

Note added on November 4, 1967.—In view of objections to our procedure, raised by Arp elsewhere in this issue, it appears desirable to emphasize the following points explicitly.

All our significance tests were calculated on the basis of statistical properties as determined empirically. There is no reliance on any assumption about some uniform or random distribution of radio sources and/or peculiar galaxies.

Arp's hypothesis (that peculiar galaxies are being found between radio sources with a frequency significantly greater than the expectation of random coincidences) refers specifically to Nos. 100–150 of his catalogue. We have not confined our attention to this

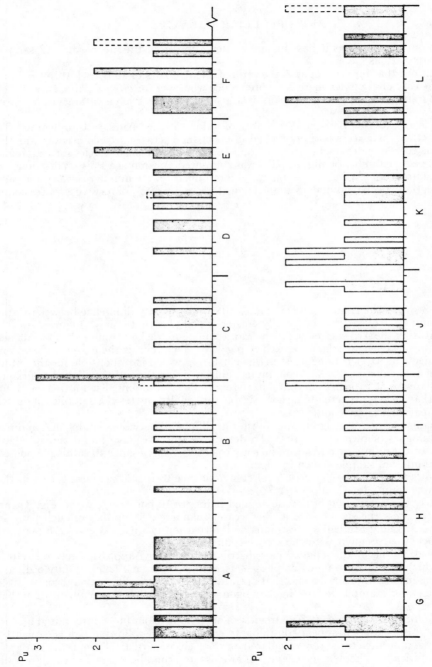

Fig. 3.—Histogram of unique aligned pairs per 10 catalogue positions for Catalogues A–L. Note that the common portions of Figs. 1 and 3 are not identical in every detail. The reason is that in Fig. 1 a poorly aligned or badly centered pair within 4° may be rejected in favor of a well-aligned and centered one that shares one of the former's sources and has the second source in a larger annulus.

TABLE 3

SOURCE COUNTS, OBSERVED AND CALCULATED PAIR COUNTS,
FOR THE 4° CIRCLES

CATALOGUE	SOURCE COUNTS PER POINT	NUMBER OF ALIGNED PAIRS		$P_c - P_o$	UNIQUE ALIGNED PAIRS P_u	P_o/P_u
		Calculated P_c	Observed P_o			
A.......	1.01	21	23	−2	16	1.44
Aa.......	0.79	12	13	−1	8	1.63
B.......	0.76	10	7	+3	7	1.00
Bb.......	0.72	8	6	+2	6	1.00
C.......	0.87	10	12	−2	8	1.50
Cc.......	0.85	10	4	+6	4	1.00
D.......	0.69	6	5	+1	5	1.00
Dd.......	0.80	7	6	+1	5	1.20
E.......	0.88	8	5	+3	4	1.25
Ee.......	0.78	7	8	−1	7	1.14
F.......	0.88	8	9	−1	7	1.29
Ff.......	0.76	7	9	−2	6	1.50
G.......	0.86	7	8	−1	6	1.33
Gg.......	0.86	10	13	−3	8	1.63
H.......	0.79	8	6	+2	5	1.20
Hh.......	0.84	9	11	−2	6	1.83
I.......	0.80	8	13	−5	8	1.63
Ii.......	0.95	12	12	0	8	1.50
J.......	0.76	8	12	−4	8	1.50
Jj.......	0.78	9	11	−2	8	1.38
K.......	0.90	14	13	+1	11	1.18
Kk.......	0.77	11	10	+1	7	1.43
L.......	0.89	15	10	+5	9	1.11
Ll.......	0.74	11	12	−1	9	1.33

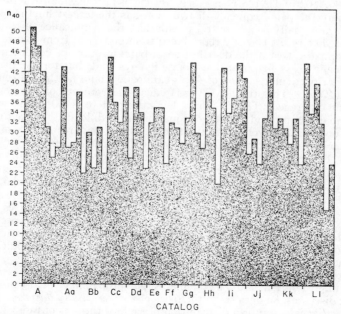

FIG. 4.—Radio-source counts per 40 catalogue positions for Catalogue A and comparison Catalogues Aa–Ll.

subset but instead included the entire catalogue in our test. We did this for two reasons: (1) For Arp's hypothesis to arise at all, there must, of course, be an excess of aligned pairs associated with A100–A150. Source counts and pair frequencies for the *whole* catalogue were necessary to establish how the subset is related to the context from which it came to be selected. (2) Reliable *empirical* determinations of statistical properties of the radio-source–peculiar-galaxies population required as large a sample as possible. This was achieved by rotating the entire set of *Atlas* positions.

Our data do, of course, contain all the information required for the similar but more limited test of A100–A150 and its rotated counterparts only. In this case we find the source counts of several controls to exceed that of A100–A150 for all annuli except from 3° to 4°, where A100–A150 has the maximum density. This maximum is only 1.4 standard deviations above the mean. The pair frequency in A100–A150, in view of its relation to source number density and in comparison with the controls, is normal (cf. appropriate segments of Figs. 1, 3, and 4 with the aid of Table 2).

IV. CONCLUDING DISCUSSION

In § I we mentioned three studies of the Arp effect published prior to the appearance of Arp (1967). Some brief comments on each are in order.

1. *Lynden-Bell, Cannon, Penston, and Rothman (1966).*—These authors' empirical results are similar to our own. Arp's peculiar galaxies are concluded to be very clustered and to be correlated with 3C radio sources to "a borderline significance" of 2 or 3 per cent. Since Lynden-Bell *et al.* used only one control group for comparison, their significance tests do not have the high confidence level which results from our much larger control population. This is particularly important when the significance of the peak which includes A102 and A145 is assessed. By choosing this small fraction of the Arp catalogue, where the pair-frequency distribution has the largest fluctuation, and comparing it with an equally small control population, the significance of the difference is overestimated. The result is easily reproduced from our data; from Figure 1 or 3 select a peak and compare it with the distribution of pairs in only the congruent segment of another catalogue. This significance estimate ignores that, but for the presence of the fluctuation, that particular section of Arp's catalogue would not have been selected.

2. *Holden (1966).*—The study by Miss Holden is a prototype of our work. There are two important differences: (1) she limits her attention to A100–A149, which gives a statistically small sample and, for reasons just mentioned, loads the dice from the start; (2) she does not generate comparison catalogues but compares counts of sources and aligned pairs with expected values computed on the basis of a uniform distribution of radio sources. The marginal significance for the frequency of aligned pairs thus found is due to selection and can in fact be reproduced in our comparison catalogues by selecting a prominent peak from such histograms as those given in Figures 1 and 3.

3. *Wagoner (1967).*—The part of Wagoner's investigation that has bearing on our work concerns the two-dimensional distribution of 3CR sources with respect to Arp's peculiar galaxies. Comparing the distribution of angular distances from peculiar galaxies of the nearest and second-nearest radio source, Wagoner finds the slight excess of sources within 4° also detected here. A χ^2 test on Wagoner's data yields a significance of 5 per cent, in agreement with our result.

We have repeatedly referred to § VI in *Arp (1967)*. One additional remark must be made. When in § VI of Arp (1967) he calculates the "probability" of any of the configurations that he discusses, he does so on the basis of a *differentially uniform* distribution of radio sources, that is, a uniform distribution of each of the sets of radio sources in successive flux-density intervals. Since radio sources are randomly rather than uniformly distributed, the calculated values of "probability" can become very small without being in any sense indicative of an anomalous configuration. The inverse of that "probability" we have called "improbability coefficient," and it proved useful occasionally when deciding which one of two interfering aligned pairs to retain for purposes of the counts.

In summary, we have compared the distribution of radio sources with respect to Arp's peculiar galaxies and with respect to 23 equally clustered but arbitrarily positioned sets of points and attempted to find where the original catalogue *differs most* from the control catalogues and what that difference is. Upon investigating circular areas centered on the catalogue points, going from 1° to 10° circles in 1° steps, the maximum difference is found to occur not at 6°, as Arp claims, but at 4°, as Miss Holden found in a less extensive test. The nature of that difference is not an excess of aligned pairs (which form very nearly one-sixth of all pairs) but a slight excess in counts of radio sources otherwise randomly distributed.

The formal, statistical significance of *that* excess is 4 per cent. It would be deceptive, however, to interpret this figure as suggestive of a genetic relation between the two classes of objects. Only in the 3°–4° annuli does the averaged source count for the Arp catalogue exceed that of *all* our comparison catalogues. For the smaller and larger annuli, several of the comparison-catalogue counts exceed that of the Arp catalogue, and there the Arp catalogue is statistically undistinguished.

This state of affairs serves to emphasize the chance character of this fluctuation. If, as Arp suggested, radio sources were ejected from peculiar galaxies, the individual, observed angular separations would be determined by several independent factors, such as projection angle, age, ejection speed, and distance from the Sun. Such a population would give a source-count maximum at small angular separations and could scarcely be expected to peak at annuli with angular diameters of 8°.

We suspect that the steep dependence of the number of pairs on the source counts (cf. equation in § IIIc) conspired with the source-count fluctuation and the usual frequency fluctuations within one catalogue to give rise to the hypothesis which has now lost its basis.

The authors express their appreciation to Russell Bosserman for his assistance in the data analysis.

REFERENCES

Arp, H. 1966a, *Science*, **151**, 1214.
——. 1966b, *Atlas of Peculiar Galaxies* (Pasadena: California Institute of Technology).
——. 1967, *Ap. J.*, **148**, 321.
Bennett, A. S., 1962, *Mem. R.A.S.*, **68**, 163.
Holden, D. J. 1966, *Observatory*, **86**, 229.
Lynden-Bell, D., Cannon, R. D., Penston, M. V., and Rothman, V. C. A. 1966, *Nature*, **211**, 838.
Pauliny-Toth, I. I. K., Wade, C. M., and Heeschen, D. S. 1966, *Ap. J. Suppl.*, **13**, 65.
Wagoner, R. V. 1967, *Nature*, **214**, 766.

Astrophysical Letters, 1972, Vol. 10, pp. 147–152   © Gordon and Breach, Science Publishers Ltd. Printed in Glasgow, Scotland

Are Quasars Associated with Bright Galaxies?

JOHN N. BAHCALL   *Institute for Advanced Study, Princeton, New Jersey, USA*

CHRISTOPHER F. McKEE   *Harvard College Observatory, Cambridge, Massachusetts, USA*

NETA A. BAHCALL   *Princeton University Observatory, USA, and Tel Aviv University, Israel*

A complete sample of 222 quasi-stellar radio source (QSRS) identifications and 166 radio quiet quasi-stellar object (QSO) identifications has been studied to look for possible correlations with the positions of bright galaxies. The results are consistent with a random correlation between the positions of quasars and bright galaxies for angular separations less than 30 arc min.

INTRODUCTION

The possibility that quasars may be associated with bright galaxies has been the subject of much discussion in recent years (e.g., see Arp (1971) and the references listed therein). Most recently Burbidge *et al.* (1971) (hereinafter referred to as BBSS) have found that 4 out of a sample of 47 identified 3C and 3C R quasars are much closer to bright galaxies listed in the *Reference Catalogue of Bright Galaxies* of de Vaucouleurs and de Vaucouleurs (1964) than would be expected if the quasars were randomly distributed with respect to the galaxies. BBSS estimated the probability of any one of their four close pairs occurring by chance to be $\sim 10^{-3}$. Since the quasars involved in these close pairs have redshifts between 0.5 and 1.4, while the bright galaxies all have redshifts less than 0.01, BBSS suggested that some quasars may have substantial redshift components that are not due to the expansion of the universe.

The use of a catalogue of radio-selected quasars for studying correlations with bright galaxies is, as BBSS have stressed, particularly useful since all that is required for statistical validity is that the sample be well defined and contain no bias with respect to bright galaxies. There remains of course the problem of deciding how to interpret formal probabilities that are assigned to 'unlikely' events whose existence was not predicted before they were noticed and for which probability arguments were only subsequently applied. [Note that the typical correlation separations discussed previously by Arp (1971) and, e.g., Sharp (1967) were several degrees while BBSS discuss correlations of several minutes of arc.] In such a case, the investigator presumably would find any of some ill-defined class of 'unlikely' events to be significant; since the size of this class is usually unknown, the probabilities of the observed 'unlikely' events become ambiguous.

One indication of whether 'unlikely' events are a statistical fluctuation or not is the predictive power of the relations (or correlations) for other observations. This also avoids any ambiguity in calculating the probability of the outcome. We have examined a complete sample of 222 QSRS identifications (from four radio catalogues) and 166 QSO identifications (from the Bologna catalogue of optically-identified quasars). Since the Bologna QSO's are not radio sources, in contrast to the quasars in the BBSS sample, we shall treat these objects separately. Our results are consistent with a random correlation of quasar (both QSRS and QSO) positions with those of bright galaxies for separations of less than thirty minutes of arc. We find fewer pairings by almost an order of magnitude than would have been expected on the basis of the BBSS results.

The details of our analysis are described in the next two sections. A brief discussion of the results is then given.

PROCEDURE AND RESULTS

The hypothesis that we wish to test is that quasars are closer to bright galaxies than would be expected if the positions of quasars and bright galaxies were uncorrelated. Unlike BBSS, we make no requirement that the quasar redshifts be already measured. In fact, one can easily imagine situations in which a quasar redshift might be measured just because the quasar seems to be near a bright galaxy or in a rich field the observer is

A.L.   AI

JOHN N. BAHCALL *et al.*

TABLE 1

Quasar–galaxy pair distribution

Sample	Definite and highly probable quasars				All quasars			
	Number	$\Delta\theta$(arc min)			Number	$\Delta\theta$(arc min)		
		0–7.5	7.5–15	15–30		0–7.5	7.5–15	15–30
3C(BBSS)[a]	38	4	0	2	38	4	0	2
3C(other)[b–f]	8	0	0	2	12	0	1	2
4C(Olsen)[d]	25	0	0	0	48	0	0	1*
Parkes[e–o]	70	0	0	1	111	1	0	1
4C(WB)[p]	0	0	0	0	51	1	0	3
Total QSRS	141	4	0	5	260	6	1	9
Bologna QSO[q]	32	0	1	4	166	2	8	12

(*a*) Burbidge *et al.* (1971); (*b*) Schmidt (1969); (*c*) Schmidt, private communication; (*d*) Olsen (1970); (*e*) Burbidge (1967); (*f*) Burbidge (1970); (*g*) Bolton and Wall (1970); (*h*) Ekers (1969); (*i*) Bolton *et al.* (1968); (*j*) Burbidge (1968); (*k*) Bolton *et al.* (1968); (*l*) Merkelijn (1969); (*m*) Hunstead (1971); (*n*) Bolton and Ekers (1967); (*o*) Shimmins (1968); (*p*) Wills and Bolton (1969); (*q*) Braccesi *et al.* (1970). *Subsequently confirmed by Schmidt (private communication): cf. Table 3.

otherwise studying. Thus using only objects with measured redshifts could introduce a bias: see the historical discussion by BBSS of the identification of the quasar 3C 275.1 near NGC 4651; also the measurement of the redshift of 3C 455 (Arp *et al.* 1972) near NGC 7413, which is not a 'bright' galaxy, however. In order to guard against other possible biases with respect to bright galaxies, we have required of all quasars in our sample that: (1) they lie at galactic latitude of 30° or greater ($|b| \geqq 30°$) and (2) they be at least as bright as 19.5 mag ($m_v \leqq 19.5$ mag). The first requirement was imposed because galaxies suffer obscuration below a galactic latitude of 30°, and the second requirement was set because a special effort is required to identify quasars fainter than 19.5 mag. These requirements reduce the size of the BBSS sample from 47 to 38, but they do not eliminate any of the galaxy–quasar pairs.

The numbers of quasars that are paired with bright galaxies from the *Reference Catalogue of Bright Galaxies* at various separations are shown in Table 1 for: (1) the reduced BBSS sample of 3C quasars; (2) the twelve 3C quasar identifications not considered by BBSS (definite and highly probable quasars 3C 37, 39, 43, 93, 216, 281, 311, and 463; possible quasars 3C 44, 90, 217, and 242);

(3) the 4C quasar identifications given by Olsen (1970); (4) the quasar identifications of sources in the Parkes Catalogue (Ekers 1969); (5) the 4C quasar identifications proposed by Wills and Bolton (1969); and (6) the QSO identifications given by Braccesi *èt al.* (1970). Sources that occur in more than one identification list (e.g., both 3C and Parkes) are counted only once in Table 1; each source was assigned to that identification list that appears nearest the top of Table 1. Identifications were regarded as 'possible' for the purposes of compiling Table 1 if the only basis for the identifications was that Sky Survey plates (or prints) show a blue stellar object near the radio position; 'highly probable' and 'definite' identifications were required to have additional confirmatory evidence such as redshifts or accurate photometry.

Incorrect quasar identifications are expected to constitute less than 40 per cent of the 'possible' identifications; thus they can only dilute any physical correlation that may exist between quasars and bright galaxies by a factor of less than two and cannot qualitatively change the tiny probabilities found by BBSS. For QSRS the estimate of the fraction of incorrect identifications is based on the work of Olsen (1970), who remarks that Schmidt spectroscopically confirmed 25 (60 per cent) out of

ARE QUASARS ASSOCIATED WITH BRIGHT GALAXIES?   149

the 38 'possible' QSRS he observed, and of Bolton and Wall (1970), who mention that two-thirds of their 'possible' QSRS were subsequently confirmed on the basis of 2-color plates showing a stellar object with a clear UV excess. The accuracy of the radio positions in these surveys (20 arc sec for Olsen, 10 arc sec for Bolton and Wall) is typical of the identifications we have used. The fraction of Bologna QSO's which have been confirmed is higher: out of the 16 objects in the Bologna catalog examined by Braccesi *et al.* (1968), 15 were found to be QSO's. Braccesi *et al.* (1970) report that unpublished work by Lynds, Grueff and Braccesi confirmed an additional 17 out of 24 candidates. Altogether 32/40, or 80 per cent of the 'possible' Bologna QSO's which have been studied spectroscopically have been confirmed.

Incorrect quasar identifications have been approximately taken into account in our probability calculations by assigning a weight of 0.5 to the 'possible' quasars. (This is a conservative procedure, since we expect at least 60 per cent of the 'possible' quasars to be real.) For example, the number of possible quasars in the total sample of QSRS is taken at full value (119) in Table 1, which describes the samples, but is counted at half value (60) in the probability calculations summarized in Table 2. Note, however, that pairs comprising a 'possible' quasar and galaxy are counted with full weight so as not to omit any possible correlation. Despite the inaccuracies inherent in using a

sample containing 'possible' quasars, we feel that this procedure is necessary because it eliminates the bias that may be introduced by requiring that the redshift already have been measured.

The probability that there should be at least as many galaxy–quasar pairs as we found is given in Table 2 for (i) the 38 3C quasars in the BBSS sample; (ii) the 103 definite and highly probable QSRS we have studied (excluding the 38 BBSS objects); (iii) all the QSRS we studied, with the 'possible' QSRS counted with a weight of 0.5 as explained above; (iv) the combined sample of QSRS, containing both our objects and the BBSS objects; and (v) the 99 Bologna QSO's. (The probability calculations are described in the following section.)

It is apparent from Tables 1 and 2 that the 3C sample studied by BBSS is somehow special: 4 out of 38 quasars in their sample are paired with bright galaxies having an apparent angular separation less than 7.5 arc min. In our sample, the numbers of quasars within 7.5 arc min of bright galaxies are: none of the 103 definite and highly probable QSRS; 2 of the 60 possible QSRS (one of which is a questionable identification: see below); and 2 of 99 Bologna QSO's. If the fraction of quasars paired with bright galaxies had been the same for our sample as it was for that of BBSS, we would have found $(4/38) \times (163 + 99) = 28$ close pairs instead of the 4 we did find. The implications of this result will be discussed below.

TABLE 2

Probability of observed quasar–galaxy pair distribution

Sample	Number	Probability P for $\Delta\theta$(arc min)[a]		
		0–7.5	7.5–15	15–30[b]
3C(BBSS)	38	8.7×10^{-5}	(0.51)[c]	0.39
QSRS (without BBSS)				
Definite and	103	(0.60)[c]	(0.22)[c]	0.58
highly probable				
All	163	0.19	0.91	0.21
QSRS (combined samples)	201	5.5×10^{-4}	0.95	0.14
Bologna QSO	99	0.43	0.08	0.51

(*a*) Here $P = \sum\limits_{k = k_{observed}}^{n} P(k;\ n;\ p)$.

(*b*) The effect of galaxy clustering has been included for $\Delta\theta = 15$–30 arc min by reducing λ by 2 for the QSRS and 1.5 for the Bologna QSO's.

(*c*) This is the probability of obtaining the observed number (zero) of pairs.

150

JOHN N. BAHCALL *et al.*

TABLE 3

Quasar–galaxy pairs

Sample	Source	Nearest-neighbor galaxy (plus additional nearby galaxies)	$\Delta\theta$ (arc min)	Quasar m_v	z	Comments
3C(BBSS)	232	NGC 3067	2	16	0.53	
	245	NGC 3351	27	17	1.03	Not mentioned by BBSS
	268.4	NGC 4138(+2)	3	18.5	1.40	
	275.1	NGC 4651	4	19	0.56	
	309.1	NGC 5832	6	17	0.90	
	351	NGC 6306(+2)	18	15.5	0.37	
3C(other)[a]	37	NGC 467	25	17.5	0.67	
	39	NGC 474(+1)	24	18	0.76	
	242	IC 610	14.5			Identification questionable
4C(Olsen)	23.24	NGC 3162	25	17.5		Confirmed by Schmidt[c]
Parkes[b]	0317−02	NGC 1305	20	19.5		
	1107+10	NGC 3547	2	18.5		Identification questionable
4C(WB)[d]	10.34	NGC 4067	28	17.5		Possible identification
	11.70	NGC 7348	17	15.5		"
	13.46	NGC 4193(+4)	5	18		"
	18.34	NGC 4382	24	18.5		"

(a) There is some uncertainty in equating sources found in the low resolution 3C survey with those found in the higher resolution 4C and Parkes surveys. We take 3C 37 = PKS 0115+02 = 4C 02.4 and 3C 39 = PKS 0118+03 = 4C 03.2. (b) PKS 1248+05 is 5 arc min from NGC 4734, but is excluded from our sample because it is faint, $m_v \sim 20$. (c) Private communication. (d) Wills and Bolton (1969).

Although the distribution of galaxy–quasar pairings at separations greater than 7.5 arc min exhibits some peculiarities, there is no evidence for any correlation. In particular, there are remarkably few QSRS within 7.5 to 15 arc min of bright galaxies (indicated by the high probability of 0.95 of getting at least the observed number of pairings; see row 4 of Table 2).

A list of the pairs we have found and their nearest-neighbor galaxies are given in Table 3 for all the optical pairs we have found except the (numerous) Bologna objects. We note that the identification as quasars of two of our paired objects, 3C 342 and PKS 1107+10, has been seriously questioned; see Bolton *et al.* (1967), Shimmins (1968), Windram and Kenderdine (1969), Olsen (1970). Also four out of the twelve listed

quasars with apparently nearby bright galaxies are in the general area of the Virgo cluster (i.e., 3C 275.1, 4C 10.34, 4C 13.46, and 4C 18.34).

PROBABILITY CALCULATIONS

We here describe briefly how the probabilities given in Table 2 were calculated (cf. BBSS). Let

$$P(k; n; p(\Delta\theta)) \equiv \binom{n}{k} p^k (1-p)^{n-k}$$

be the probability that for n quasars one finds exactly k pairs with angular separations in the specified range $\Delta\theta$, where $p(\Delta\theta)$ is the probability that an average quasar has a nearest-neighbor bright galaxy in the range $\Delta\theta$. The individual probabilities $p(\Delta\theta)$ can be written

$$p(\Delta\theta) = \pi[(\Delta\theta)^2_{max} - (\Delta\theta)^2_{min}]\,\lambda$$

ARE QUASARS ASSOCIATED WITH BRIGHT GALAXIES?   151

where $\lambda \equiv$ the number of bright galaxies per square degree and $\Delta\theta_{max}$, $\Delta\theta_{min}$ are the maximum and minimum angular separations considered. Following BBSS, we estimate λ directly from the *Reference Catalogue of Bright Galaxies*. Over the whole sky (excluding $|b| < 30°$) we find λ (average) $= 0.1$ galaxies per square degree and, for the 3C region studied by BBSS, $\lambda_{3C} = 0.12$. The above values of λ are in good agreement with the average values given by BBSS. For the largest angular separations we consider ($15' < \Delta\theta \leq 30'$), there is some evidence that galaxy clustering is a significant factor. We have made a crude correction for clustering by using λ (effective) $= 0.05$ galaxies per square degree for $15' < \Delta\theta < 30'$. This correction (which of course *lowers* our calculated probabilities for accidental pairings) is based on the fact that for the 9 quasars with nearest-neighbor galaxies in the above-mentioned angular range there were on the average two galaxies per quasar. The bright galaxy density in the region of the Bologna survey is higher than the average over the whole sky above thirty degrees galactic latitude. There are 11 bright galaxies in the *Reference Catalogue* in the 36 square deg of the Bologna survey, i.e., λ (Bologna) $\simeq 0.3$ per square deg; and λ (Bologna, corrected) $\simeq 0.2$ per square deg for $15' < \Delta\theta < 30'$ (where the correction for galaxy clustering in the Bologna field was made in a similar manner as the previously described clustering correction for the other fields).

DISCUSSION

Our results show that, of all the quasar samples we have considered, only the 3C objects studied previously by BBSS exhibit an unexpectedly close angular correlation between quasars and galaxies. This correlation is significant for separations $\Delta\theta \leq 7.5$ arc min; no correlation is apparent for $7.5 < \Delta\theta \leq 30$ arc min.

An application of the χ^2 test based on the data for $\Delta\theta \leq 7.5$ arc min shows that the probability that the BBSS sample of QSRS and our sample of definite and highly probable QSRS were drawn from the same population is only 6×10^{-3}. Hence, either (1) the BBSS sample of QSRS differs from our sample in some significant manner; or (2) one or both of the samples represent a rare statistical fluctuation.

Consider the first possibility. The only obvious difference between the samples is that the QSRS in the BBSS sample were chosen from the 3C catalogue and hence have a higher average radio flux. However, this is not a valid explanation of the difference in results: Quasars from the 3CR catalogue generally have a radio flux at 178 MHz $S_{178} \gtrsim 9$ fu. Eighteen quasars in our sample satisfy this criterion (17 definite or highly probable identifications; 2 possible identifications × weight of 0.5) based on the assumption of a mean spectral index of -0.7. None of these 18 sources is within 30 arc min of a bright galaxy. Thus there is no evidence that the difference between our results and those of BBSS can be explained by supposing that only quasars with the largest radio fluxes are correlated with bright galaxies.

The second (and more likely) possibility is that one or both of the samples represent a large statistical fluctuation. In this case one would like to determine which sample is more representative of the set of all QSRS. Although our sample is somewhat larger than theirs, it is not large enough to show that their sample, and not ours, is a statistical fluctuation. Correspondingly, the probability that the six pairs in the combined sample are due to chance (6×10^{-4}), while larger than that for the four pairs in the BBSS sample (9×10^{-5}), is still extremely small. Of course, on the basis of all the data presently available, the most *probable* situation in a purely statistical sense is that the mean fraction of QSRS closely paired with bright galaxies is intermediate between the number due to chance (4.9×10^{-3}) and the number obtained from the B^2S^2 data ($4/38 = 0.105$). In this 'most probable case', both samples are extreme fluctuations, whence the low value of χ^2 mentioned above. Because of the disparity between the two samples, however, this case does not seem to be significant.

We conclude that the available evidence on the existence of a close ($\Delta\theta \leq 7.5$ arc min) angular correlation between galaxies and quasars is not consistent. All that one can say at present is that the BBSS result that $4/38$ of quasars are closely paired with bright galaxies was not confirmed with our larger, independent sample.

We thank E. M. Burbidge, G. R. Burbidge, P. M. Solomon and P. A. Strittmatter for communicating their results prior to publication and H. C. Arp, M. Schmidt,

152   JOHN N. BAHCALL *et al.*

P. A. Strittmatter, and S. P. McKee for helpful conversations. This work was supported in part by grants GP-16147 and GP-23780 from the National Science Foundation and grant SFG-0-3005 from the Smithsonian Astrophysical Observatory. JNB also acknowledges partial support by Tel Aviv University, Israel. CFM began this work while at the Hale Observatories, Pasadena, California.

REFERENCES

Arp, H. C., 1971, *Science*, **174**, 1189.

Arp, H. C., Burbidge, E. M., Mackay, C. D., and Strittmatter, P. A., 1972, *Astrophys. J.*, **171**, L 41.

Bolton, J. G., and Ekers, J., 1967, *Austral. J. Phys.*, **20**, 109.

Bolton, J. G., Kinman, T., and Wall, J., 1968a, *Astrophys. J.*, **154**, L105.

Bolton, J. G., Shimmins, A., and Merkelijn, J., 1968b, *Austral. J. Phys.*, **21**, 81.

Bolton, J. G., and Wall, J., 1970, *Austral. J. Phys.*, **23**, 789.

Braccesi, A., Formiggini, L., and Gandolfi, E., 1970, *Astron. Astrophys.*, **5**, 264.

Braccesi, A., Lynds, R., and Sandage, A., 1968, *Astrophys. J.*, **152**, L105.

Burbidge, E. M., 1967, *Annu. Rev. Astron. Astrophys.*, **5**, 399.

Burbidge, E. M., 1968, *Astrophys. J.*, **154**, L109.

Burbidge, E. M., 1970, *Astrophys. J.*, **160**, L33.

Burbidge, E. M., Burbidge, G. R., Solomon, P. M., and Strittmatter, P. A., 1971, *Astrophys. J.*, **170**, 233.

Ekers, J., Ed. 1969, *Austral. J. Phys., Astrophys. Suppl. No. 7.*

Hunstead, R. W., 1971, *Monthly Not. Roy. Astron. Soc.*, **152**, 277.

Merkelijn, J., 1969, *Austral. J. Phys.*, **22**, 237.

Olsen, E. T., 1970, *Astron. J.*, **75**, 764.

Schmidt, M., 1969, *Annu. Rev. Astron. Astrophys.*, **7**, 527.

Sharp, J., 1967, M.S. thesis, San Diego State College.

Shimmins, A., 1968, *Austral. J. Phys.*, **21**, 65.

Vaucouleurs, G. de, and Vaucouleurs, A. de, 1964, *Reference Catalogue of Bright Galaxies* (University of Texas Press, Austin).

Wills, D., and Bolton, J. G., 1969, *Austral. J. Phys.*, **22**, 775.

Windram, M. D., and Kenderdine, S., 1969, *Monthly Not. Roy. Astron. Soc.*, **146**, 265.

Received 26 January 1972

Astrophysical Letters, 1972, Vol. 11, pp. 77–82   © Gordon and Breach, Science Publishers Ltd. Printed in Glasgow, Scotland

3C Quasi-Stellar Objects and Bright Galaxies

C. HAZARD and N. SANITT   *Institute of Theoretical Astronomy, Cambridge, England*

The suggestion by Burbidge *et al.* that the correlation between the positions of 3C quasi-stellar objects and bright galaxies provides evidence of a local class of QSO's is examined by the use of other source and galaxy lists. These studies failed to reveal any similar correlation. It is shown the results of the two investigations are difficult to reconcile on the assumption that the 3CR–QSO galaxy correlation has any physical significance.

INTRODUCTION

In a recent paper, Burbidge *et al.* (1971) have shown that of the 47 quasi-stellar objects in the 3C catalogue (Edge *et al.* 1959) and the 3CR catalogue (Bennett 1962), four with redshifts in the range 0.5–1.4 are closer to bright galaxies (mag $\lesssim 13$) in the revised Shapley–Ames catalogue (de Vaucouleurs and de Vaucouleurs 1964) than would be expected if the 3C QSO's were randomly distributed. With the observation of Arp (1971) that the QSO Markarian 205 ($z = 0.070$) appears to be physically connected to the nucleus of NGC 4319 ($z = 0.006$) it appears that a local origin for QSO's can no longer be lightly dismissed. At the same time, the similarity in radio structure between the physical properties of double radio galaxies and double quasi-stellar radio sources (QSRS's), calculated on the assumption of a cosmological origin of QSO redshifts, and also the recent demonstrations that at least some QSO's have the same redshift as apparently associated faint clusters (Gunn 1971, Oke 1971), leave little doubt that for many QSO's the redshifts are cosmological.

If the evidence of Arp and of Burbidge *et al.* is accepted, it follows that QSO's comprise at least two classes of object with similar apparent properties, but in the cosmological case having an intrinsic luminosity up to some 3 mag brighter than the brightest cluster galaxy (i.e., M(cosmological) $= -25$) and in the local case up to 7 mag fainter than a normal spiral (i.e., M (local) $= -11$). The rather low redshift of Markarian 205 may indeed indicate that we are dealing with a new class of object, perhaps a Seyfert-nucleus without the usual embedding galaxy; the absence of the surrounding galaxy would then have to be taken as somehow related to the discrepant redshift (Rees and Sargent 1971). The 'local' QSO's found by Burbidge *et al.*,

however, have redshifts typical of cosmological QSO's, and can in no way be distinguished by their apparent properties.

While a local origin for *all* QSO's has the attraction of avoiding the serious energy-problems posed by the assumption of a cosmological origin, a local origin for only *some* QSO's not only leaves intact the requirements as to energy but also demands an explanation of two types of object identical in all their apparent properties but differing by about 14 mag in absolute luminosity, and also demands a new explanation of the large redshifts of the local QSO's.

It is therefore important to assess the significance of the correlation between 3C QSO's and bright galaxies. There is always some doubt as to the significance of a correlation based on such small numbers, and presumably in this case first brought to attention by the chance notice of the existence of one or two close QSO-galaxy pairs. A true assessment of the significance would take into account all the possible types of correlation that could have been noticed in this way. However, the suggestion of Burbidge *et al.* can be tested either by providing additional evidence that the QSO-galaxy pairs are associated, perhaps by the detection of connecting material, or by testing its implications against other lists of QSO's and galaxies.

SELECTION EFFECTS

Before carrying out any analysis we attempted to find a selection effect which would explain the observed correlation. The non-random distribution of galaxies has been shown by Burbidge *et al.* not to affect their conclusions. Since by now the regions around all the 3CR sources have been extensively studied, and accurate positions and structures are available, there would not appear to

C. HAZARD AND N. SANITT

be any selection in the identification procedure that would lead to QSO's near bright galaxies being preferentially noted. In this respect it may be noted that in the early 3C identification programmes based on positional accuracies of no better than ± 1 arc min the bias would rather be against detecting close QSO-bright galaxy pairs, since the bright galaxy would be easily noticed (and identified), while in such a large search area a QSO would be difficult to observe. With present positional accuracies measured in arc sec this problem does not arise.

It would appear that any selection effect must be one which would lead to the listing of a QSO-galaxy pair in a given catalogue, when the same QSO if isolated would not be included. Such a situation can arise when a QSO below the nominal limiting flux lies close to a bright galaxy which is also a radio emitter. Hence, if a galaxy of 4.5 fu lies close to a QSO also of 4.5 fu it will appear as a source of 9 fu and appear in the 3CR list while, if separated, neither source would be included. A radio flux of this order is to be expected for many of the galaxies brighter than 13 mag. The separations of ≤ 3.5 arc min found by Burbidge et al. for three of their QSO galaxy pairs are themselves suspicious since the lobe separation of the 3C interferometer was about 7.5 arc min and two sources separated by 3.5 arc min could not be resolved. However, we have examined all four suggested pairs and can find no evidence of emission from the galaxy; nor does the present knowledge of the emission from normal galaxies and the expected numbers of QSO's at different flux levels suggest that the effect would be significant. We did find, however, that one of the pairs should not be considered as a 3C source, namely that listed as 3C 232, which is close to the galaxy NGC 3067. In the original 3C catalogue this source is listed as doubtful with a flux of 8.5 fu, but it does not appear in the 3CR catalogue. The 4C survey (Pilkington and Scott 1965) shows it to consist of two sources (4C 32.32 and 4C 32.33), each of 4 fu, and separated by about 20 arc min. It is the source 4C 32.33 which lies close to NGC 3067. However, even if this source is ignored and attention is confined to the remaining three sources, which are all found in the 3CR catalogue, the conclusions of Burbidge et al. are not affected. Thus, out of the total of 40 3CR QSO's, there are three QSO-galaxy pairs

with a separation of ≤ 7 arc min. The density of galaxies brighter than mag $\approx +13$ is about 0.1 per square degree, giving a probability of three or more close pairs occurring by chance of approximately 7×10^{-4}, compared with an estimated probability of 6×10^{-5} for four or more pairs from the sample of 47 3C QSO's. This is much smaller than the very conservative estimate of $\leq 5 \times 10^{-3}$ given by Burbidge et al. Even if 3C 232 is excluded the observed correlation of 3CR QSO's and galaxies appears inconsistent with a random distribution of galaxies and QSO's, and warrants further investigation.

SOUTHERN-HEMISPHERE QSO's AND BRIGHT GALAXIES

Since the 3CR QSO's are restricted to a region north of declination $-5°$ the most direct check of the significance of the results of Burbidge et al. would be to consider an equivalent sample of southern QSO's. We have therefore selected from the 85-MHz catalogue of Mills et al. (1958) and the 408-MHz Parkes catalogue (Ekers 1969, Bolton et al. 1964) all QSO's and possible QSO's between declinations $-5°$ and $-40°$, the southern declination limit of the Palomar sky survey, which have a flux at 178 MHz equal to about 9 fu, corresponding to the lower flux limit of the 3CR catalogue. Redshifts were available only for QSO's north of declination $-25°$. More southerly objects, where we have to rely only on the positional agreement, can be considered only as possible QSO's. In this way 30 definite QSO's and 8 possibles were selected.

The positions of the selected QSO's were compared with the positions of galaxies in the revised Shapley–Ames catalogue. Following Burbidge et al., we restricted the search to a radius of 7 arc min around each QSO. Although the number of QSO's was similar to that in the 3CR catalogue we did not find one case of a QSO-galaxy coincidence; this is consistent with a random distribution of QSO's relative to galaxies and gives no support to the suggested correlation of QSO's and bright galaxies.

COMPARISON OF 3C QSO's WITH ZWICKY GALAXIES

In the previous section we have shown that the

unexpectedly large number of QSO-galaxy pairs found for 3CR QSO's is not found in a similar sample of southern-hemisphere QSO's. Can we explain this discrepancy by assuming that the 3CR QSO's are somehow special, possibly because the region of sky in which they lie covers the central regions of the Virgo cluster? To investigate this possibility we consider the correlation with galaxies in this region fainter than +13 mag. We first consider whether a large fraction of the 3CR QSO's can be associated with nearby galaxies. If so, this would imply that of the 43 QSO's in the sample of Burbidge *et al.* not noted to be within 7 arc min of a bright galaxy, a large fraction must be associated either with galaxies at even greater distances or with galaxies fainter than the lower limit of the revised Shapley–Ames catalogue. Two possibilities for the nature of the association must be considered: (a) the galaxies and associated QSO's are separating, possibly because of the ejection of the QSO's from the galaxies, or vice versa; or (b) the QSO's are bound to the galaxies with a physical separation corresponding to an angular separation ≦ 7 arc min for galaxies with optical magnitudes between +12 and +13. The four associated galaxies and QSO's noted by Burbidge *et al.* are consistent with a constant galaxy-QSO separation of the order of 10 kpc, the value of the angular separation ($\Delta\theta$) increasing approximately linearly with decreasing redshift. An apparent restriction to small physical separations could also arise if the optical luminosity of a QSO decreases with increasing distance from the associated galaxy, but the apparent luminosities of the suggested QSO-galaxy pairs show no evidence for such a relationship.

It was shown by Burbidge *et al.* that extending the search area from 7 to 20 arc min produces only one more possible QSO-galaxy association, compared with an expected increase of between 1 and 2 if the QSO's and the revised Shapley–Ames galaxies are randomly distributed. This is inconsistent with an ejection hypothesis involving the majority of 3CR-QSO's unless (a) the QSO's are accelerating away from their associated galaxies or (b) the ejection of QSO's from galaxies is non-random but occurs in bursts correlated among the galaxies.

To investigate whether a large fraction of the 3C QSO's can be associated with fainter galaxies we have compared the positions of the 43 non-paired QSO's with the positions of the galaxies in the catalogue of Zwicky *et al.* (1961). This catalogue lists about 31,000 galaxies brighter than 15.7 mag with a mean space density $\rho \approx 1.6$ per square degree north of declination $-3.5°$, while the revised Shapley–Ames catalogue lists 2600 galaxies brighter than 13 mag with $\rho \approx 0.1$. Again following Burbidge *et al.*, we considered QSO's within the radius of 7 arc min of a galaxy; but to take into account that QSO's might be bound to galaxies we also considered radii of 1.6 to 3.5 arc min, thus allowing for the greater mean distances of the Zwicky galaxies. The assumption of bound orbits is not essential, although we note that the correlation found by Burbidge *et al.* is most obvious if for galaxies brighter than +13 mag the search radius is restricted to ≦ 7 arc min.

The results are listed in Table 1 and show no evidence of any correlation. We conclude that only a small fraction of 3CR QSO's can be associated with nearby galaxies and that this correlation is evident only if the search area around a QSO is restricted to ≦ 7 arc min. Table 2 lists the positions and QSO-galaxy separations for all QSO's found within 7 arc min of a galaxy.

NON-3C QSO's AND GALAXIES

To exclude the possibility that we have been unfortunate in our choice of southern QSO's and to investigate the QSO-galaxy correlation for QSRS's fainter than the lower limit of the 3CR catalogue we have also carried out an analysis of QSRS's with a flux less than 9 fu.

For the QSO's in the 3CR catalogue there is now no doubt as to the reliability of the identifications, but for other QSO lists the optical studies and particularly the redshift measurements are not so complete. We have selected five lists of identifications (see Table 1 notes), all of which contain numbers of suggested QSO's comparable with the numbers in the 3C catalogue, and for which we consider the majority of the suggested identifications are likely to be genuine.

Since we wished to compare these lists with the galaxies of Zwicky *et al.* we confined our attention to QSRS's north of declination $-3.5°$. Further-

80   C. HAZARD AND N. SANITT

TABLE 1

Expected and observed numbers of QSRS's close to galaxies in the catalogue of Zwicky *et al.*

Source list (see notes)	Number of QSRS in catalogue of Zwicky *et al.*	Expected number of QSRS within			Actual number of QSRS within			QSRS within 7 arc min of galaxies. See Table 2 for details.
		1.6	3.5 arc min	7.0	1.6	3.5 arc min	7.0	
1	43	0.2	0.7	2.8	0	2	4	3C 268.4, 3C 275.1, 3C 309.1, 3C 345
2	52	0.2	0.9	3.4	1	2	3	PKS 0106 + 01[b], PKS 1514 + 00[a], CTA 102
3	58 QSO } 34 QSO?	0.3	1.5	5.9	1	4	10	PKS 0225 − 014 = 4C − 01.11, PKS 0106 + 01[b], PKS 1514 + 00[a], PKS 0317 − 02 = 4C − 02.15, PKS 0906 + 01 = 4C + 01.24, PKS 0950 + 00, PKS 1055 + 01 = 4C + 01.28, PKS 2318 + 02, PKS 1618 + 007, PKS 1021 − 00.
4	20 QSO } 4 QSO?	0.1	0.4	1.5	0	0	0	
5	31 BSO	0.1	0.5	2.1	0	0	2	PKS 0317 − 02 = 4C − 02.15 4C − 03.38 (4C − 01.11 is listed as a blank field)
6	25 QSO } 43 QSO?	0.2	1.1	4.4	0	1	3	4C 30.13, 4C 32.33, 4C 30.23
Total 7	144 QSO } 93 QSO?	0.8	4.0(0.03)	15.3(0.52)	1	7(2)	19(3)	

1. 3C QSRS.
2. List of QSRS from Burbidge and Burbidge (1967), excluding 3C and PHL objects.
3. Parkes 207-MHz survey, which is complete down to 0.35 fu (Wall *et al.* 1971).
4. Parkes 207-MHz survey of selected regions, complete down to 0.1 fu (Wall *et al.* 1971).
5. BSO's from 4C catalogue in the declination range −3.5° to 0° from the 408-MHz Mills Cross Survey (Munro 1971a, b).
6. QSO's and possible QSO's from 4C catalogue, declination range +20° to +40°. The 25 confirmed QSO's are out of a total of 38 of the suggested identifications which have been studied optically (Olsen 1970).
7. The total numbers of QSO's and possible QSO's covered by lists 1 to 6. A source appearing in more than one list is counted only once in this row and the BSO's in list 5 have been listed as QSO. The expected and observed numbers of QSRS near to more than one galaxy are shown in brackets, the significant excess of observed over expected being due to the effect of galaxy clustering.
(a) PKS 1514 + 00 has 2 galaxies within 3.5 arc min.   (b) PKS 1021 − 00 has 3 galaxies within 7.0 arc min.

more, as the 3CR QSO's found close to bright galaxies are all coincident with the radio sources, when no optical confirmation of the suggested identification was available we therefore selected only blue stellar objects (BSO's) for which the optical and radio positions agree within the limits of the positional errors. We have also not considered BSO's noted close to galaxies where the radio source seems certainly to be associated with the galaxy. Radio-quiet QSO's close to galaxies are a separate problem and more difficult to investigate, although of considerable interest in view of the observations of Arp (1971) on Markarian 205 and NGC 7603. The numbers of confirmed

and possible QSO's selected from each list are given in Table 1 except for the identifications of Munro (1971a, b) where the positional accuracy of about ±5 arc sec in each coordinate ensures that the majority of the suggested identifications are genuine.

As in the previous section we have considered radii around the selected QSRS's of 1.6, 3.5 and 7 arc min to allow for the greater distances of the Zwicky galaxies and for a possible dependence of distance on radio flux. The results are listed in Tables 1 and 2, together with the numbers expected on the assumption of a random distribution of galaxies and QSO's. Also given are the corresponding

3C QUASI-STELLAR OBJECTS AND BRIGHT GALAXIES   81

TABLE 2: List of QSRS's found within 7 arc min of a galaxy in the catalogue of Zwicky *et al.*

Radio source	RA (1950)	Dec (1950)	Nearest galaxy NGC number	Nearest galaxy Magnitude (M_{pg})	Angular separation of QSRS and galaxies (arc min)
PKS 0106 +01	01ʰ 06ᵐ 04ˢ	01° 19′		14.8	3.5
PKS 0225 −014 =4C −01.11	02 25 35	−01 29 06		14.4	5.2
PKS 0317 −02 =4C −02.15	03 17 57	−02 19 06	NGC 1298	14.2 15.2	4.2 4.4
4C 30.13	08 01 35	30 20 57		15.4	5.9
PKS 0906 +01 =4C +01.24	09 06 36	01 33 48		15.7	5.2
4C −03.38	09 45 23	−03 28 02		14.5	6.6
PKS 0950 +00	09 50 13	00 14 30		15.4	3.5
4C 32.33	09 55 25	32 38 23	NGC 3067*	12.8	1.9
PKS 1021 −00	10 21 57	−00 37 36		⎧15.5 ⎨15.6 ⎩15.5	2.2 3.4 5.2
PKS 1055 +01 =4C +01.28	10 55 56	01 50 03		14.9	5.1
4C 30.23	11 41 45	30 14 50		15.4	6.1
3C 268.4	12 06 42	43 56	NGC 4138*	12.2	2.9
3C 275.1	12 41 28	16 39 19	NGC 4651*	11.8	3.5
3C 309.1	14 58 58	71 52 19	NGC 5832*	13	6.2
PKS 1514 +00	15 14 15	00 26 01		⎧15.6 ⎨15.7	2.0 1.6
PKS 1618 +007	16 18 16	00 43 54		15.5	5.3
3C 345	16 41 18	39 54 11	NGC 6212	15.0	4.6
CTA 102	22 30 08	11 28 23	NGC 7305	15.1	5.1
PKS 2318 +02	23 18 13	02 40 36		15.6	6.2

* These galaxies are listed in the revised Shapley–Ames catalogue.

numbers for the QSO's covered by all the lists, since many QSO's appear in more than one list.

Provided the density of QSRS's is much less than that of galaxies, the number N of QSRS's out of a total q which are expected to lie within r arc min of n galaxies is given by,

$$N = qp^n e^{-p}/n!$$

where $p = \rho \pi r^2/3600$ and ρ is the number of galaxies per square degree.

Since $p \approx 6.7 \times 10^{-2}$ the expected number of correlations involving more than one galaxy is always small, and is only 0.5 even for the total of 237 QSRS's and $r = 7$ arc min. However, three QSRS's were noted within 7 arc min of two or more galaxies, which is clearly a result of and consistent with the known clustering of galaxies. Because of this clustering the numbers of QSRS's close to galaxies is more relevant than the number of galaxies close to QSRS's.

The results given in Table 1 show good agreement between the numbers of expected and observed correlations for all source lists. In no instance is there an excess of observed over expected exceeding 2.0 standard deviations. Considering the confirmed and suggested QSO's separately, we found no evidence of a significant difference between the two groups. Note that no coincidences were observed with galaxies listed in the revised Shapley–Ames catalogue apart from those already noted by Burbidge *et al.* Hence a comparison of our selected QSO lists with both bright galaxies and galaxies in the catalogue of Zwicky *et al.* provides no confirmation of the correlation noted by Burbidge *et al.* with the 3CR QSO's. The elimination of spurious QSO's from lists 3 to 6 cannot affect this conclusion even if a large fraction of the unconfirmed identifications are spurious.

82   C. HAZARD AND N. SANITT

DISCUSSION

The absence of a QSO-galaxy correlation for southern-hemisphere QSRS's with a flux range similar to those in the 3CR catalogue shows that the correlation found by Burbidge *et al.* must be peculiar to the region of sky covered by the 3CR survey. Even in this region the correlation is evident only for 3CR sources and galaxies in the revised Shapley–Ames catalogue, and is most apparent for QSO-galaxy separations of $\leqq 7$ arc min. The preference for separations $\leqq 7$ arc min can be explained by assuming that the suggested local QSO's tend to lie at distances from the associated galaxy ≈ 10 kpc, a typical separation of a globular cluster. This could suggest that the local QSO's are associated with collapsed globular clusters while the cosmological QSO's are associated with collapsed galactic nuclei; the possibility of the collapse of globular clusters has been discussed by Peebles (1972). The difference of about 14 mag between their intrinsic luminosities would then be roughly proportional to the masses involved in the two cases. However, we have to account for the failure to find a QSO-galaxy correlation for (a) those Zwicky *et al.* galaxies not listed in the revised Shapley–Ames catalogue and (b) QSRS's below the lower limit of the 3CR catalogue. The Zwicky galaxies have a magnitude limit of about $+15.7$ mag compared with a limit of about $+13$ mag for the revised Shapley–Ames galaxies. Their associated local QSO's could therefore be fainter by approximately 2.5 mag than those associated with Shapley–Ames galaxies. If associated with fainter Zwicky galaxies, two of the three 3CR pairs found by Burbidge *et al.* would have QSO magnitudes $> +20$ mag and would therefore not have been detected. The failure to detect any correlation for source lists fainter than the 3CR is more difficult to explain, since at least in the 3CR region of the sky we would expect a percentage of QSO-galaxy pairs similar to that found for the 3CR sources. Typically the majority of sources in these lists are less than 1 mag fainter than the majority of the 3CR sources, which appears too small a difference to explain plausibly the lack of correlations as due to a correlation of optical and radio luminosities. Since the majority of fainter QSRS's in Table 1 lie in the same general region of sky as the 3CR sources it appears that it is not so much the region of sky

which is somehow special but rather the 3CR sources themselves.

CONCLUSIONS

The failure to confirm the correlation found for 3CR sources and bright galaxies by using other source and galaxy lists must be weighed against the high formal significance of this correlation. This weighting must to some extent be subjective, but it is our opinion that the restrictions which must be placed on the distribution and properties of 'local QSO's' to reconcile the results of the two investigations make it highly improbable that the 3CR-QSO galaxy correlation has any physical significance, and that this investigation rather demonstrates the problems inherent in establishing a rare phenomenon from a limited body of data.

We thank Dr M. J. Rees for helpful discussions.

REFERENCES

Arp, H. C., 1971, *Astrophys. Letters*, **7**, 221.
Bennett, A. S., 1962, *Mem. Roy. Astron. Soc.*, **68**, 163.
Bolton, J. G., Gardner, F. F., and Mackey, M. B., 1964, *Austral. J. Phys.*, **17**, 340.
Burbidge, G. R., and Burbidge, E. M., 1967, *Quasi-Stellar Objects* (W. H. Freeman & Co., San Francisco).
Burbidge, E. M., Burbidge, G. R., Solomon, P. M., and Strittmatter, P. A., 1971, *Astrophys. J.*, **170**, 233.
Edge, D. O., Shakeshaft, J. R., McAdam, W. B., Baldwin, J. E., and Archer, S., 1959, *Mem. Roy. Astron. Soc.*, **68**, 37.
Ekers, J. A., 1969, *Austral. J. Phys. Astrophys. Suppl.* No. 7.
Gunn, J. E., 1971, *Astrophys. J.*, **164**, L113.
Mills, B. Y., Slee, O. B., and Hill, E. R., 1958, *Austral. J. Phys.*, **11**, 360.
Munro, R. E. B., 1971a, *Austral. J. Phys.*, **24**, 263.
Munro, R. E. B., 1971b, *Austral. J. Phys.*, **24**, 617.
Oke, J. B., 1971, *Astrophys. J.*, **170**, 193.
Olsen, E. T., 1970, *Astron. J.*, **75**, 764.
Peebles, P. J. E., 1972, *General Relativity and Gravitation*, in press.
Pilkington, J. D. H., and Scott, P. F., 1965, *Mem. Roy. Astron. Soc.*, **69**, 183.
Rees, M. J., and Sargent, W. L. W., 1972, *Comments Astrophys. Space Phys.*, **4**, 1.
Vaucouleurs, G. de, and Vaucouleurs, A. de, 1964, *Reference Catalogue of Bright Galaxies* (University of Texas Press, Austin).
Wall, J. V., Shimmins, A. J., and Merkelijn, J. K., 1971, *Austral. J. Phys. Astrophys. Suppl.* No. 19.
Zwicky, F., Herzog, E., Wild, P., Kowal, C. T., and Karpowicz, M., 1961, *Catalogue of Galaxies and Clusters of Galaxies* (California Institute of Technology, Pasadena).

Received in final form 20 March 1972

REPRINTS OF PAPERS
SELECTED BY
HALTON ARP

COMMENTS ON REPRINTED PAPERS

In the following pages a selection of papers dealing with the possible need for new interpretations of extragalactic redshifts is presented. The purpose of this section is to give actual examples of the different approaches of various authors to this problem. The reader of this book is also enabled to conveniently read some of the original papers which are sometimes not easily accessible.

The reprints are arranged in a roughly chronological order. In order to give a flavor of the starting years of the controversy, four early papers are reprinted here. They start with Terrell (*Science*, 145, 1964), who after the original discussion by Greenstein and Schmidt was the first to put forth a serious proposal that the quasars might not be at cosmological distances. Two years later Hoyle, Burbidge, and Sargent introduced the idea of the physical difficulty of having very luminous, condensed synchrotron sources. Almost at the same time, Arp published the first evidence for association of quasars and high-redshift galaxies with nearby objects (*Science*, 151, 1966) and Hoyle and Burbidge wrote a long, critical discussion of the physical consequences of the "local" and cosmological interpretation of quasar distances. The

last mentioned article (*Astrophys. J.*, 144, 1966) is almost pro-
phetic in discussing, pro and con at an early stage, most of the
questions that are currently being debated. (Because of length, a
technical discussion of the difficulties of synchrotron sources sup-
plying energy for quasars at cosmological distances, *Journal*
pages 542 to 550, are omitted from the article as reprinted here.)

From 1966 to 1971 only Arp published material on the
association of quasars and high-redshift objects with relatively
nearby galaxies. Long papers in *The Astrophys. Journal* (148,
321, 1967) and in *Astrofizika* (4, 59, 1968) are not reprinted here.
A short paper (*Nature*, 223, 1969) gave the first general suggestion
of a nonvelocity, nongravitational redshift cause. Then papers in
1970 (*Astron. J.*, 75 and *Nature*, 225) gave associations of quasars
and companion galaxies with low-redshift galaxies which were
supported by the best quantitative and statistical evidence to that
date.

In 1971 a number of papers began to appear on the subject
of discordant redshifts. Hoyle and Narliker (*Nature*, 233) made
the specific suggestion that matter with low-mass electrons could
give nonvelocity redshifts. Burbidge, Burbidge, Solomon, and
Strittmatter showed that four quasars in the *3C Cambridge Radio
Catalog* fell closer than 7 arc minutes from large spiral galaxies,
a number much in excess of that expected by chance. Tifft had
been independently investigating redshifts in clusters of galaxies
and adduced the presence of nonvelocity redshifts in the Virgo
cluster (*Bull. Amer. Astron. Soc.*, 3, 1971) and the Coma cluster
(*Bull. Amer. Astron. Soc.*, 4, 1972, and *Astrophys. J.*, 175,
1972). In a current paper (*Astrophys. J.*, 181, 15 April), Tifft
has applied the multiple redshift-magnitude band system of the
galaxy clusters to the known quasars and concludes the quasar redshifts are also quantized. If the latter result can be substantiated,

it is a very powerful insight into the nature of the intrinsic redshift mechanism. (The latter two Tifft papers, because of their
length, are not reprinted here.)

In a paper evaluating all the known data on groups and
clusters of galaxies, Jaakkola gave impressive evidence for systematically higher redshifts in Sc galaxies than E galaxies at the
same distance (*Nature*, 234, 1971). Shortly thereafter,
de Vaucouleurs (*Nature*, 236, 1972) supported this conclusion from
his own extensive expertise on the Virgo cluster. (The latter
author is currently preparing a more extensive analysis of the
Virgo cluster from this standpoint.) In 1971 Arp published an extensive, nontechnical review paper of the subject in *Science* (174)
(not reprinted here), which brought more notice of the problem
outside strictly astronomical quarters. Kellermann, in his Warner
Prize lecture in Puerto Rico in December 1971, summarized the
evidence for many radio sources being local that had hitherto been
assumed to be at cosmological distances.

In 1972 and 1973 Rowan-Robinson brought attention to the
evidence for flat-spectrum compact quasars being local and proposed compromise resolutions where some quasars would be local
and some cosmological. In April of 1972 at the Seattle meeting of
the American Astronomical Society, Arp (*Bull. Amer. Astron.
Soc.*, 4) presented a paper with new evidence for redshift discrepancies in Stephan's Quintet and in other chains and groups of
multiple interacting galaxies. At the same meeting, Hoyle summarized in his Russell Lecture the evidence for discordant redshifts. Hoyle's Russell Lecture, previously available only in
preprint form, is published in the following section for the first
time.

Papers are now appearing rapidly on this subject, and by
publication date of the present book much new information will

undoubtedly be available. The controversy may or may not be resolved by that date, but there seems to be some value to sum-
marizing the debate at a peak of interest. Looking back from some
future stage of enlightenment it may be possible to discern more
accurately the course by which we can wend our way through confusing data, claims and counterclaims, to important progress in
scientific knowledge.

LIST OF PAPERS REPRINTED

Terrell, J. Quasi-stellar diameters and intensity
 fluctuations. *Science*, 145, 1964.

Hoyle, F., Burbidge, G. R. and Sargent, W. L. W.
 On the nature of the quasi-stellar sources.
 Nature, 209, 751, February 19, 1966.

Arp, H. Peculiar galaxies and radio sources. *Science*,
 151, 1214, 1966.

Hoyle, F. and Burbidge, G. R. On the nature of the
 quasi-stellar objects. *Astrophys. J.*, 144,
 534 – 541 and 551 – 552, 1966.

Arp, H. Redshifts of very young objects. *Nature*, 223,
 386, 1969.

Arp, H. Distribution of quasi-stellar radio sources on
 the sky. *Astron. J.*, 75, 1, 1970.

Arp, H. Redshifts of companion galaxies. *Nature*, 225,
 1033, 1970.

Hoyle, F. and Narliker, J. On the nature of mass.
 Nature, 233, 41, 1971.

Burbidge, E. M., Burbidge, G. R., Solomon, P. M. and
 Strittmatter, P. A. Apparent associations be-
 tween bright galaxies and quasi-stellar objects.
 Astrphys. J., 170, 233, 1971.

Tifft, W. G. The correlation of redshift with magnitude
        in the Coma cluster. *Bull. Amer. Astron. Soc.*,
        3, 391, 1971.

————. Redshift, morphology, and integrated magni-
        tude relations in the Coma cluster. *Bull. Amer.
        Astron. Soc.*, 4, 238, 1972.

Jaakkola, T. On the redshifts of galaxies. *Nature*, 234,
        534, 1971.

de Vaucouleurs, G. Non-velocity redshifts in galaxies.
        Nature, 236, 166, 1972.

Rowan-Robinson, M. Are quasars local or cosmological?
        Nature, 236, 112, 1972.

Kellermann, K. I. Radio galaxies, quasars, and
        cosmology. *Astron. J.*, 77, 531, 1972.

Arp, H. Morphology and redshifts of galaxies. *Bull.
        Amer. Astron. Soc.*, 4, 397, 1972.

Hoyle, F. The developing crisis in astronomy. Russell
        Lecture, A.A.S. Seattle, 1972. Caltech preprint
        OAP286.

Reprinted from Science, August 28, 1964, Vol. 145, No. 3635, pages 918-919

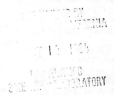

Quasi-Stellar Diameters and Intensity Fluctuations

Abstract. *It is shown that there are relativistic limits on the fluctuations in brightness which may be observed for a large spherical source, and also for more general sources, so that quasi-stellar objects are probably no more than light-days in size. There is thus a possibility that they may be close to our galaxy.*

The relatively rapid fluctuation (1–4) of photographic magnitude has been one of the most interesting and puzzling features of quasi-stellar objects. These fluctuations indicate that the objects cannot be larger than a few light-years, or perhaps even light-days, in diameter. These size limitations are based on the finite speed of light, and have been discussed in several papers (1, 2, 5, 6).

Hoffmann has recently suggested (6) that rapid fluctuations and a large size may still be consistent, provided that a central source in a roughly spherical object triggers simultaneous brightening of the entire surface. This interesting suggestion is, unfortunately, not consistent with the special theory of relativity. A very large spherical surface brightening simultaneously at all points would not be seen as such by an observer, because of the finite speed of light and the different distances to various points of the surface. The brightening would be seen to begin at the nearest part of the surface and to spread out from there to the edge in a time given by R/c, where R is the radius of the object and c the velocity of light. If the surface were oscillating in brightness with period T small compared to R/c, a concentric set of bright rings would be seen, expanding continually from the center.

For an object sufficiently far away that no surface details could be seen, there could thus be a large relativistic reduction in the visible fluctuations of magnitude. It is possible to calculate the exact relation between the visible brightness fluctuation and the true fluctuation of Hoffmann's spherical surface. An integration over the sur-face, using Lambert's cosine intensity law, yields the fraction $F(\alpha)$ of the sinusoidal fluctuation which is visible at a distance:

$$F(\alpha) = (2/\alpha^2)(2 + \alpha^2 - 2\cos\alpha - 2\alpha\sin\alpha)^{\frac{1}{2}} \quad (1)$$

in which α is the ratio of circumference to period, in appropriate units:

$$\alpha \equiv 2\pi R/cT \quad (2)$$

This relativistic effect vanishes, so that $F(\alpha)$ approaches unity, for small sources or long periods:

$$F(\alpha) = 1 - \alpha^2/36 + \cdots , \quad \alpha \ll 1 \quad (3)$$

For large values of α, however, the visible fraction of the fluctuation is inversely proportional to α:

$$F(\alpha) \cong 2/\alpha = cT/\pi R, \quad \alpha \gg 1 \quad (4)$$

Since the intensity at the source cannot become negative, the oscillating component at the source must have a smaller amplitude than the average brightness. Therefore, at a distance, the ratio of maximum to minimum visible intensity must be less than

$$[1 + F(\alpha)]/[1 - F(\alpha)].$$

This immediately establishes a maximum value α_m of α which can be associated with a given oscillation in brightness. Letting Δm be the change in stellar magnitude from maximum to minimum brightness, we have

$$\Delta m = 2.5 \log_{10}\{[1 + F(\alpha_m)]/[1 - F(\alpha_m)]\} \quad (5)$$

This may be approximated by

$$\alpha_m/\pi = D_m/cT \cong 10(\log_{10} e)/\pi\Delta m = 1.382/\Delta m, \quad \alpha \gg 1 \quad (6)$$

giving the maximum diameter D_m which could produce sinusoidal fluctuations Δm in visual magnitude of period T.

This asymptotic relation approximates the exact relation, Eq. 5, in a very satisfactory way, giving an error in D_m of the order of ± 10 percent or less for $\Delta m \leqq 0.7$, which is the range of interest for quasi-stellar sources. These relations mean, to give an example, that for a spherical source with $\alpha = 2\pi$, or $D = 2cT$ (for example, 1-year period and 2-light-year diameter), the apparent fluctuations would be reduced to 31.8 percent of the true fluctuations. The maximum and minimum visible brightness could not differ by as much as a factor of 1.93, or 0.71 magnitudes ($\Delta m = 0.69$ from Eq. 6).

Although Eq. 6 was derived on the assumption of a spherical surface with sinusoidally oscillating brightness (7), it can be put on a more general basis. Consider, for example, an astronomical source composed of a number n of adjacent spherical oscillating sources, all alike but with completely random relative phases. The fluctuation of the total light intensity I from the average, I, may be described in terms of the relative standard deviation, $\sigma(I)/I$. It may readily be shown that the combination of n sources has a fluctuation which is reduced from that of a single source by the factor $n^{1/2}$, the square root of the ratio of total source areas (2). Similar considerations have been discussed previously (2, 5). Thus, in this case also, the visible fluctuation Δm is inversely proportional to an effective diameter, so that Eq. 6 should still hold for n independent sources.

Still more generally, the source could merely be assumed to be composed of a number of independently flashing surface areas. This still leads to Eq. 6 if the independent surface areas are taken to have area c^2T^2/π. This exact choice is by no means forced on us, but is at least a reasonable choice for the maximum area for which the oscillating brightness may remain in phase, considering the finite speed of light signals between portions of the area.

Thus we may use Eq. 6 for quasi-stellar objects with some confidence that it will give a physically reasonable maximum diameter. The precision is greatest, of course, for sinusoidal light fluctuations of a spherical source, but the result is as accurate, in less simple cases, as the meanings of "period" and "effective diameter" permit. The fluctuations observed (1–4) for the quasi-stellar objects 3C48, 3C196, and 3C273 are of the order of 0.3 magnitudes, or a factor of 1.32 in brightness. A period of 2 years would mean, according to Eq. 6, a source diameter less than 9.2 light-years (8.6 light-years, from Eq. 5). If the night-to-night variations and bright flashes observed for 3C273 and 3C48, amounting to as much as 0.7 magnitude with a period of days (1–4), are real and general phenomena, the objects cannot be more than a few light-days in size, as Smith and Hoffleit (1) have stated. In the case of a multiple-source disc produced by gravitational contraction, as suggested by Hoyle and Fowler (5, 8), the effective diameter would have to be considered to be essentially that of the one source emitting most of the light at a given time, rather than that of the entire assemblage.

It should be kept in mind that Eq. 6 yields an upper limit for the source diameter, not an average, and that the light source would perhaps be considerably smaller than this size. Thus the quasi-stellar objects—at least the optically visible parts—are probably less than a few light-days in size and could be of the order of the solar system in size, or less (9). These objects are then many orders of magnitude less than galactic size, simply on the basis of the light fluctuations.

Since the quasi-stellar objects are not of galactic size, it is not absolutely necessary to suppose that they are at the distances of the order of 1000 Mparsec which follow from the application of Hubble's law. The large red shifts observed for these objects (10–13) may be due to the same cause as that of the most distant galaxies, or they may be gravitational in origin, or they may simply be due to relativistic velocities of nearby objects. The first explanation requires optical power outputs perhaps 100 times larger than those of the largest and brightest gal-axies (1–3, 5) and is so far the favorite explanation, although the energy source is not clear. The gravitational explanation is usually ruled out (5, 10, 12) primarily on spectroscopic grounds.

The third possibility, that the objects might be much closer to our galaxy, with the red shift due only to the relativistic Doppler shift, has not been much discussed. A minimum distance is determined by the lack of observed proper motion. W. H. Jefferys (14) has established the proper motion of 3C273 as unobservably small (0.001 ± 0.0025 second of arc per year), and W. J. Luyten (15) has obtained a similar result. If it is assumed, for example, that 3C273 originated near the center of our galaxy, about 8 kparsec (26,000 light-years) from the sun, a proper motion of 0.002 second per year and a recessional velocity of 0.146 c (10) would correspond to 190 kparsec observed distance, and to an origin about 5 million years ago. This distance is a few galactic diameters away, further than the clouds of Magellan, but not so far as the Andromeda nebula which is about 500 kparsec distant.

How such highly relativistic objects could be produced by an explosion or explosions in our galaxy is not known, and this is probably the principal objection to the idea. Considerable kinetic energy alone would be required, amounting to 1.1 percent of the rest mass for 3C273, and to 9.6 percent for 3C147, the fastest such object yet observed, with $v/c = 0.410$, and $\Delta\lambda/\lambda = 0.545$ (13). The optical energy requirements are reduced by a factor of 10^7 at this distance, however, amounting to about 10^5 times the output of our sun. A possible source of high velocities, if the objects are assumed to be local, would be a gravitational collapse (5, 8) in our galaxy.

Whatever their distance, it is clear that the quasi-stellar objects are not of galactic size, but are, at least optically, probably less than light-days in size (16).

James Terrell
University of California, Los Alamos
Scientific Laboratory, Los Alamos,
New Mexico

References and Notes

1. H. J. Smith and D. Hoffleit, *Nature* **198**, 650 (1963).
2. T. A. Matthews and A. R. Sandage, *Astrophys. J.* **138**, 30 (1963).
3. A. Sandage, *ibid.* **139**, 416 (1964).
4. A. S. Sharov and Yu. N. Efremov, *Astron. Zh.* **40**, 950 (1963); *Soviet Astron.* (English transl. *Astron. Zh.*) **7**, 727 (1964).
5. J. L. Greenstein, *Sci. Am.* **209**, No. 6, 54 (1963); H. Y. Chiu, *Phys. Today* **17**, No. 5, 21 (1964).
6. B. Hoffman, *Science* **144**, 319 (1964).
7. It is interesting to note that a somewhat different case, that of a luminous disc viewed perpendicularly, with waves of brightness spreading out from the center, has precisely the same solution, Eq. 5. In either case an assumed recession of the source would not change the result, but the period *T* should be as measured in the rest frame of the source.
8. F. Hoyle, W. A. Fowler, G. R. Burbidge, E. M. Burbidge, *Astrophys. J.* **139**, 909 (1964).
9. The solar system is about ½ light-day across. These considerations would not rule out an extremely small, dense, region in the center of a galaxy or cluster as a quasi-stellar object, as suggested by S. M. Ulam and W. E. Walden [*Nature* **201**, 1202 (1964)].
10. M. Schmidt, *Nature* **197**, 1040 (1963).
11. J. B. Oke, *ibid.*, p. 1040.
12. J. L. Greenstein and T. A. Matthews, *ibid.*, p. 1041.
13. M. Schmidt and T. A. Matthews, *Astrophys. J.* **139**, 781 (1964).
14. W. H. Jefferys, III, *Astron. J.* **69**, 255 (1964).
15. W. J. Luyten, *Minnesota University Astron. Observatory Publ.* **3**, No. 13 (1963).
16. A letter by A. T. Young has just been published [*Science* **145**, 72 (1964)], giving some similar considerations on the size of quasi-stellar objects. Young's results are based on square-wave modulation of a light source. If a large spherical source of radius R is "turned off" for a time interval t, it will undergo a maximum reduction in apparent brightness amounting to $(2ct/R) - (ct/R)^2$. This is, as in Young's example, 19 percent decrease for a 10-light-year-radius source turned off for 1 year. If such a fluctuation were seen, the correct maximum radius would be obtained from Eq. 4 if the period were taken to be $T = \pi t$. This is the period corresponding to the amplitude of the visible fluctuation and its maximum rate of change, for a sinusoidal oscillation. Thus the results of this paper are as applicable to monotonic and single-pulse fluctuations as to sinusoidal oscillations, if the "period" is suitably estimated. Young's formulation and the one given here both give a maximum rate of change of luminosity equal to $2c/R$ logarithmically, or to 1.0857 $(2c/R)$ in units of stellar magnitude.
17. Supported by the AEC. Some of the matters considered have been clarified in interesting discussions with G. R. Burbidge, A. N. Cox, and S. M. Ulam, but they should not be held responsible for the opinions expressed here.

22 June 1964

(Reprinted from Nature, Vol. 209, No. 5025, pp. 751–753, February 19, 1966)

ON THE NATURE OF THE QUASI-STELLAR SOURCES

By Prof. F. HOYLE, F.R.S.
University of Cambridge, and California Institute of Technology
AND
Prof. G. R. BURBIDGE and
Prof. W. L. W. SARGENT
University of California, San Diego

IN a recent paper, under the same title as this article[1], two of us have discussed both the usual cosmological interpretation of the red-shifts of the spectrum lines of the quasi-stellar objects and also the possibility that these objects are comparatively 'local' to our Galaxy, with distances of 1–10 Mpc. The work was begun with the conventional prejudice toward the cosmological interpretation, the initial aim being to disprove the local hypothesis. In the outcome, however, we were not able to satisfy ourselves that the local hypothesis must be discarded—in fact we arrived at a balanced position between the two hypotheses. In this communication we do not propose to repeat the former arguments but to put forward new considerations which seem to us to point toward the local hypothesis.

The new argument turns on the Compton process in which relativistic electrons are inelastically scattered by low-energy photons. The importance of this process was realized by Greenstein and Schmidt[2] and also discussed by Ginzburg, Ozernoi and Syrovatsky[3]. Like synchrotron radiation, the Compton process depends on the presence of electrons of relativistic energies. The electrons lose energy by the synchrotron process and by the Compton process in accordance with the ratio:

$$\frac{\text{Loss by Compton scattering}}{\text{Loss by synchrotron scattering}} = \frac{U_{\text{rad}}}{H^2/4\pi} \quad (1)$$

where U_{rad} is the energy density of the radiation field. In order to apply the relation (1) to the Compton process it is necessary to discuss two important questions: (a) What is the intensity of the synchrotron emission from quasi-stellar objects? (b) What are the dimensions of the emission regions? We begin with a discussion of (a).

The observed radiation from quasi-stellar objects is thought to be derived from two processes: (i) synchrotron radiation and (ii) radiative recombination of an ionized gas. The most detailed studies have been carried out for 3C 273 and 3C 48. There is no question but that the radio frequency radiation from the quasi-stellar objects is generated entirely by the synchrotron process. However, as will become clear from the later discussion, it is the high-energy density of radiation which comes almost entirely from the flux emitted in the optical and infra-red spectrum which is of critical importance here.

Matthews and Sandage[4] ascribed the whole of the continuous radiation from 3C 48 in the wave-length range $3.0-8.3 \times 10^{14}$ c/s to the synchrotron process. Greenstein and Schmidt[2] and Oke[5] have both discussed the optical spectrum of 3C 273 in terms of radiative recombination of a hot gas. In Fig. 1 we compare the observed continuous energy distribution in the optical and near-infra-red regions published by Oke and by Johnson[6] with the theoretical radiative recombination spectra of the two models discussed. The two theoretical spectra have been given an arbitrary vertical shift to facilitate comparison of the forms of the continua.

Oke chose a value of $T_e = 160,000°$, which gives the observed Balmer discontinuity, B; however, the Paschen continuum must then be ascribed to a component of unknown origin (which Oke supposed to be synchrotron radiation) which rises steeply into the near infra-red. As Oke pointed out, Greenstein and Schmidt's model with $T_e = 16,000°$ requires Balmer and Paschen discontinuities much larger than those observed; thus in their case also a spectrum rising steeply into the infra-red must be added to the recombination spectrum. Searle et al.[7] have shown that in each case the spectrum which must be added has the same slope as those of the Crab nebula and 3C 48. This is very strong evidence that the near infra-red continuum of 3C 273B is dominated by synchrotron radiation; Dibai and Pronik[8] have also given strong arguments for this conclusion. The presence of a small Balmer discontinuity together with the Balmer emission lines shows that a hydrogen recombination spectrum is also present and begins to dominate below λ4,000.

Extension of the observations of 3C 273 into the infra-red ($\nu = 3 \times 10^{13}-1.4 \times 10^{14}$ c/s) by Johnson and Low[9] and into the millimetre range ($\nu = 10^{11}-3 \times 10^{11}$ c/s) by Low[10] and Epstein[11] has shown that this object is an exceedingly powerful emitter in these regions. For other quasi-stellar objects there are no published observations of fluxes in the infra-red, but indications that the fluxes are rising in the near infra-red suggest (because of the effect of the red-shift) that a synchrotron contribution may be important in the optical region in some objects.

An integration of the energy distribution curve of 3C 273 shows that the total flux received at the Earth in the frequency range $10^{11}-5 \times 10^{14}$ c/s is 5.3×10^{-9} erg/cm²/ sec. The bulk of this radiation is emitted in the frequency range $10^{11}-10^{13}$ c/s. For 3C 48, Sandage[12] obtained a flux of 1.1×10^{-11} erg/cm²/sec for this frequency range.

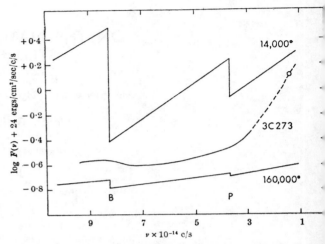

Fig. 1. The continuous energy distribution of 3C 273 in the optical and near infra-red is compared with the theoretical hydrogen recombination continua appropriate to electron temperatures of 14,000° and 160,000° given by Oke (ref. 5). The continuous line represents Oke's observations of 3C 273. The circle represents Johnson's (ref. 6) measurement of the flux at 2·2μ. The ordinate is the logarithm of the observed flux per unit frequency interval from 3C 273 at the Earth's surface. The units are 10^{-24} ergs cm⁻² sec⁻¹ (c/s)⁻¹. The abscissa is the frequency in units of 10^{14} c/s. Note that the theoretical continua have been given arbitrary vertical shifts

It has often been pointed out that detection of a significant degree of polarization in the radiation might add weight to the circumstantial evidence for the flux being synchrotron radiation. In the case of 3C 48, Hiltner found that the degree of polarization in the optical region is less than 2 per cent. Recently, Schmidt[13] has reported that in 3C 273B there is no measurable polarization in the frequencies near 7.5×10^{14} c/s, but at 6×10^{14} c/s about 4 per cent polarization is present. This is consistent with a steadily increasing proportion of synchrotron radiation at lower frequencies.

We conclude that the bulk of the energy emitted from 3C 273 and probably from the quasi-stellar objects in general is synchrotron radiation. We consider next the dimension of the source from which this radiation is emitted.

It is well known that it is the flux variations from the source which set a much stronger upper limit for the size than the 'quasi-stellar' characteristic. 3C 273B has been shown to vary in the optical region in time scales of years and months[11]. Recent observations in frequencies near 3×10^{11} c/s (ref. 9) suggest that there have been large variations in periods of months. In other quasi-stellar objects variations in the optical flux of similar ar

sometimes shorter time scales are seen. In the case of
3C 345 significant optical and spectroscopic variations on
a time scale of weeks have recently been reported[15,16]
though no observations of the infra-red flux in this object
are available. Thus a reasonable scale to take for the size
of the region in which this synchrotron flux is generated
is $\sim 10^{17}$ cm or about a light month.

We return now to our discussion of the inverse Compton
process. In attempting to estimate the relative importance
of the synchrotron process and the Compton scattering
process, we first considered that the dimension of a
'coherent' emission region is such that a relativistic electron
can travel in its lifetime across the whole region. For a
sphere of radius R this requires $\tau > R/c$, where τ is the
electron lifetime against synchrotron radiation. Calculating τ from the synchrotron process alone, we have:

$$\tau = 10^{12}\nu^{-1/2}H^{-3/2} \text{ sec} \qquad (2)$$

where H is the magnetic intensity in gauss and ν is the
critical frequency in c/s of the electron in question. From
the inequality $\tau > R/c$, we obtain:

$$H^{3/2} < 3 \times 10^{22}\nu^{-1/2}R$$

giving:

$$\frac{H^2}{4\pi} < \sim 10^{29}\nu^{-2/3}R^{-4/3} \qquad (3)$$

R being in cm.

Suppose first that the whole emission comes from just
one such coherent region. Then, if d is the distance to the
source and s is the flux in ergs cm^{-2} sec^{-1} at the Earth,
$L = \pi R^2 c \, U_{\text{rad}} \simeq 4\pi d^2 s$, neglecting a red-shift factor, so
that:

$$U_{\text{rad}} \approx \frac{4d^2 s}{R^2 c} \qquad (4)$$

Taking the ratio of (4) to (3):

$$\frac{U_{\text{rad}}}{H^2/4\pi} = \frac{4d^2 s}{10^{29}c}\left(\frac{\nu}{R}\right)^{2/3}$$

If we substitute for s the value of $5 \cdot 3 \times 10^{-9}$ erg/cm^2/sec,
then:

$$\frac{U_{\text{rad}}}{H^2/4\pi} = 63 \cdot 6 \, D^2 \left(\frac{\nu}{R}\right)^{2/3} \qquad (5)$$

where D is measured in megaparsecs (Mpc) ($d = 3 \times 10^{24} D$): Putting $D = 470$ Mpc, the cosmological distance
for 3C 273, we have:

$$\frac{U_{\text{rad}}}{H^2/4\pi} = 1 \cdot 3 \times 10^7 \left(\frac{\nu}{R}\right)^{2/3}$$

Thus, if $R = 10^{17}$ cm and $\nu = 10^{14}$ c/s (the near infra-red):

$$\frac{U_{\text{rad}}}{H^2/4\pi} = 1 \cdot 3 \times 10^5 \qquad (6)$$

It is of some interest to apply a similar argument to the
case of 3C 345. This quasi-stellar source which has a redshift $z = 0 \cdot 595$ (ref. 17) has varied significantly in a time
scale of $\sim$ light week or $\sim 10^6$ sec (ref. 15). In this case
we do not have such strong evidence that a major part of
the flux is synchrotron radiation. The flux that is seen
is ultra-violet radiation and we have no knowledge concerning the flux in the frequency range discussed for
3C 273 because of the much larger red-shift. However,
we shall assume that the flux is synchrotron in origin.
Taking an apparent magnitude of $17^{\text{m}}0$ for 3C 345, we
find that for this object $s = 8 \cdot 4 \times 10^{-12}$ erg/cm^2/sec.
Thus for this source, going through the same steps as
before:

$$\frac{U_{\text{rad}}}{H^2/4\pi} = 1 \cdot 0 \times 10^{-1}D^2\left(\frac{\nu}{R}\right)^{2/3}$$

The cosmological distance, D, is equal to 1,770 Mpc,
and thus:

$$\frac{U_{\text{rad}}}{H^2/4\pi} \approx 3 \cdot 1 \times 10^5 \left(\frac{\nu}{R}\right)^{2/3}$$

Putting $\nu = 6 \times 10^{14}$ c/s, $R = 3 \times 10^{16}$ cm:

$$\frac{U_{\text{rad}}}{H^2/4\pi} \approx 2 \cdot 3 \times 10^4 \qquad (7)$$

Thus, provided that the ultra-violet radiation from 3C 345
is synchrotron in origin, we arrive at much the same ratio
as for 3C 273.

The results (6) and (7) give a manifest contradiction
because U_{rad} in (6) and (7) was calculated from the
synchrotron process alone. Our results would therefore
require the energy density of radiation from the Compton
process to be greater than that from the synchrotron
process by the large factor $\sim 10^5$, yielding a divergence
in which the inverse Compton process acts on itself,
giving a repeated stepping up of the energies of quanta
initially generated by the synchrotron process. When
U_{rad} is estimated from the synchrotron process, the
right-hand side of (1) must not be greater than unity.
It may be noted that this requirement prevents one from
arguing that the optical and infra-red emission arises
from the Compton process acting on radio-frequency
quanta. For, since the energy of the former much exceeds
the latter, it would be necessary on this idea for the Compton process to multiply the energy of the synchrotron
process by a large factor. The same process would then
multiply itself by a similar factor, giving an impossible
divergence.

Reference to (4) shows that these difficulties are avoided
if the distances of the quasi-stellar objects are reduced by

a factor $\sim 10^2$, that is to say, from characteristic distances of 1,000 Mpc to characteristic distances of 10 Mpc. It seems to us rather unlikely that the cosmological interpretation of the distances can be saved by any modification of our argument except perhaps by modifying our assumption that the emission comes from just one coherent region—in the sense we have used coherence (compare equations (2) and (3)). We continue our discussion by examining the consequences of denying this assumption.

Suppose there are n coherent sub-regions each of radius R. Proceeding in the same way as before, but now for a single sub-region, we find that $4\pi U_{rad}/H^2$ is reduced by n. Evidently, for n large enough this ratio can be reduced to a value of order unity—even when D is a cosmological distance. However, since the fluctuation data require all n sub-regions to be fitted into an overall dimension of $\sim 10^{17}$ cm, it is clear that for $n \gg 1$ the radius R of each region must be taken substantially less than 10^{17} cm. Thus the factor $(v/R)^{2/3}$, which appears in the analysis, cannot now be taken much greater than unity. It follows that n must be still larger than the numerical values appearing in equations (6) and (7)—$n \gtrsim \sim 10^6$ is needed.

The problem is then to explain how so many sub-regions can vary in phase with each other, as they must do if their combined radiative output is to vary at all appreciably in a time scale of a few weeks. A correlation between the sub-regions is needed—they cannot be autonomous. We might think of a magnetic field maintaining a connexion from one region to another, but for the connexion to be strong enough to co-phase the regions the magnetic intensity could not be significantly less in the inter-region zone than it is within the regions themselves. The concept of separate sub-regions then becomes of doubtful validity and it seems better to return to a single coherent region of dimension $\sim 10^{17}$ cm. However, we can now regard the 'coherence' as being maintained by a magnetic field rather than by a diffusion from place to place of the relativistic particles. In terms of our previous analysis, we can increase the value of H^2 until the ratio $4\pi U_{rad}/H^2$ falls to unity. The effect of such an increase is to give extremely short lives to the relativistic electrons. Instead of a few weeks, the lifetimes are only $\sim 10^3$ sec—this is for the electrons which radiate the main energy, at $v \simeq 10^{14}$ c/s.

It would seem then that the assumption of cosmological distances for the quasi-stellar objects leads to the requirement that relativistic particles are generated *in situ*. From what? Perhaps from oscillations in the magnetic field. In order to explain the optical fluctuation data such oscillations must propagate with essentially the speed of light—the oscillations would need to be electromagnetic, not hydromagnetic.

A mechanism for generating electromagnetic waves has been discussed by Hoyle, Narlikar and Wheeler[18]. The waves were regarded as being emitted in the dynamical oscillations of a highly collapsed star, the wavelength being a few kilometres. Such waves do not propagate unless the electron density is very low. Hence it would be necessary for an inner region of dimension $\sim 10^{17}$ cm surrounding a collapsed object to be largely devoid of matter. To prevent the waves simply radiating freely into space it would also be necessary in such a model to trap the waves within the inner region, by surrounding the region with a gas cloud of comparatively high density. It may be added that the distinction between synchrotron emission and the Compton process is now lost; the two processes are the same in a purely electromagnetic field.

Difficulties remain, however. The most immediate is that the outer gas cloud must have a characteristic dimension of the same order as the radius of the inner region $\sim 10^{17}$ cm. For a dimension as small as this it is hard to obtain sufficient line emission (for example, for Hβ in 3C 273) without requiring the electron density to be so high that the cloud is very opaque in the continuum.

We conclude that either the quasi-stellar objects are at distances of ~ 10 Mpc or less, or the physical model associated with these objects must be substantially different from the theories at present in vogue.

This work has been supported in part by grants from the National Science Foundation, in part by the National Aeronautics and Space Administration through contract *NsG–357*, and also by the Office of Naval Research (*Nonr–220(47)*).

[1] Hoyle, F., and Burbidge, G. R., *Astrophys. J.* (in the press).

[2] Greenstein, J. L., and Schmidt, M., *Astrophys. J.*, **140**, 1 (1964).

[3] Ginzburg. V. L., Ozernoi, L. M., and Syrovatsky, S. I., *Sov. Phys. Dokl.*, **9**, 3 (1964).

[4] Matthews, T. A., and Sandage, A. R., *Astrophys. J.*, **138**, 30 (1963).

[5] Oke, J. B., *Astrophys. J.*, **141**, 6 (1965).

[6] Johnson, H. L., *Astrophys. J.*, **139**, 1022 (1964).

[7] Searle, L., Rogers, A. W., Sargent, W. L. W., and Oke, J. B., *Nature*, **208**, 1190 (1965).

[8] Dibai, E. A., and Pronik, V. I., *Astr. Tsirkulyar Akad. Nauk S.S.S.R.*, Nos. 282, 286 (1964).

[9] Low, F. J., and Johnson, H. L., *Astrophys. J.*, **141**, 336 (1965).

[10] Low, F. J., *Astrophys. J.*, **142**, 1287 (1965).

[11] Epstein, E. E., *Astrophys. J.*, **142**, 1285 (1965).

[12] Sandage, A. R., *Astrophys. J.*, **139**, 416 (1964).

[13] Schmidt, Th., *Z. Astrophys.*, **62**, 217 (1965).

[14] Smith, H. J., *Quasi-Stellar Sources and Gravitational Collapse*, edit. by Robinson, I., Schild, A., and Schucking, E. L., 221 (Univ. Chicago Press, 1965).

[15] Goldsmith, D. W., and Kinman, T. D., *Astrophys. J.*, **142**, 1693 (1965).

[16] Burbidge, E. M., and Burbidge, G. R., *Astrophys J.*, **143**, 271 (1965).

[17] Burbidge, E. M., *Astrophys. J.*, **142**, 1674 (1965).

[18] Hoyle, F., Narlikar, G., and Wheeler, J. A., *Nature*, **203**, 914 (1964).

Printed in Great Britain by Fisher, Knight & Co., Ltd., St. Albans.

Reprinted from Science, March 11, 1966, Vol. 151, No. 3715, pages 1214-1216

Reports

Peculiar Galaxies and Radio Sources

Halton Arp

Abstract. *Pairs of radio sources which are separated by from 2° to 6° on the sky have been investigated. In a number of cases peculiar galaxies have been found approximately midway along a line joining the two radio sources. The central peculiar galaxies belong mainly to a certain class in the recently compiled Atlas of Peculiar Galaxies. Among the radio sources so far associated with the peculiar galaxies are at least five known quasars. These quasars are indicated to be not at cosmological distances (that is, red shifts not caused by expansion of the universe) because the central peculiar galaxies are only at distances of 10 to 100 megaparsecs. The absolute magnitudes of these quasars are indicated to be in the range of brightness of normal galaxies and downward. Some of the radio sources which have been found to be associated with peculiar galaxies are galaxies themselves. It is therefore implied that ejection of material took place within or near the parent peculiar galaxies with speeds between 10^2 and 10^5 kilometers per second. After traveling for times of the order of 10^7 to 10^9 years, the luminous matter (galaxies) and radio sources (plasma) have reached their observed separations from the central peculiar galaxy. The large red shifts measured for the quasars would seem to be either (i) gravitational, (ii) collapse velocities of clouds of material falling toward the center of these compact galaxies, or (iii) some as yet unknown cause.*

A 4-year study of peculiar galaxies has been recently completed (*1*). This *Atlas of Peculiar Galaxies* was compiled in order to systematically study physical processes in galaxies and relationships between different kinds of galaxies and to gain a more realistic picture of the contents of space. At the end of the project, positions of *Atlas* objects were compared to positions of radio sources. Aside from a few well-known identifications, there was no significant number of coincidences in position. Shortly after the *Atlas* had been submitted for publication in January 1966, however, a remark by J. L. Sérsic (*2*) caused me to reexamine the relation between radio sources and peculiar galaxies.

At first, only radio sources of greater than 9 flux units in the *3CR Catalogue* (*3*) were considered. The numbers of these radio sources are such that if they were distributed at random in the northern sky (excluding the area of the Milky Way) we would have to draw a circle of 4.°6 radius, on the average, in order to include one radio source within it. Peculiar galaxies from the *Atlas*, however, fell significantly closer than this to the radio sources on the average. It was also noticed that radio sources of similar flux densities tended to form pairs separated by from 2° to 6° on the sky. A certain class of peculiar galaxy, Nos. 100 through 150 in the *Atlas*, often fell approximately midway along a line joining these pairs. To date, out of 27 peculiar galaxies in the *Atlas* which have so far been assigned as probable or possible origins of radio sources, 23 are numbered between 100 and 200, and 18 of these are numbered between 100 and 150. Other kinds of galaxies of this brightness do not fall between pairs of radio sources. This can be demonstrated by referring to *Atlas* galaxies Nos. 1 through 50 and 200 through 338. Of these, only three fall between radio-source pairs or groups. Peculiar galaxies numbered 102 through 145 in the *Atlas* are characterized as elliptical galaxies with either disturbed spiral galaxies nearby or disturbed material which appears in many cases to be ejected from these ellipticals. The typical distance from one of these kinds of central peculiar galaxies to the nearest radio source is about 2°. The distance to the second radio source is 4° on the average. The direction from the peculiar galaxy to the second source agrees with the direction of the line joining the two sources to within ±5°. The probability that any single configuration like this could be accidental is less than about one part in 1500.

In the peculiar galaxies Nos. 145, 127, and 142 there are elongated shreds of material in the vicinity which point in the direction of the line joining the pair of radio sources. In Nos. 148, 125, 139, and 160 there is a third nearby radio source which is in an opposite direction to the material which appears to be being ejected from the central elliptical galaxy. In general, there is a tendency for filaments and axes within the peculiar galaxies to point toward the radio sources. Some peculiars have four neighboring radio sources tending to be paired oppositely, and No. 108 in the Southern Hemisphere is surrounded by five Mills, Slee, and Hill sources (*4*).

Systematic spectroscopic and photometric observations on the central peculiars remain to be made. Fragmentary data presently available, however, indicate that these galaxies have apparent magnitudes in range $m_{pg} = 13$ to 15. Five red shifts indicate recession velocities of the order of 2 to 10×10^3 km/sec. The velocity-distance relation would give distances from 10 to 100 megaparsecs and absolute magnitudes for the peculiar galaxies themselves of $M_{pg} = -18$ to -20. Such magnitudes would place these central peculiar galaxies among the brighter known galaxies.

As radio observations continue to improve their resolution and positioning accuracy, some radio sources will probably continue to remain optically unidentified, whereas other sources will become identified with optical galaxies of varying degrees of compactness. As far as is known now, the list of radio sources associated here with peculiar galaxies (Table 1) contains at least five quasars in the probable category and three others in the possible category. Association of these quasars with the class of peculiar galaxies just discussed would indicate that the quasars are not at cosmological distances (that is, expansion of the universe only accounts for a small part of their observed red shifts). Since the apparent magnitudes of quasars is in the $m_{pg} = $ 15 to 19 range, their absolute magnitudes would appear to be in the range of that of normal galaxies and fainter. The quasar 3C273 is identified here as belonging to the bright Virgo-cluster elliptical NGC 4472. Using a distance modulus of $m - M = 30.3$ magnitudes gives $M_R = -17.4$ magnitude for 3C273 (average over light variations).

At the distances estimated for these peculiar galaxies, the radio sources spread out in space for a distance of

Fig. 1 (right). A giant elliptical galaxy, NGC 2937, and strongly disturbed matter around it. The galaxy is No. 142 in the *Atlas of Peculiar Galaxies (1)*. This type of galaxy seems now to be preferentially located between pairs of radio sources widely spaced in the sky. Note the shred of luminous matter pointing up and to the right—within 8° of the direction to the radio source 3C222, which is 1°.7 away.

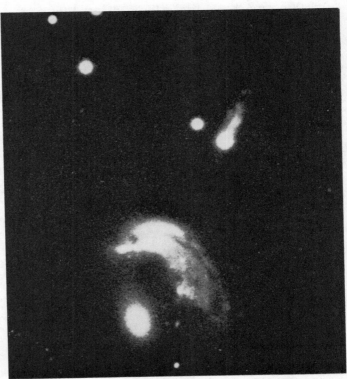

from 1 to 10 megaparsecs around them. The implication is that the radio sources were ejected from within or near the central galaxy at some time in the past. This has been the model for previous radio sources identified with galaxies (radio sources which show a strong tendency to pair across the center galaxy). Maltby, Matthews, and Moffat (5) have indicated that for radio sources heretofore identified with optical galaxies the sources probably originate near the nucleus with diameters of the order of 10^3 parsecs. As the two components of the source move away from the nucleus out to distances of the order of 200×10^3 parsecs, they reach diameters of the order of 100×10^3 persecs. The results here open up a whole new class of identifications, similar in the pattern of opposing pairs of radio sources but with separations ranging from 5 to 50 times larger. The diameters of the radio sources in the new, wider identifications, however, are smaller and sometimes associated with luminous matter of varying degrees of compactness, including quasars. The question is what velocities and what flight-travel times are involved for the much larger angular separations here identified. There are two ways I have estimated the time since the ejection event. One is that the central galaxies are not exactly at the point of inferred origin. They lie very roughly between 0.1 and 1 megaparsec from that origin. At 300 km/sec, a reasonable peculiar velocity for a galaxy to have, they would have to have been traveling for about 10^9 years. Another estimate comes from the rotation period of the galaxies, which is characteristically a few times 10^8 years. In order to have approximate, but still not exact, alignment of shreds and filaments and axes, and opposition of ejected material, then the time since the event must be over 10^8 years from this estimate also. In order to reach the observed distances, the ejected material would then have to be traveling around 3000 km/sec. These kinds of velocities of ejected material have been observed from exploding galaxies (6).

Among the radio sources listed in

Table 1. Probable associations of peculiar galaxies and radio sources.

Central galaxy (*Atlas* Nos., NGC Nos., and other designations and sources)	Radio sources (Cambridge Survey Nos. and Mills, Slee, and Hill Nos.)
......, NGC 62, ...	MSH 00–16, MSH 00–17
No. 127, NGC 191, ...	MSH 00–07(NGC 157), MSH 00–01*1*
No. 140, NGC 274, 75, VV 81	MSH 00–0*13*, MSH 00–0*10*
No. 157, NGC 520, VV 231	3C39, 3C44, MSH 01+03, MSH 01+06 (last two may be redundant)
No. 166, NGC 750, 51, VV 189	3C46, 3C67
No. 145, ..., Minkowski, Gates, Reaves	3C65, 3C66 (galaxy) (3C73 ?)
No. 131, ..., VV 336	MSH 02–1*11*, MSH 02–1*14*, MSH 02–1*16* MSH 02–1*13* (perhaps more to south)
No. 141, ..., VV 123	3C173. 1, 3C184
No. 108, ..., VV 346	MSH 02–2*18*, MSH 02–2*19*, MSH 02–2*20* MSH 03–2*1*, MSH 03–22
No. 143, NGC 2444, 45, VV 117	3C183, 3C186, (3C189), 3C194
No. 55, ..., VV 155	3C216 (QSS), 3C219 (two galaxies)
No. 142, NGC 2936, 37, VV 316	3C222, MSH 09+*04*, MSH 09–07, MSH 09+06
No. 148, ..., VV 32	3C247, 3C252, 3C254 (QSS)
No. 197, IC 701, VV3	(3C256 ?), 3C258, 3C263.1, 3C264 (galaxy)
No. 160, NGC 4194, ...	3C266, 3C277, 3C277. 1
No. 134, NGC 4472, ...	3C273 (QSS), 3C274 (M87)
Nos. 139, 196, ..., Herzog	3C277.3, 3C284, 3C287 (QSS)
No. 125, ..., A. Wilson	3C337, 3C338 (NGC 6166), 3C345 (QSS)
No. 102, ..., Zwicky, VV 10*	3C352, 3C356
No. 130, , VV 263	3C467, 3C1, 3C9 (QSS)
Nos. 35, 201, ..., ..., VV 257, VV 38	3C2 (QSS), 3C15, 3C17 (galaxy)
No. 282, NGC 169, A. Wilson	3C14, 3C19, 3C28 (galaxy)
Pec., ..., ...	3C16, 3C18
No. 121, ..., A. Wilson	Nearby MSH sources
......, NGC 5223, 28, 33, ...	3C286 (QSS), 3C288
No. 117, IC 982, 83, ...	3C293.1, 3C300
No. 111, NGC 5421, VV 120	3C293, 3C294
Pec., , ...	3C437, 3C437.1, 3C442
No. 110, ..., ...	Numerous MSH sources between declination −10° and −20° and right ascension 22h20m and 23h24m

* Below No. 102 the associations are classed as possible rather than probable.

Table 1 are nine previously identified with optical galaxies (5). Since they are at approximately the same distances from the parent peculiar galaxies as the radio sources in general, there is no reason to believe that the luminous material is traveling with any different velocity than the plasma which emits the radio signals. In particular, the quasars in Table 1 are essentially the same distance as the radio sources and luminous matter and, therefore, must be traveling with spatial velocities of the order of 10^3 km/sec.

Two very approximate direct checks are available. First, M87, the opposite particle to 3C273, is traveling away in our line of sight, relative to the central elliptical NGC 4472, with a velocity of 300 km/sec. Symmetry implies that 3C273 is approaching with this component of velocity in the line of sight. Its true space velocity must be somewhat higher, depending on the projection factor. The second piece of evidence comes from a triplet of peculiar galaxies studied by Zwicky (7). Two ellipticals traveling at +7000 km/sec are seen attached to a spiral traveling at +100 km/sec. It appears now that this spiral is being ejected with a velocity of about 7×10^3 km/sec. Order of magnitude limits to the ejection velocities of the radio sources then seem to lie between 10^2 and 10^4 km/sec.

The nature of the large red shifts observed in the quasars then remains unexplained. We have shown here that neither cosmological recessional velocities nor large Doppler velocities can be the answer (the latter is also doubtful because of the failure to observe blue shifts). As discussed by Greenstein and Schmidt (8), line shifts in the gravita-tional field of a very dense body should be the only remaining possibility. I would like to add one other possibility: that of high collapse velocity of material toward the center of these very compact objects. The region radiating the emission lines would have to be opaque to obscure the blue shifts from the back side. A hotter surface below should produce absorption lines on the red side of the emission lines (reverse P Cygni effect). That surface might, however, be red-shifted out of visibility—or possibly the emitting material might actually form a hollow shell collapsing from a larger spherical shell distribution. If only parts of that shell were luminous at a given time, fairly narrow emission lines could possibly result. Since there are objections to both the gravitational and collapse explanations of the large red shifts, however, the true explanation may lie in a direction so far not considered.

The results so far open up a number of possibilities. Among them are: (i) If it is possible to observe large spectral red shifts which are noncosmological, then the velocity-distance relationship for ostensibly normal galaxies should be regarded with slightly more caution. (ii) If material in galactic amounts can be ejected from other galaxies, the possibility is raised that certain kinds of galaxies, particularly spiral galaxies, can be much younger than other kinds of galaxies. (iii) If ejection from a center can take place more slowly, with differential rotation, then spiral galaxies themselves may be a manifestation of this phenomenon. (iv) Very small, compact dwarf galaxies with emission lines recently discovered (9) appear to be characteristically double. It may be possible to show that these also represent fairly recent ejections from nearby galaxies. (v) The cause and mechanism of the ejection of material from galaxies becomes an even deeper puzzle which may be connected with the formation of galaxies. (vi) The ejection of material into intergalactic space must affect the composition of that space. If the ejected plasma is ultimately controlled by the conditions it finds there, then we have experiments which connect the galaxies with the conditions in the space in which they are imbedded.

HALTON ARP

Mount Wilson and Palomar Observatories, Carnegie Institution of Washington, California Institute of Technology, Pasadena

References

1. H. Arp. *Atlas of Peculiar Galaxies* (California Institute of Technology Bookstore, Pasadena, 1966); *Astrophys. J., Suppl.*, in press.
2. J. L. Sérsic, National University, Cordoba, Argentina, privately remarked that a peculiar galaxy he had been studying in the southern hemisphere had three radio sources within 1½° (NGC 6438).
3. "The revised 3C (3d Cambridge) catalogue of radio sources." A. S. Bennett, *Mem. Roy. Astron. Soc.*, vol. 58, p. 163 (1962).
4. B. Y. Mills, O. B. Slee. E. R. Hill, *Australian J. Phys.* 11, 360 (1958); *ibid.* 13, 676 (1960); E. R. Hill and B. Y. Mills, *ibid.* 15, 437 (1962).
5. P. Maltby, T. A. Matthews, A. T. Moffett, *Observations of the California Institute of Technology Radio Observatory* 4, IV. "A discussion of 24 identified sources" (1962).
6. C. R. Lynds and A. Sandage, *Astrophys. J.* 137, 1005 (1963); E. M. Burbidge, G. R. Burbidge, V. C. Rubin, *ibid.* 140, 942 (1964); E. M. Burbidge and G. R. Burbidge, *ibid.* 142, 1351 (1965).
7. F. Zwicky, *Ergeb. Exakt. Naturw.* 29, 344 (1956).
8. J. L. Greenstein and M. Schmidt, *Astrophys. J.* 140, 9 (1964).
9. H. Arp, *ibid.* 142, 402 (1965); F. Zwicky, *ibid.* 143, 197 (1966).

25 February 1966

ON THE NATURE OF THE QUASI-STELLAR OBJECTS

F. HOYLE

University of Cambridge, Cambridge, England

AND

G. R. BURBIDGE

University of California at San Diego, La Jolla, California

Received July 6, 1965; revised October 4, 1965

ABSTRACT

In this paper we discuss the origin of the quasi-stellar objects from two different points of view: (i) that they are objects at cosmological distances as has been commonly supposed, and (ii) that they are local objects situated at distance $\sim$1–10 Mpc. In the introductory section the optical properties of the quasi-stellar objects are compared with the optical properties of galaxies associated with strong radio sources and Seyfert nuclei from both points of view. Section II is devoted to a discussion of (i), and on this basis it is argued that they are most probably the nuclei of galaxies which have reached a high-density phase at which time the formation of massive objects and their subsequent evolution has occurred.

Apart from the suggestion of Terrell, little attention has been paid until now to the possibility that they are local objects, and so we have considered this in considerable detail. A plausible case can now be made for supposing that they are coherent objects which have been ejected at relativistic speeds from the nuclei of galaxies at times when they erupt to give rise to strong radio sources and other phenomena. On this basis a likely candidate to give rise to the objects in our vicinity is NGC 5128 which is a powerful radio source in which at least two outbursts appear to have occurred. In this case some objects with blueshifts may be present. The fraction of such objects and the solid angle about NGC 5128 in which they lie is given as a function of the distance of NGC 5128, average speed of ejection, and time which has elapsed since the explosion. Calculations are also made of the redshift-magnitude relation to be expected in the local theory, and comparison is made with the relation in orthodox cosmology.

Some of the problems associated with the hypothesis that the objects lie at cosmological distances have arisen through the observation of Dent of the radio emission from 3C 273B. This shows that at 8000 Mc/s the intensity has increased by about 40 per cent over the last 2 years, and that the spectrum is flat over the range 200–8000 Mc/s and may be flat over a wider range. It is shown that a model based on the synchrotron process is able to explain the form of the spectrum with the object at a cosmological distance provided that the following conditions prevail: (1) the electron distribution is optically thin; (2) the magnetic-field intensity depends on distance r from the center with the form $H = H_0(a/r)^n$; (3) the electron energy spectrum is everywhere the same; (4) the energy density of the electron distribution is of the form $W = W_0(b/r)^m$; (5) $n + m = 3$. Even with this model there is considerable difficulty in explaining a flat spectrum beyond 10^4 Mc/s. The difficulties associated with the model are somewhat reduced in the local theory.

In the concluding sections a number of programs are outlined which may enable us finally to determine whether the objects are at cosmological distances or are local.

I. INTRODUCTION

Of the sample of rather more than one hundred radio sources for which optical identifications have been made, more than thirty have turned out to be associated with starlike objects. Because optical identifications are confined to sources for which good radio position measurements have been made, and because good position measurements so far exist only for the brighter sources, it was thought until recently that data concerning these starlike objects could only be accumulated rather slowly. However, Sandage (1965) has begun to identify starlike objects which are not associated with the brighter radio sources. Sandage has described the objects as quasi-stellar galaxies, but in this paper we shall prefer the term quasi-stellar object since at this stage we do not think enough is known about their nature for a definitive name to be chosen. Indeed it is the purpose of this paper to discuss the ambiguities of interpretation of the quasi-stellar objects.

QUASI-STELLAR OBJECTS   535

The new work of Sandage follows pioneering investigations by Humason, Zwicky, Haro, and Luyten on faint blue stars at high latitudes. It gives support to the view that quasi-stellar objects may be rather common—the at present rather poor statistics indicate as many as $\sim$4 objects per square degree of sky, giving a total of $\sim$1.5 $\times$ 10^5. This implies that if the quasi-stellar objects are at cosmological distances their spatial density is about 0.01 per cent of the density of galaxies. If the objects are closer than cosmological distances, their density is of course correspondingly greater. The new objects are radio quiet in the sense that if they are radio sources the flux at the Earth must be less than 10^{-25} W/m^2/(c/s) at 178 Mc/s.

The quasi-stellar objects have been associated with galaxies for two reasons. Lines in their spectra show redshifts similar to the cosmological redshift for galaxies, and those objects which are strong radio sources turn out to have intrinsic radio luminosities that are comparable with the intrinsic luminosities of strong radio galaxies—provided the redshifts are interpreted cosmologically. However, both of these pieces of evidence are circumstantial. There is little or no direct evidence to connect quasi-stellar objects with galaxies. Indeed there is some evidence for an anticorrelation, for so far no quasi-stellar object has been found in a cluster of galaxies. The optical object associated with 3C 273 has an apparent magnitude of about 12.8. If 3C 273 were in a cluster of galaxies, the galaxies would have apparent magnitude about 18 and would easily be observed. The redshifts for 3C 48, 3C 47, and 3C 147 are 0.367, 0.425, and 0.545, and at the cosmological distances indicated by these shifts galaxies would probably have been detected if these objects were in clusters. The shifts for 3C 254, 3C 245, CTA 102, 3C 287, and 3C 9 (Schmidt 1965) are so great that if any galaxies were associated with these objects they would be beyond the plate limit. Of the three new objects reported by Sandage, two have small redshifts and associated galaxies, if there were any, would be readily observed. On the whole, therefore, the evidence is against quasi-stellar objects being correlated spatially with galaxies. It is to be anticipated that any remaining ambiguity in this question will soon be eliminated.

It is our purpose in this paper to discuss the further consequences of supposing (a) that the quasi-stellar objects are at cosmological distances, (b) that they are extragalactic but of local origin. In §§ II and III we discuss the qualitative implications of (a) and (b), respectively, while in §§ IV and V we shall be concerned with more quantitative problems, in particular in § IV with the question of whether the observations of Dent (1965) of a variation in the intensity of the high-frequency radio emission from 3C 273B throws light on the nature of this object.

II. QUASI-STELLAR OBJECTS AT COSMOLOGICAL DISTANCES

To avoid pedantry, we shall drop the conditional "if at cosmological distances" clause and will write in this section as if the quasi-stellar objects were known to be at cosmological distances—the conditional clause obviously applies to the whole of this section. However, certain of the properties of quasi-stellar objects, briefly reviewed below, apply also to the conditions of § III. These will be indicated by an asterisk. Following this review we shall consider three theories which have been proposed to explain the nature of the objects.

First, we note three properties which distinguish the quasi-stellar objects from *normal* galaxies.

1. Optically, the quasi-stellar objects are up to forty times brighter than the most luminous galaxies. All of the quasi-stellar radio sources for which redshifts are available have absolute magnitudes in the range -24 to -26, compared to -22 for the most luminous galaxies. Two of the three new objects observed by Sandage have small redshifts and the absolute magnitudes are near -21. Hence the quasi-stellar objects cover a range extending upward by 5 mag. from normal galaxies.

2*. The objects are all exceedingly compact. In the majority of cases they are indistinguishable from stars on direct plates, although in one of Sandage's new cases the object shows a fuzziness that distinguishes it from a stellar image.

On the cosmological hypothesis, this sets a limit to the size of the optical object at about 1 kpc for the nearer systems and about 3 kpc for the more distant objects.

A more powerful limitation on the sizes of the objects comes from variation in the optical brightness. In all cases in which repeated measurements have been made of the objects associated with radio sources, variations in light have been seen or suspected. This means that a major part of the luminosity in such objects must come from a source that is at most only a few parsecs in extent.

3*. The optical radiation emitted by these objects is not radiation from ordinary stars. The continuum is almost certainly a mixture of radiation from a hot diffuse gas with a non-thermal component, possibly synchrotron radiation. The emission lines are also those characteristic of a rather highly excited gas (Greenstein and Schmidt 1964; Schmidt 1964; Sandage 1965).

In contrast to these differences, certain interesting similarities emerge when quasi-stellar objects are compared with *abnormal* galaxies.

4*. The emission lines in the spectra of quasi-stellar objects have widths which indicate large random motions in the emitting gas, ≥ 1000 km/sec. This is also characteristic property of the spectra of the nuclei of Seyfert galaxies. In a general way, the latter also have the properties 3*, although the emission lines are stronger relative to the continuum than is the case in quasi-stellar objects.

The nuclei of Seyfert galaxies are known to be compact. Their stellar appearance on direct plates sets a limit of about 50 pc to their sizes. Evidently, there is more than a superficial resemblance between such nuclei and the quasi-stellar objects. We can perhaps venture the predictions (i) that the colors of these nuclei will turn out similar to the quasi-stellar objects, (ii) that light variations will be found.

5*. A jet emerges from 3C 273B which is similar to the jet emerging from the nucleus of M87.

6*. The radio source MSH 14-121 has two components separated by $\sim 37''$. The optical object associated with MSH 14-121 lies on the line connecting the two components (Véron 1965). This is similar to the situation in many radio galaxies. In this respect also the centers of galaxies appear to play the same role as quasi-stellar objects.

7. The linear size associated with 3C 47 is of order 200 kpc or more, which is comparable with the sizes of large radio galaxies.

8. The radio emission from the quasi-stellar sources is $\sim 10^{44}$ ergs sec^{-1}, comparable with the strongest radio galaxies.

9*. The N-type radio galaxies have starlike nuclei outside which faint features can be seen (Matthews, Morgan, and Schmidt 1964). Like the fuzziness associated with 3C 48 these features could be due to a jet, or a series of jets, emerging from the nucleus.

Possibly Cygnus A should also be included in this category—i.e., of galaxies having features in common with quasi-stellar objects. The extent of the region giving emission lines in Cygnus A has dimensions of ~ 6 kpc, the absolute magnitude is about -21, and the emission lines in this system are very strong.

We are strongly impressed by items 4*–9*, which seem to us to indicate a close connection between the physical processes in quasi-stellar objects and those which take place in the nuclei of some galaxies. It must be emphasized, however, that item 1 remains a major and critical difference. The absolute magnitudes of the nuclei of Seyfert galaxies are about -18, a thousand times fainter than the most luminous quasi-stellar objects. This difference is unavoidable so long as we accept the cosmological interpretation of the redshifts. We are then dealing with a phenomenon in which the optical output ranges from $\sim 10^{42}$ ergs sec^{-1} to $\sim 10^{46}$ ergs sec^{-1}, and in which the emission comes from a volume of only a few cubic parsecs. This and the fact that there is no evidence of any stellar features in the optical spectra place important constraints on any theory which interprets the quasi-stellar redshifts as cosmological. If normal stars exist in the quasi-stellar objects they are completely overborne in luminosity by the continuum of thermal and non-thermal origin emitted by a hot plasma.

We turn now to three theories which have been put forward. Field (1964) has suggested that the objects are galaxies in the process of formation, and Sandage (1965) has supported this point of view. The difficulty in this theory is to understand the similarities between the newly forming galaxies and the nuclei of old, established galaxies. Although we are not unsympathetic to the idea of newly forming galaxies (Burbidge, Burbidge, and Hoyle 1963), we have not so far been able to understand how items 4*–9* can be understood in this theory.

A far more radical theory is that both the quasi-stellar objects and the nuclei of some galaxies are relics of a high-density phase of the whole Universe. Many cosmologists are attracted by an oscillating model for the Universe. The critical problem in such a model is to explain why the Universe switches from contraction to expansion. So far no strict mathematical explanation has been given within the usual framework of cosmology; the switch is simply assumed. Granted, however, that a switch takes place in some fashion, it is reasonable to argue that if the whole Universe can "bounce" so can a localized object. The time scale for the bounce of a localized object is not the same for a distant observer as it is for an observer moving with the object. The former is greater than the latter by a dilatation factor, $(1 - \xi)^{-1/2}$, where ξ is the largest value of the relativistic parameter, $2\,GM/R$, for a mass point in the usual Schwarzschild theory, attained during the oscillation. If ξ comes exceedingly near unity, this dilatation factor can greatly exceed the oscillation time scale for a comoving observer. Indeed, the time scale for the external observer could be as long as 10^{10} years, so it would be possible for us to see localized objects emerging from a highly relativistic situation.

Turning now to the third theory, Ulam and Walden (1964), Gold (private communication), and Woltjer (1964) have all noticed that, if the star density is high near the centers of galaxies, star collisions may be frequent enough to produce appreciable optical emission. Acceleration of particles to cosmic-ray energies in a rapidly moving gaseous assembly might be responsible for the radio emission. Similar ideas have also been suggested by these authors for the quasi-stellar objects.

The absence of any detectable stellar component in the spectra of the quasi-stellar objects casts doubt on this theory, at any rate on the idea that the optical emission arises from star collisions. It was also pointed out by Hoyle (1964) that the variations in the optical emission of 3C 273 cannot be explained in terms of star collisions. It would seem, therefore, that a more hopeful line of attack would be to argue that star collisions are responsible for producing a massive object of the kind first discussed by Hoyle and Fowler (1963a, b), and that both the optical and radio properties are controlled by the massive object. The further evolution of such an object has been studied by Hoyle and Fowler (1965) and by Fowler (1966).

This latter form of the third theory is perhaps the most conservative attack on the problem. The main difficulty in the theory is to understand how the star density can become high enough to produce any appreciable development through the mechanism of star collisions. A qualitative discussion of this question has been given by Gold, Axford, and Ray (1965). Preliminary calculations by Ulam and Walden (1964) indicate that collisions will increase rather rapidly if the star density exceeds $10^6/pc^3$, and that the process becomes essentially catastrophic at a density of $10^9/pc^3$. These values may be compared with the star density at the center of M31, $\sim 10^3/pc^3$. Nothing is known about the star density at the centers of other more distant galaxies.

The difficulty can be understood in more general terms in the following way. Divide the inner regions of a galaxy into a large number of small volumes, say, the region within 100 pc of the center into a million equal cells. Take the time average of the stars, and of their motions, within each cell—i.e., attach an observer to each cell and let him observe the stars that pass through his individual cell. Why should one particular observer, the one associated with a cell at the geometric center of the galaxy, obtain a result substantially different from any other observer? What distinguishes the center as a singular point? We believe these questions to be unanswerable, and the theory to be consequently untenable, if the nuclei of galaxies have condensed by contraction from a diffuse gas, in accordance with the usual picture of their origin. In such a picture we would expect the stars to have sufficient angular momentum about the geometric center for one cell in our imaginary model to be indistinguishable from another, at any rate over the first 10–100 pc from the center.

An alternative suggestion for the origin of the nuclear regions of galaxies, and for the elliptical galaxies as a whole, has recently been put forward by Hoyle and Narlikar (1965). In their oscillating model the expansion phase takes place nearly as in the Einstein–de Sitter cosmology. The latter is a limiting case in the sense that comparatively small inhomogeneities can restrain expansion over limited volumes. It was suggested that elliptical galaxies are such restrained volumes possessing mass concentrations at their centers, the mass concentration at the center of a massive elliptical being $\sim 10^9 \, M \odot$. On this basis it is possible to derive theoretically the form of the light distribution within ellipticals. This turns out in very good agreement with the observed distribution, suggesting that the expansion picture may well be correct. If so, the center is singular from the origin of a galaxy; it does not have to develop. The center can have an initial density comparable to the mean density of the Universe at its most compact state. The most suitable criterion for determining this density numerically is from a comparison with observation of the formula

$$ 2 \, \frac{G \mathfrak{M}}{c^2} \simeq c \, (G\rho)^{-1/2}, \tag{1} $$

where $\mathfrak{M}$ is an upper limit to the total mass of clusters of galaxies. Here $(G\rho)^{-1/2}$ is the unit of time associated with the required mean density ρ. The right-hand side of formula (1) is therefore of the order of the dimensions of a length, L, say, and formula (1) expresses the theoretical result that the relativistic parameters associated with $\mathfrak{M}$ and L, $2 \, G\mathfrak{M}/c^2 L$, must be of order unity. Numerically, equation (1) leads to

$$ \mathfrak{M} \simeq 4 \times 10^8 \rho^{1/2} M\odot . \tag{2} $$

If we set $\mathfrak{M}$ equal to the mass of a typical cluster of galaxies, $\sim 10^{13} \, M\odot$, (2) gives $\rho \simeq 10^{-9}$ gm cm^{-3}, which is close to the value used by Hoyle and Narlikar. However, it is possible that $\mathfrak{M}$ should be set equal to the masses of the largest clusters, $\sim 10^{15} \, M\odot$, in which case $\rho \simeq 10^{-13}$ gm cm^{-3}. The latter value is close to what would be required to give $\sim 10^9$ stars per cubic parsec. Although there would be some reduction of density due to expansion, an *initial value* equivalent to between 10^6 and 10^9 stars/pc^3 is entirely possible.

On this basis elliptical galaxies, and perhaps some spirals, are born with their nuclei already at the critical density necessary for the development of massive objects. It is likely that violent events in which material is thrown out of the nucleus early in the history of such galaxies lead to a quasi-equilibrium in which the nucleus is always close to instability—i.e., to a further outburst. The occurrence of an outburst would be expected to stabilize the situation for a while, until further evolution eventually brings on a new outburst, or until the whole inner dense nucleus has been dissipated and the galaxy becomes finally inactive.

Of the three theories mentioned or discussed above the third seems to us in many ways the most attractive for the case of radio galaxies. For the quasi-stellar objects the third theory raises an awkward problem, however. Because of the similarities between quasi-stellar objects and radio galaxies, we are loath to accept a quite different theory for the quasi-stellar objects. Yet if we suppose the latter to be massive objects situated at the centers of dense star systems we are obliged to ask what star systems? In particular, what star systems can we have that are not associated with clusters of galaxies? A possible answer would be the dwarf elliptical galaxies which probably have a high spatial density, existing in profusion as field galaxies. The mystery then is why dwarf ellipticals can set up objects with a far greater optical output than the objects which develop at the centers of *massive* ellipticals. The natural expectation would be to have things the opposite way around.

III. QUASI-STELLAR OBJECTS AS LOCAL PHENOMENA

There is no question in our minds but that the line shifts which have been measured in the quasi-stellar radio sources and the objects studied by Sandage are Doppler shifts. The case against them being gravitational in origin has been made in detail for 3C 273 and 3C 48 by Greenstein and Schmidt (1964) and in our view it is overwhelming. If the objects are local it is therefore necessary to explain how velocities nearly up to c, relative to the usual standard of rest, have been derived. This is the immediate problem which any local theory has to face.

The minimum total energy necessary to explain the emission from strong radio galaxies is of order 10^{60} ergs. This value is calculated on the basis of equipartition between the total energy of the magnetic field and the total energy of the synchrotron electrons, protons being assumed absent. Allowance for a deviation from equipartition and for protons making a dominant contribution to the energy could readily raise the requirement to 10^{62} ergs. Hence we already know that 10^{60}–10^{62} ergs is involved in the outbursts of strong radio galaxies. The energy distribution of the relativistic particles, if it is at all like that of normal cosmic rays, is such that the main contribution to the total energy comes from particles with individual energies not much above 1 BeV $\simeq 10^{-3}$ erg. It seems then as if we are involved in 10^{63}–10^{65} particles moving at speeds comparable to c. This corresponds to a total mass of 10^{6}–10^{8} $M_\odot$ moving at relativistic speed. It is clearly permissible therefore to argue that a mass of the general order of 10^{7} $M_\odot$ is ejected at relativistic speed from a strong radio galaxy, and it is on this that a local theory for the origin of the quasi-stellar objects must turn.

It has been customary to think of the matter ejected from radio galaxies as a diffuse cloud of separated particles. What we have now to consider is the possibility that in addition to a diffuse emission there may also be an ejection of compact objects, and that such objects make up an appreciable fraction of the ejected material. In several respects it is easier to understand the observational data in these terms. If a single object breaks explosively into two objects, the two objects must fly apart in opposite directions, agreeing immediately with the characteristic property of radio galaxies, that they tend to be double and that the join of the two sources tends to pass through the center of the associated galaxy. A phenomenon such as the jet of M87 would seem to be more readily understood if a series of compact objects exists along the line of the jet. Otherwise it is hard to see why radial lines of force of a magnetic field should be confined to a jet. Also numerous condensation knots appear to exist within the jet.

Rather than suppose the nucleus of a galaxy to eject a large number of compact objects, it is possible that the number of objects grows by repeated subdivision. First there are two major objects, then each of these objects breaks into two, and so on in a cascade process. If, at each subdivision, the direction of separation of the two components is arbitrary, an approximation to isotropy develops as the cascade proceeds. This approximation evidently becomes closer as the multitude of fragments expands away from the parent galaxy.

It is important in the local theory that the parent galaxy be able to provide sufficient energy to explain the observed properties of the many ejected quasi-stellar objects. The problem would appear most severe in the case where all the blue "stars" of the Haro-

Luyten catalogue are taken to be quasi-stellar objects. We shall therefore discuss the problem in these terms without prejudice to the question of whether this is really so (cf. the discussion by Kinman 1965). Our aim is to show that sufficient energy can be made available even in the most difficult case.

Write $N(m)$ for the number of objects brighter than magnitude m. Sandage finds $d \log N/dm = 0.383$. We obtain substantially this relation from the following postulates:

i) The objects have expanded out from a local source and are now approximately isotropically distributed with respect to the Galaxy.

ii) The total mass of the objects within unit logarithmic interval of mass is constant.

iii) Postulate (ii) applies not only to the total distribution of objects but at every ejection speed.

iv) The optical output of an object is proportional to its mass.

The second postulate requires the number of objects with masses between M and $M + dM$ to be proportional to dM/M^2. Using (iv) the number with intrinsic luminosities between L and $L + dL$ is proportional to dL/L^2. If all the objects are at the same distance, as those with a particular ejection speed are in view of (i), the number with apparent luminosities between S and $S + dS$ is proportional to dS/S^2 and the number brighter than S is proportional to $1/S$. If this is true for every ejection speed it is true for the total distribution of objects. Writing $N(S)$ for the number brighter than S, we get

$$\log N = -\log S + \text{const.} = 0.4m + \text{const.} \quad \text{and} \quad d \log N/dm = 0.4 . \quad \text{(2a)}$$

From postulate (ii) the total mass requirement is

$$\sim M_{\max} \ln M_{\max}/M_{\min} , \quad \text{(3)}$$

where $M_{\max}$ is the maximum mass to be found among the objects and $M_{\min}$ is the minimum mass. Since the logarithmic factor is unlikely to be much greater than 10, the mass requirement is not more than $\sim 10\, M_{\max}$, so that the most massive and brightest object can contain as much as 10 per cent of the total mass. With $\sim 10^7\, M_\odot$ for the latter, we can have a maximum object mass of $\sim 10^6\, M_\odot$. Since our distribution requires the number of objects more massive than M to be proportional to $M_{\max}$, we have

1 object with mass $M_{\max}$,

10 objects with masses $> 0.1\, M_{\max}$,

100 objects with masses $> 0.01\, M_{\max}$,

and so on.

This completes our discussion on the basis that *all* blue "stars" of the Haro-Luyten catalogue are quasi-stellar. In this extreme case the requirement is for an emitted mass no greater than the amount involved in the outburst of a radio galaxy. Should only a minor fraction of the Haro-Luyten "stars" turn out to be quasi-stellar, modification of the above discussion of the slope of the $N(m)$-curve might well be needed. Our point, however, is that a reduction in the number of quasi-stellar objects must ease the energy requirement.

The advantage of the local theory is that it relates the properties of quasi-stellar objects immediately and directly to the radio galaxies. They are of the same stuff, with a similar structure, to the objects giving rise to the properties of the radio galaxies. Similarities such as items 4*–9* of § II become much more readily understandable.

Two possibilities arise in the local theory for the source of the quasi-stellar objects. Terrell (1964 and private communication) has suggested that the objects have been ejected from the nucleus of our own Galaxy, and he has pointed out that Burbidge and Hoyle (1963) proposed an explosion in the galactic nucleus in order to explain the existence of a transient halo and the outflow of gas in the plane of the Galaxy. According to Burbidge and Hoyle the explosion occurred about 10 million years ago. Hence, if we take

$c/3$ as the characteristic ejection speed of the objects, their present distances should be about 1 Mpc. Since the characteristic distance of the objects on the cosmological picture is $\sim 10^3$ Mpc, energy requirements are reduced by a factor $\sim 10^6$. The optical emission of 3C 273, instead of being $\sim 10^{46}$ ergs sec^{-1}, becomes $\sim 10^{40}$ ergs sec^{-1}, and the total emission over 10 million years is $\sim 10^{55}$ ergs. Because 3C 273 is probably one of the brightest of the quasi-stellar objects (intrinsically) the total energy requirement is greater than 10^{55} ergs by only one or two powers of 10, say 10^{57} ergs, which is not much different from the energy output suggested by Burbidge and Hoyle.

The second possibility is that the objects have emerged from a powerful radio galaxy in the neighborhood of the Galaxy. The galaxy NGC 5128 is an immediate suggestion, because NGC 5128 is known to have undergone two major outbursts in the last few million years. Taking 10 Mpc as the characteristic distance in this case, the total energy requirement is increased to $\sim 10^{59}$ ergs, equivalent to the rest-mass energy of $\sim 10^5 \, M\odot$. This also is consistent with what is thought to have been involved in the outbursts of NGC 5128.

If the objects come from the Galaxy, no cases showing a Doppler blueshift are to be expected. If the objects come from NGC 5128, there is the possibility that some slowly moving objects still lie between the Galaxy and NGC 5128. These would show a blueshift. Consider objects to have been emitted isotropically from NGC 5128 with speed v a time τ ago, and let D be the distance between NGC 5128 and the Galaxy. Evidently $v\tau/D$ is dimensionless. The fraction of objects showing blueshift is 0.5 $(1 - v\tau/D)$ if $v\tau/D < 1$ and is zero otherwise, and in the case $v\tau/D < 1$ the blueshifted objects are found in a solid angle

$$2\pi\{1 - [1 - (v\tau/D)^2]^{1/2}\}$$

centered on NGC 5128. For $v\tau/D$ small, approximately half of the objects have blueshifts, but on the sky they are concentrated closely around NGC 5128, e.g., $v\tau/D = 0.1$ gives 45 per cent blueshifts, but the solid angle about NGC 5128 is only 0.031 steradian. As $v\tau/D$ increases, so does the solid angle, but the blueshifted fraction decreases, e.g., $v\tau/D = 0.6$ gives a solid angle of ~ 1.25 steradian but the fraction of blueshifted objects has fallen to 20 per cent.

Taking 10 Mpc as the characteristic distance, say, for $v = c/2$, we require $\tau \simeq 60$ million years for the time that has elapsed since the relevant explosion in NGC 5128. The distance D is rather uncertain; 4 Mpc is the current estimate. With these values we have $v\tau/D = 0.6$ for $v \simeq 0.1 \, c$. Blueshifts of this amount in directions toward NGC 5128 could confirm this theory. Absence of blueshifts would go a long way toward disproving it, although it may be possible to increase τ/D sufficiently for only very small shifts to be permitted and there could be a paucity of slowly moving objects.

It is a point against objects from NGC 5128 that τ must be taken at least as great as ~ 30 million years if the objects are to appear approximately isotropic when viewed from the Galaxy. This is longer than the time which has elapsed since the first of the two known explosions in NGC 5128, assuming the latter to be given by dividing the dimension of the extended radio source around NGC 5128 by the velocity of light. Possibly the extended source is no longer expanding at appreciable speed, in which case the elapsed time since the first outburst could be ~ 30 million years in consonance with the present requirement. Alternatively it seems possible that NGC 5128 has undergone a succession of explosions before the one which gave rise to the extended radio source which we see at present.

Finally, we notice that the characteristic distances given above, ~ 1 Mpc for objects from the Galaxy and ~ 10 Mpc for objects from NGC 5128, are so great that proper motions must be very small. For example, an object at 10 Mpc with transverse motion $0.1 \, c$ would have a proper motion less than $0''.001$/yr. Estimates by Luyten (1963) and by Jeffreys (1965) place the proper motion of 3C 273 at less than $0''.01$/yr (Luyten) and less than $\sim 0''.0025$ (Jeffreys).

VII. CONCLUSIONS

This paper has been concerned with the possible origins of the starlike objects which are neither stars nor normal galaxies. Of the large number of objects which are probably present down to 19 mag., spectra in which Doppler shifts can be measured have so far been obtained for fourteen objects, and redshifts have been obtained in all of these. The situation as to their origin is rather similar to that which existed 50 years ago when the spiral nebulae were also a great mystery. About the same number of Doppler shifts had been measured, largely by Slipher, and there was considerable confusion as to whether they were of galactic or extragalactic origin. As is well known, conclusive proof of their extragalactic nature came in the next decade.

With the discovery of the redshifts of the quasi-stellar radio sources, the most natural theory was to assume that these were also objects at cosmological distances and with the exception of the proposal by Terrell this is what has been generally assumed. However, in this paper we have attempted to discuss the physical nature of the objects, assuming either that they are at cosmological distances or that they are extragalactic but local at distances typically of 1–10 Mpc.

There are a number of observational programs which may eventually indicate which of these hypotheses is correct. In concluding we shall list some of these.

1. The model that we have proposed to account for the form of the spectrum and the variability at high frequency observed by Dent in 3C 273B is just able to give rise to a flat spectrum out to about 10^4 Mc/s if the object is at a cosmological distance. Detailed observations out into the infrared will enable this model to be tested further.

2. As has been emphasized by many authors, detailed and accurate studies of 3C 273 and other starlike objects in all possible frequency ranges are greatly needed to determine the time scales over which they vary.

3. The angular diameter of the radio source 3C 273B is obviously of critical importance in deciding whether it is a very distant object. This question has been discussed at the end of § V. If the object can be proved to have an angular diameter $\sim0\overset{''}{.}5$ and also is variable indicating a dimension of a few light-years and is truly a single compact object, then it must be local.

4. Identification of more bright starlike objects may enable a significant test to be made as to whether or not they are associated with clusters of galaxies.

5. If the objects are in general at cosmological distances, then the bulk of them fainter than 16 mag. should have redshifts $z > 1$. If this is found to be the case, the local origin hypothesis will not be disproved. However, if many of the faint starlike objects are found to have small redshifts $z \leq 0.1$ the model proposed by Sandage cannot be retained. The compact object discovered by Arp (1965) which is distinguishable from a star on a good direct plate has an apparent magnitude of 17.9 and $z = 0.004$. On the local hypothesis this would probably be an object ejected from the Galaxy.

6. The detection of objects with blueshifts would establish the correctness of the local explanation of such objects. On the picture described here NGC 5128 is a probable source, while some may come from our own Galaxy. We should not expect blueshifts from objects of galactic origin, but as was discussed in § III a search for such objects bearing in mind that they may have come from NGC 5128 is urgently required. Since this is a southern galaxy ($\alpha = 13^{\mathrm{h}}22^{\mathrm{m}}4$, $\delta = -42°46'$ [1950]), searches in its vicinity must be carried out from the southern hemisphere.

7. On the local theory depending on the time which has elapsed since objects were ejected from NGC 5128, we shall expect to see some asymmetry in the distribution of the objects on the sky. While it may be difficult to detect such asymmetry by optical methods, it is important that the distribution of the radio sources of small angular diameter over the sky be investigated, since on the local theory that fraction of the radio sources associated with quasi-stellar objects are local.

8. If the ejection of coherent objects from the nuclei of galaxies is commonplace it

552   F. HOYLE AND G. R. BURBIDGE

may be possible to detect them about galaxies such as M82 in which explosive events are known to have taken place comparatively recently. It is interesting that an optical identification of a quasi-stellar object of 19 mag. with a radio source very close to NGC 4651 (which was originally identified as the source) has recently been made (Sandage, Véron, and Wyndham 1965). As these authors have pointed out, NGC 4651 has a very peculiar jetlike structure and on the local hypothesis the 19-mag. object has been ejected from that galaxy.

Note added in proof, March, 1966: Since this paper was written, data on flux variations in QSO's with time scales as short as months or weeks (e.g., D. W. Goldsmith and T. D. Kinman, *Ap. J.*, **142**, 1693, 1965) have been published. These data lead to a paradoxical situation which may be resolved by supposing that the objects are local. We have given a discussion of this elsewhere (F. Hoyle, G. R. Burbidge, and W. L. W. Sargent, *Nature*, **209**, February 19, 1966).

On the local hypothesis it is our present view that 10 Mpc may be a better estimate of distance than the range 1–10 Mpc. We now think it may be preferable to consider the observed QSO's as being ejected, not just by our Galaxy or by NGC 5128, but by a wider population of galaxies—perhaps with an average frequency of ∼10 QSO's per galaxy. On this basis the QSO's would have luminosities comparable to, although probably somewhat less than, the nuclei of Seyfert galaxies.

We are indebted to Allan Sandage for giving us a copy of his manuscript in advance of publication and also for the use of his house, where the bulk of this paper was written. We also wish to acknowledge the many interesting discussions we have had with Margaret Burbidge, William Fowler, and Maarten Schmidt. This work has been supported in part by a grant from the National Science Foundation and by NASA through contract NsG-357.

REFERENCES

Arp, H. C. 1965, *Ap. J.*, **142**, 402.
Burbidge, E. M., Burbidge, G. R., and Hoyle, F. 1963, *Ap. J.*, **138**, 873.
Burbidge, G. R., and Hoyle, F. 1963, *Ap. J.*, **138**, 57.
Dent, W. A. 1965, *Science*, **148**, 1458.
Field, G. B. 1964, *Ap. J.*, **140**, 1434.
Fowler, W. A. 1966, *Ap. J.*, **144**, 180.
Gold, T., Axford, W. I., and Ray, E. C. 1965, *Quasi-stellar Sources and Gravitational Collapse*, ed. I. Robinson, A. E. Schild, and E. L. Schucking (Chicago: University of Chicago Press), p. 93.
Greenstein, J. L., and Schmidt, M. 1964, *Ap. J.*, **140**, 1.
Hazard, C., Mackey, M. B., and Shimmins, A. J. 1963, *Nature*, **197**, 1037.
Hoyle, F. 1964, *Nature*, **201**, 804.
Hoyle, F., and Fowler, W. A. 1963a, *M.N.*, **125**, 169.
————. 1963b, *Nature*, **197**, 533.
————. 1965, *Quasi-stellar Sources and Gravitational Collapse*, ed. I. Robinson, A. E. Schild, and E. L. Schucking (Chicago: University of Chicago Press), p. 17.
Hoyle, F., and Narlikar, J. V. 1965, *Proc. R. Soc.* (in press).
Jeffreys, W. H. 1965, *Quasi-stellar Sources and Gravitational Collapse*, ed. I. Robinson, A. E. Schild, and E. L. Schucking (Chicago: University of Chicago Press), p. 219.
Kinman, T. D. 1965, *Ap. J.*, **142**, 1241.
Luyten, W. J. 1963, *Pub. Astr. Obs. Minnesota III*, No. 13.
Matthews, T., Morgan, W. W., and Schmidt, M. 1964, *Ap. J.*, **140**, 35.
Oke, J. B. 1965, *Ap. J.*, **141**, 6.
Sandage, A. R. 1965, *Ap. J.*, **141**, 1560.
Sandage, A. R., Véron, P., and Wyndham, J. 1965, *Ap. J.*, **142**, 1307.
Scheuer, P. A. G. 1965, *Quasi-stellar Sources and Gravitational Collapse*, ed. I. Robinson, A. E. Schild, and E. L. Schucking (Chicago: University of Chicago Press), p. 373.
Schmidt, M. 1964, *Ap. J.*, **141**, 1.
————. 1965, *ibid*, p. 1295.
Shklovsky, I. S. 1964, *Astr. Zh.*, **41**, No. 5.
Terrell, J. 1964, *Science*, **145**, 918.
Ulam, S., and Walden, W. 1964, *Nature*, **201**, 1202.
Véron, P. 1965, *Ap. J.*, **141**, 332.
Woltjer, L. 1964, *Nature*, **201**, 803.

386

NATURE. VOL. 223. JULY 26. 1969

Red-shifts of Very Young Objects

I WOULD like to call attention to one of the consequences of our current belief that there is a red-shift–distance relation for galaxies. It is accepted that as we observe increasingly distant galaxies we see them at earlier and earlier stages of their life history because of the finite velocity of light. If we could therefore observe sufficiently faint galaxies, we would see galaxies very close to their moment of creation, or even, conceptually, close to the moment of creation of the matter which would make up the galaxies. Empirically we find that the red-shifts of more distant galaxies increase and we must conclude with a high degree of certainty that, if we could see distant enough matter, it would be very young and have an extremely high red-shift. We can summarize this reasoning in the following statement: If we observe the universe near age zero, then it has a very large red-shift.

Now suppose some matter were to be created locally (that is, at a distance at which normal galaxies do not have a very large red-shift). In the initial stages following the creation of this matter it would have an age very near age zero. We can then use the axiom that things equal to the same thing are equal to each other. We can say that the age zero universe is identically equal to itself, and, whether we see it at great distances or close by, that it will have a very high red-shift. I suggest that we have proved the following theorem: In the limit, observing the red-shift of matter at a great distance which was created at the same epoch as ours is equivalent to observing the red-shift of nearby matter which was created at a very recent epoch.

We can discuss this idea in a little greater detail by remarking that the curved space of general relativity predicts that, as we observe increasingly distant galaxies, their diameters will reach a point where they appear to enlarge again, until, if we could observe extremely distant galaxies, their diameters would begin to merge, and going further, in the limit, we would expect to see the age zero universe in every direction we look. The operational definition which we must give to the "creation" of local matter is that before the event instruments could detect no matter, but afterwards they could. Because a logical definition of the universe would include all matter, past or present, we would have to say that the new matter did not come into our universe from elsewhere. Rather we would have to consider that it existed, perhaps as a diffuse or virtual potential, and then started to localize at the time which we have called "creation". (Phenomenologically this may be indistinguishable from volumes in the initial universe which have had retarded expansion as discussed by a number of authors[1].) If these objects are eventually to be detected by normal observational methods, then the localization process must initially consist of decreasing the object's apparent diameter and lowering its red-shift, both effects naturally enhancing its visibility. There is no reason to expect discontinuity in this process so that we would expect some kind of continuous disengagement from the age zero universe toward the state of its ultimate character. On the other hand, there is no reason to expect the special case of linearity in this process and we can point to the physics on the edge of our observable universe which affects the red-shift in a nonlinear fashion at great time–distance (the deceleration curvature in the linear Hubble relationship on the macroscopic scale).

Because we know that the presence of matter defines the properties of space, it has been previously suggested[2] that, if new matter did materialize, it would form regions where large amounts of matter were already in a high density state. This would, of course, suggest the nuclei of galaxies as a possible materialization point, with the more compact nuclei being most favourable.

The reason for discussing the theorem at this time is to point out that we should deductively expect, if our present models of red-shift distances and finite light signal velocities are correct, if we were ever directly to observe an object recently created, in the sense defined here, it would have a very high red-shift regardless of the distance at which the object was located from us.

HALTON ARP

Mount Wilson and Palomar Observatories,
Carnegie Institution of Washington,
California Institute of Technology,
Pasadena, California 91106.

Received May 19, 1969.

[1] Ne'eman, Y., Astrophys. J., 141, 1303 (1965); Novikov, I. D., Sov. Astron., 8, 857 (1965); Harrison, E. R., Astron. J., 73, S182 (1968).

[2] Hoyle, F., Eleventh Solvay Conf., La Structure et l'Evolution de l'Univers, 57 (1958). Hoyle, F., Galaxies, Nuclei and Quasars, 125 (Harper and Row, New York, 1965).

From the ASTRONOMICAL JOURNAL
75, No. 1, 1970, February—No. 1376
Printed in U. S. A.

Distribution of Quasistellar Radio Sources on the Sky

HALTON ARP

Mount Wilson and Palomar Observatories, Carnegie Institution of Washington and California Institute of Technology
(Received 16 June 1969; revised 12 November 1969)

All QSR's currently known that have measured redshifts are analyzed as a function of their apparent magnitude V. It is shown that a minimum of 40% of the intermediate-brightness QSR's ($16.2 \leq V \leq 17.0$ mag) form pairs which are separated by from 4° to 13° on the sky. This precludes the possibility that these QSR's can be at the great distances given by the usual assumption that they obey the galaxy redshift–distance relation. These intermediate-brightness QSR's seem instead to be associated with very bright galaxies, such as M81, NGC 1068, and the Virgo Cluster. Faint QSR's ($V > 17.0$ mag) are shown to be distributed on the sky in the same characteristic way that slightly less bright galaxies are distributed ($9.0 \leq m_{pg} \leq 12.5$ mag). Computer analysis demonstrates that these faint QSR's fall much closer to galaxies in this magnitude range than could randomly positioned QSR's. Statistically significant alignment of the nearest QSR's across just those galaxies with the most strongly correlated distances is also demonstrated. The most conspicuous grouping of faint QSR's in the sample is shown to be associated with the exploding galaxy NGC 520. The four faint QSR's in this region lie with extreme accuracy along a previously established ejection line from NGC 520. Similarity of redshifts further demonstrates the physical reality of the association, and the behavior of radio fluxes and indices agrees with previously published relationships for ejected radio sources. Evidence that the highest redshifts are associated with low intrinsic luminosities is used to explain the cutoff in numbers of QSR's with $z \gtrsim 2$ and the anomalous distribution of high-redshift QSR's first pointed out by Strittmatter *et al*. Finally, the brightest QSR's are indicated to be for the most part associated with M31 and the Local Group of galaxies.

I. INTRODUCTION

PREVIOUS evidence (Arp 1966b, 1967, 1968b) has indicated that the high redshifts of quasistellar sources are not reliable indicators of their distance. In the search for some parameter that might be more strongly correlated with distance, one might consider the flux strength of the radio emission. Evidence from galaxies indicates, however, that periods of active radio emission can be relatively short-lived and the intensity of the activity not necessarily correlated with the luminosity of the galaxy. We also have, of course, the existence of radio-quiet quasistellar objects (QSO's), which in all other aspects resemble the quasistellar radio sources (QSR's) and make it difficult to see how their radio strengths could be a measure of their distance. The remaining candidate for a distance criterion is the optical apparent magnitude. If the range in the intrinsic absolute magnitude of the QSO's is less than their range in apparent magnitude due to differing distance, then the apparent magnitude would be a usable distance criterion. The following paper analyzes the distribution of QSR's on the sky as a function of apparent magnitude. The inter-associations between QSR's themselves and QSR's and bright galaxies which result indicate, in this observer's opinion, that this approach is successful and enables QSR's in general to be identified with their galaxies of origin.

II. LISTS OF KNOWN QSR'S

There are two major radio surveys that have been published for a long enough time to enable reasonably complete optical identifications of the entries to be attempted. One is the 3C survey at 178 MHz (Edge *et al.* 1959; Bennett 1962), which covers the northern hemisphere from Dec. $= +90°$ to about $-5°$. The other is the Parkes catalogue, which surveys at 635 MHz from Dec. $= +27°$ to $+20°$ (Shimmins and Day 1968), surveys at 408 MHz from Dec. $= +20°$ to 0° (Day *et al.* 1966), and surveys at 1410 MHz from Dec. $= 0°$ to $-20°$ (Shimmins *et al.* 1966).

The QSR's that have been identified from radio sources in these catalogues are listed in Tables I, II, and III of the present paper. It is quite clear that these QSR identifications are not limited by faintness of optical apparent magnitude. That is, even though the Sky Survey Schmidt plates, for example, go easily to $V = 20$ or 21 mag, there are very few radio sources in these catalogues that have been or will be identified with stellar-appearing objects fainter than apparent

1

2   HALTON ARP

TABLE I. QSR: $V<16.2$, Dec.$\geq -20°$, all flux strengths, $|b^{II}|\geq 20°$.

QSR	V	z	S(408)	S(750)	α
1. 2128−12	15.98	0.50	1.5	...	+0.2
2. 2135−14	15.54	0.20	10.0	...	−0.8
3. 2251+11	15.80	0.32	3.7	...	−0.8
4. 2344+09	15.92	0.68	2.7	...	−0.5
5. 0405−12	14.79	0.57	9.3	...	C
6. 0837−12	15.76	0.20	5.7	...	−0.9
7. 3C232	15.78	0.53	...	2.1	−0.57
8. 1004+13	15.15	0.24	3.3	...	−0.8
9. 3C249.1	15.72	0.31	...	3.7	−0.71
10. 3C273	12.80	0.16	...	46.0	−0.24
11. 3C345	15.96	0.59	...	7.9	−0.34
12. 3C351	15.28	0.37	...	5.2	−0.73
13. 3C454.3	16.10	0.86	...	13.4	−0.16

TABLE II. QSR: $16.2\leq V\leq 17.0$ mag, Dec.$\geq -20°$.

Radio source	V	z	S(408)	S(750)	α
NGH ($b^{II}\geq +20$)		S(408)$\geq$1.7 f.u.			
1. 0859−14†	16.59	1.33	3.1	...	−0.4
2. 1049−09	16.79	0.34	5.8	...	−0.8
3. 1127−14	16.90	1.19	5.0	...	+0.2
4. 3C263*	16.32	0.65	...	5.1	−0.81
5. 1217+02	16.51	0.24	1.7	...	−0.9
6. 1229−02	16.75	0.39	4.6	...	−0.5
7. 1252+11	16.64	0.87	2.8	...	−0.1
8. 3C281	17.02	0.60	5.4	...	−1.1
9. 3C298	16.79	1.44	24.4	...	−1.1
10. 3C309.1*	16.78	0.90	...	11.5	−0.48
11. 1510−08	16.52	0.36	3.0	...	0.0
12. 1354+19	16.20	0.72	6.0	...	−0.7
13. 3C323.1	16.69	0.26	...	4.0	−0.77
14. 3C334	16.41	0.56	...	3.5	−0.77
15. 3C380*	16.81	0.69	...	23.2	−0.76
SGH ($b^{II}\leq -20°$)		S(408)$\geq$1.5 f.u.			
1. 2145+06	16.47	0.37	3.4	...	C
2. 2216−03	16.93	0.90	2.8	...	C
3. 0003+15	16.40	0.45	2.6	...	−0.6
4. 3C48*	16.20	0.37	...	25.5	−0.79
5. 0119−04	16.88	1.96	2.2	...	−0.4
6. 0122−00	16.70	1.07	1.5	...	−0.1
7. 0232−04	16.46	1.43	3.2	...	−0.7
8. 3C95	16.24	0.61	...	5.7	−1.09
9. 3C94	16.49	0.96	...	5.3	−0.98

* Dec. greater than Pks limit of +27°.
† This QSR is listed $V=16.59$ in Barbieri *et al.* (1967). $m_{pg}=16$ in Bolton *et al.* (1966), and $m_v=17.8$ in Day *et al.* (1966). Since it is not known to the author whether variability accounts for these discrepancies, the object has been included at its brighter magnitude, and again in Table III at its fainter magnitude.

visual magnitude $V\sim 19.5$ mag. This cutoff in apparent magnitude for the identified QSR's has been shown very clearly by Bolton (1969). The radio completeness is not uniform, however, since the various surveys go to different limiting signal strengths and survey at different frequencies. The 3C region, for example, must contain uncatalogued, fainter radio sources or sources with flat radio indices, some of which may be ultimately identified with QSR's of fairly bright optical apparent magnitude. Comparable objects have already been identified in the Parkes region. (The 4C catalogue is furnishing such QSR identifications, but QSR's from 4C identifications have not been included here because the fairly recent publication of the 4C and the large

TABLE III. QSR: $V>17.0$ mag, Dec.$\geq -20°$.

Radio source	V	z	S(408)	S(750)	α	Radio source	V	z	S(408)	S(750)	α
NGH ($b^{II}\geq +20°$)		S(408)>5.0 f.u.				SGH ($b^{II}\leq -20°$)		S(408)$\geq$1.5 f.u.			
1. 3C186	17.60	1.06	...	2.7	−1.19	1. 3C432	17.96	1.81	...	2.9	−1.08
2. 3C191	18.40	1.95	...	3.4	−1.09	2. 2146−13	(19.5)	1.80	5.6	...	−0.9
3. 3C196	17.60	0.87	...	23.9	−0.83	3. 2209+08	(18.5)	0.49	4.1	...	−0.6
4. 0812+02	(18.5)	0.40	6.0	...	−0.9	4. 3C446	18.39	1.40	...	8.7	−0.61
5. 3C204	18.21	1.11	...	2.4	−0.95	5. 2223+21	(18)	1.96	...[2]	...	...
6. 3C205	17.8	1.53	...	4.0	−0.81	6. PHL 5200	(18.2)	1.98	...[3]	...	...
7. 3C207	18.15	0.68	...	4.1	−0.69	7. CTA 102	17.32	1.04	7.1	...	C
8. 3C208	17.42	1.11	...	4.6	−1.23	8. 3C454	18.40	1.76	...	3.5	−0.80
9. 0859−14†	(17.8)	1.33	5.4	...	−0.4	9. CTD 141	17.30	1.02	...[4]	...	...
10. 3C245	17.25	1.03	...	5.1	−0.75	10. 2345−16	(18.5)	0.6	2.5	...	−0.2
11. 1055+20	17.07	1.11	...[1]	...	−0.6	11. 2354+14	18.18	1.81	3.7	...	−1.1
12. 3C254	17.98	0.73	...	5.4	−0.91	12. 3C2	19.35	1.04	...	6.0	−0.80
13. 1116+12	19.25	2.12	5.5	...	−0.6	13. 3C9	18.21	2.01	...	3.9	−1.05
14. 3C261	18.24	0.62	...	2.5	−0.56	14. 0056−00	17.33	0.72	3.9	...	−0.3
15. 1136−13	17.8	0.55	12.8	...	−0.8	15. 0106+01	18.39	2.11	3.5	...	−0.7
16. 3C268.4	18.42	1.40	...	3.6	−0.73	16. 0114+07	(18)	0.86	(3.0)	...	−1.1
17. 3C270.1	18.61	1.52	...	5.0	−1.09	17. 0115+02	(17.5)	0.67	(4.5)	...	−0.6
18. 3C275.1	19.00	0.56	...	5.0	−0.86	18. 0118+03	(18)	0.77	(5.0)	...	−1.1
19. 3C277.1	17.93	0.32	...	3.6	−0.57	19. 0123+25	(18)	2.38	...[5]	...	−0.8
20. 3C279	17.75	0.54	...	10.8	−0.35	20. 3C47	18.10	0.43	...	7.0	−0.99
21. 3C280.1	19.44	1.66	...	2.6	−0.92	21. PHL 1078	18.25	0.31	...[6]	...	...
22. 3C287	17.67	1.06	...	9.7	−0.52	22. PHL 1093	17.07	0.26	...[7]	...	...
23. 3C286	17.30	0.85	...	19.2	−0.38	23. 0159−11	(17.5)	0.67	6.5	...	−0.5
24. M13−011	17.68	0.63	10.1	...	−0.8	24. 0214+10	(17)	0.41	2.3	...	−0.7
25. 3C288.1	18.12	0.96	...	2.9	−1.08	25. 0229+13	17.71	2.07	(2.3)	...	+0.2
26. M14−121	17.37	0.94	10.2	...	−0.7	26. 0336−01	18.41	0.6	3.5	...	C
27. 1317−00	17.32	0.89	5.4	...	−0.7	27. 0403−13	17.09	0.57	8.7	...	−0.5
28. 3C336	17.47	0.93	...	4.4	−0.79	28. 0420−01	(18)	1.7	1.5	...	+0.2

[1] $S(635)=4.7$. [2] $S(635)=(2.0)$. [3] $S(178)=3.2$. [4] $S(178)=6.6$. [5] $S(635)=1.1$. [6] $S(178)=2.2$. [7] $S(178)=3.8$.
† See note to Table II.

DISTRIBUTION OF QSR's 3

number of sources contained therein make it uncertain whether very complete QSR identifications have been made as yet.)

In analyzing the distribution of QSR's in various regions of the sky, therefore, flux cutoff limits have been set in this paper so that a fairly complete sample to the same limiting radio strength is dealt with in each region. As mentioned previously, it is assumed that negligible numbers of QSR's are undiscovered because of faint optical apparent magnitude. No declinations south of Dec. $= -20°$ are considered in the present paper in order to confine the regions analyzed to those which have been searched for optical identifications from the active northern observatories. Finally, in order to restrict ourselves to QSR's with quantitative data and to avoid being involved with unconfirmed identifications, only QSR's with measured redshifts have been used in the present paper. The data for the identified QSR's have been taken from Burbidge (1967), Barbieri *et al.* (1967), Schmidt (1968), and a list of

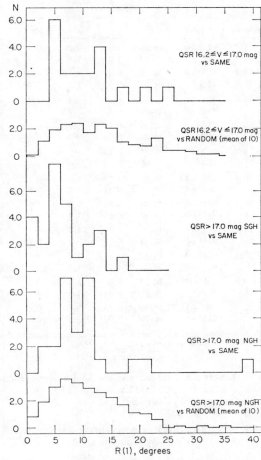

FIG. 2. Frequency distributions of QSR separations on the sky: Intermediate-brightness QSR's are shown in *top* histogram. Faint QSR's in SGH are shown in *middle* histogram and in NGH in *bottom* histogram. For the first and last cases, average distributions obtained from randomly distributed QSR's are shown for comparison.

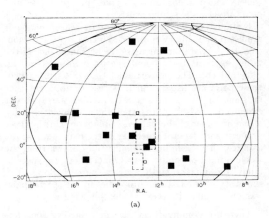

(a)

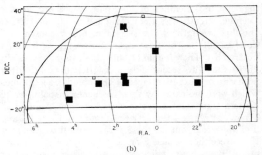

(b)

FIG. 1. (a) *Above:* Intermediate-brightness QSR's ($16.2 \leq V \leq 17.0$ mag) plotted in the NGH, from Table II. Large dotted rectangle represents approximate bounds of Virgo Cluster, smaller dotted rectangle represents southern extension of Virgo Cluster. Small open rectangles represent brightest galazies in NGH ($m_{pg} \leq 8.9$ mag). (b) *Below:* Intermediate-brightness QSR's plotted in the SGH. Open squares represent brightest galaxies ($m_{pg} \leq 10.0$ mag).

measures by Bolton and Kinman kindly supplied in advance of publication.

Division into Magnitude Intervals: Inspection of the frequency distribution (e.g., see Bolton 1969) of V magnitudes in the QSR sample indicated a deficiency of V magnitudes around $V = 17.0$ mag. This makes it convenient to divide two groups at $V = 17.0$ mag. A brighter group was rather arbitrarily divided off at $V = 16.2$ mag. As analysis proceeded, it appeared that the division at $V = 17.0$ mag was indeed physically significant for the north Galactic hemisphere, at least. For the brighter group, while there is some overlapping

4   H A L T O N   A R P

TABLE IV. Spatial pairs.

Separation	Object	V	z	$S(408)$	α
4°.3	0119−04	16.88	1.96	2.2	−0.4*
	0122−00	16.70	1.07	1.5	−0.1
5.4	1229−02	16.75	0.39	4.6	−0.5
	1217+02	16.51	0.24	1.7	−0.9
6.0	3C281	17.02	0.60	5.4	−1.1
	1252+11	16.64	0.87	2.8	−0.1
7.3	3C95	16.24	0.61	11.6	−1.09*
	3C94	16.49	0.96	9.9	−0.98
8.4	3C334	16.41	0.56	6.3	−0.77*
	3C323.1	16.69	0.26	(7.3)	−0.77
10.9	1127−14	16.90	1.19	5.0	+0.2
	1049−09	16.79	0.34	5.8	−0.8
13.1	2216−03	16.93	0.90	2.8	C*
	2145+06	16.47	0.37	3.4	C
13.9	3C298	16.79	1.44	24.4	−1.1
	1354+19	16.20	0.72	6.0	−0.7
17.8	0232−04	16.46	1.43	3.2	−0.7
	0122−00	16.70	1.07	1.5	−0.1
20.6	1510−08	16.52	0.36	3.0	· 0.0
	3C298	16.79	1.44	24.4	−1.1
25.5	0003+15	16.40	0.45	2.6	−0.6
	0122−00	16.70	1.07	1.5	−0.1

* Closely matching fluxes and indices.

in characteristics, it appears in the final analysis that a division at $V=16.4$ or 16.5 mag would have been somewhat more physically significant.

III. IS DISTRIBUTION OF INTERMEDIATE-BRIGHTNESS QSR'S RANDOM ON THE SKY?

Figure 1(a) and (b) shows the distribution of the intermediate-brightness QSR's ($16.2 \leq V \leq 17.0$ mag) in both the NGH and SGH. There are a number of aspects of these distributions which appear significant, but in this section we will first test the very obvious pairing of the QSR's which seems to characterize their distribution on the sky.

A. Pairing of QSR's

Because the region north of Dec. $= +27°$ had not been surveyed in the 3C catalogue to as faint a flux limit, particularly for flat index sources, as in the Parkes catalogue south of this limit, the distribution of the intermediate-brightness QSR's was therefore tested only in the region $-20° \leq$ Dec. $\leq 27°$, where the radio survey was uniform. A computer program gave the distance, in degrees on the sky, from each QSR to its nearest neighbor. In order to avoid boundary effects, only distances from QSR's in Fig. 1(a) and (b) with Dec.

$\geq -15°$ and $|b^{\mathrm{II}}| \geq 25°$ were tested against the distribution of QSR's in the larger area. These 19 QSR's had a distribution of nearest-neighbor distances that showed a strong excess of small separations, particularly at 4°–6°, and a less strong excess at 12°–14° separation. By testing these same 19 QSR's against generated random populations, it was possible to show that the probability of getting the observed number of separations between 4°–6° by accident is only 0.0065 (0.7%). The probability of accidentally getting a peak as big as this somewhere in this range of separations is about 4%. The middle histogram in Fig. 2 shows, however, that the faint QSR's in the SGH have peaks in the distribution of their separations at exactly the same values of 4°–6° and 12°–14°. The faint QSR's in the NGH have two peaks also, but slightly closer together at 6°–8° and 10°–12°. In the past, separations of both 3C and Parkes radio sources have been shown by Wagoner (1967) to prefer values around 5° and 4°, respectively. Therefore, independent sets of data as well as independent analyses give, quantitatively and qualitatively, the same result.

Distances between each QSR and its nearest neighbor are listed in Table IV in order of increasing separation. The small separation entries in the beginning part of the table represent the obvious spatial pairs seen in Fig. 1(a) and (b). The properties of the QSR's in each of these pairs are listed, and it is readily apparent that in some of the small-separation pairs the apparent magnitude, redshift, flux strength, and spectral index of one member of the pair resemble more closely these properties in the other member of the pair than, on the average, they resemble random values drawn from the whole range of properties. Coincidences in spectral index and flux strength are the most striking. The similarity of the last two properties in the spatial pairs was tested quantitatively in the following way:

There are 19 individual QSR's in Table IV. Considering the first member of any pair, there are, in the average, slightly less than two chances out of 18 that the flux strength of the second member, chosen randomly, will be the next nearest flux strength in an ordered list. There would be only about four chances out of 18 that the second flux strength would be either adjacent or next-to-adjacent. The angular separations which were indicated by their frequency distribution to contain significant excesses were between 4° and 14°. Therefore, we test the first eight pairs in Table IV and find that four out of these eight have flux strengths that are either adjacent or next-to-adjacent, and four of the pairs have indices that are either identical or adjacent. The probability of getting the latter numbers of similarities accidentally is about 5% for the flux strengths and about 1% for the indices.

Bartlett (private communication) has computed a slightly different test. He finds that, if 16 ordered quantities are paired arbitrarily, the chance of getting five or more pairs with adjacent or next-to-adjacent properties is 2%, and to get four or more pairs with adjacent properties it is 1%. The first computation applies to the observed pairing in the first 16 flux strengths, and the latter, approximately, to the indices.

Since the properties of flux strength and index are independent, the probability of having this result occur accidentally for both properties simultaneously is less than 0.1%. (In addition to these tests on the whole set of properties, it is striking to note that it is just the same pairs in which the indices are almost identical that the flux strengths tend to be the closest.) We find, then, that the distribution of intermediate-brightness QSR's on the sky shows a significant excess of small separations, and that the properties of some of these spatially associated QSR's resemble each other more closely than could reasonably occur by chance. Of course it is just the resemblance of flux strengths and indices which leads to the identification of commonly accepted pairs of ordinary radio sources (e.g., see Moffet 1965).

Our first major conclusion is, therefore, that an appreciable portion of intermediate-brightness QSR's are not distributed randomly on the sky but are physically associated together in pairs that cover from 4°–13° in the sky. A simple light-travel-time argument then shows that the QSR's cannot be at distances indicated by a cosmological interpretation of the redshifts if their lifetimes are less than 10^8 yr. They would be beyond each others event horizons and, therefore, could not be physically associated.

B. Association with Very Bright Galaxies

If we proceed on the conclusion that the intermediate-brightness QSR's must be closer than given by the usual assumption about their redshift, then we must ask what nearer classes of objects are at the same distance as these QSR's.

Figure 1(a) and (b) demonstrates two other distribution properties of the intermediate-brightness QSR's that appear decidedly nonrandom. One is the apparent line of QSR's that originates from the Virgo Cluster. The other aspect of the distribution, and connected in a way to the first, is that generally the intermediate-brightness QSR's fall close in the sky to the very bright galaxies. (Close, for example, to M81, NGC 4594, and NGC 4826 in the NGH, and NGC 1068, M33, and M31 in the SGH.) We are led to the hypothesis that these intermediate-brightness QSR's are associated with these bright galaxies and would then predict that the fainter (more distant) QSR's would be associated with fainter (more distant) galaxies. In the next section we test this prediction.

IV. DISTRIBUTION OF FAINT QSR's

We define faint QSR's as those with $V > 17.0$ mag, and look at their distribution in the NGH first.

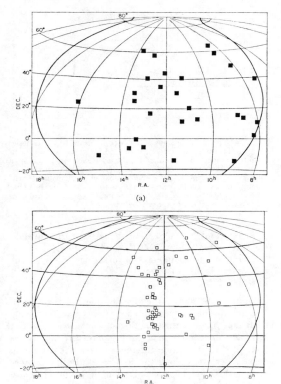

(a)

(b)

FIG. 3. (a) *Above:* All faint QSR's ($V > 17.0$ mag) in NGH are plotted (Dec. $\geq -20°$, $b^{II} \geq +20°$). Radio sources as listed in Parkes and 3C catalogues with $S(408) \gtrsim 5.0$ f.u. (b) *Below:* All bright galaxies ($9.0 \leq m_{pg} \leq 11.0$ mag) from Shapley–Ames catalogue are plotted.

Since an appreciable portion of the NGH is north of Dec. $= +27°$, where the 3C catalogue surveys to a brighter limit than the Parkes catalogue, we list only QSR's in the NGH section of Table III that have radio strengths greater than $S(408) = 5.0$ f.u. For average radio-source spectral indices, this gives about the same radio-flux cutoff in both the 3C and Parkes catalogues, and therefore gives a reasonably homogeneous sample of optically faint QSR's in the NGH. This set of QSR's is shown plotted in Fig. 3(a). It is immediately apparent that this plot is quite similar to the familiar distribution of galaxies in the NGH (see, for example, Shapley and Ames 1932; de Vaucouleurs and de Vaucouleurs 1964).

There are three obvious characteristics of the distribution of faint QSR's in the NGH:

1. a broad, strong concentration at about R.A. $= 12^h 30^m$, running in declination from Dec. $= -10°$ to $+60°$;

6   HALTON ARP

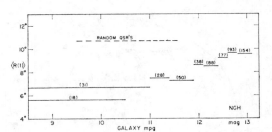

FIG. 4. The distance from a galaxy to its nearest QSR is designated $R(1)$, the mean ,$\langle R(1) \rangle$, is plotted for galaxies of different brightness ranges. The number of galaxies in each magnitude span is written in parentheses above its $R(1)$ value. When the real QSR's are replaced by the same number of random points, the average distance to the nearest galaxy is 10°.7, shown by the upper dashed line. (Plotted dashed line represents mean of total of 50 random distributions for selected classes.)

2. a sprinkling of QSR's to the west of this, over to the Galactic latitude cutoff R.A. = 8ʰ to 9ʰ;

3. an almost complete lack of faint QSR's to the east, from R.A. = 14ʰ over to Galactic latitude cutoff at R.A. = 17ʰ to 18ʰ.

Figure 3(b) shows the distribution on the sky of galaxies in the magnitude range $9.0 \leq m_{pg} \leq 11.0$ mag from the Shapley–Ames catalogue. *These bright galaxies show the same three distribution characteristics as the faint QSR's.* The major difference is that the maximum density of QSR's along the R.A. = 12ʰ30ᵐ line is more dispersed than the concentration of bright galaxies. The galaxies could be described as forming a concentrated core to the somewhat more dispersed QSR distribution.

The coincidence of QSR and galaxy distributions can be analyzed numerically by using the same computer program referred to earlier. The distance from each galaxy to the nearest three QSR's is computed, and the angles, as seen from the galaxy, between the first and second, first and third, and second and third QSR are also computed. In order to avoid edge effects, only galaxies within $-10° \leq$ Dec. $\leq 90°$ and $b^{II} > 30°$ were tested, and, in order to avoid the weighting of statistics to the strong concentration of galaxies in the Virgo Cluster (which are apparently associated with brighter QSR's, in any case), all galaxies with $0 \leq$ Dec. $\leq 20°$ and $12^h00^m \leq$ R.A. $\leq 13^h00^m$ were excluded.

The first test averaged the distance from each galaxy to the nearest QSR, designated as $R(1)$, over the 428 galaxies in this region of the NGH that were brighter than $m_{pg} = 13.1$ mag (the listed limit of the Shapley–Ames catalogue). This average distance turned out to be $\langle R(1) \rangle = 8°.8$. The same number of positions as the QSR's were then distributed randomly in the same area of the sky in which the real QSR's are distributed. Successive trials with ten different sets of random QSR's showed that the average distance from Shapley–Ames galaxies to these randomly distributed QSR's was

$\langle R(1) \rangle = 10°.7$, with a range from 9°.3–12°.6. The chance of getting a value as low as the observed $\langle R(1) \rangle = 8°.8$ is only 0.017 (1.7%; Bartlett, private communication).

The computer program, however, also enables different classes of galaxies to be compared to the QSR distribution. Figure 4 shows that the average distance from brighter and brighter classes of galaxies to the same set of faint QSR's becomes progressively smaller, until the distance computed from the 31 brightest galaxies $(8.0 \leq m_{pg} \leq 11.0$ mag) is $\langle R(1) \rangle = 6°.7$. The average distance to the 18 brightest $\Delta 8.0 \leq m_{pg} \leq 10.6$ mag) is $\langle R(1) \rangle = 5°.8$. But empirical tests of ten random QSR distributions with the former and 30 with the latter show that the average distance from these bright galaxies to a randomly distributed QSR population is still $\langle R(1) \rangle = 10°.7$. The significance level of these correlations is high, and the chance of finding any one of these brighter classes of galaxies falling accidentally this close to the QSR's is only of the order of 2–3%.

Because of previous evidence on the association of radio sources and galaxies, it is tempting to conclude that these QSR's are specifically associated with these bright galaxies. One qualification should be noted, however, and that is that it is well known that bright galaxies show strong clustering. As fainter classes of galaxies are considered, their distribution becomes more

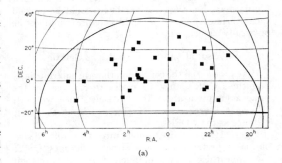

(a)

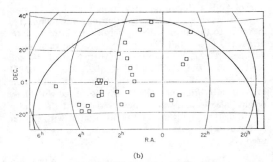

(b)

FIG. 5. (a) *Above:* All faint QSR's in SGH ($V > 17.0$ mag) are plotted (Dec. $\geq -20°$, $b^{II} \leq -20°$). Radio sources are listed in Parkes and 3C catalogues with $S(408) \geq 1.5$ f.u. (b) *Below:* All bright galaxies in SGH ($10.0 \leq m_{pg} \leq 12.0$ mag) are plotted.

DISTRIBUTION OF QSR's   7

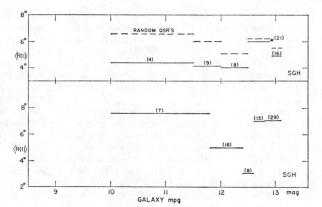

FIG. 6. (a) *Above:* Mean distances of nearest intermediate and faint QSR's ($m_{pg} \geq 16.2$ mag) to different classes of galaxies in the SGH are plotted. Dashed line represents mean for two random distributions of QSR's. (b) *Below:* Same quantity $\langle R(1) \rangle$, plotted for SGH galaxies relative to only faint ($V > 17.0$ mag) QSR's.

uniform. But some faint galaxies are associated with the bright clusters. What has been demonstrated so far is that the QSR's are associated with centers of bright galaxy clustering. As the analysis proceeds, we can try to find out which, if any, galaxies associated with these cluster centers are more strongly associated with the QSR's.

One very important further proof of the reality of this association comes from the alignment angles computed for these QSR's. Seen from the Galaxy, the angle between the first nearest and second nearest QSR is designated $A12$. If this angle is 180°, the two QSR's are exactly aligned across the Galaxy. In the area of the NGH which we have considered, the 31 brightest galaxies ($8.0 \leq m_{pg} \leq 11.0$ mag) have near alignment, $140° \leq A12 \leq 180°$, which occurs 13 times. By using generated random QSR positions, again it is possible to find empirically that the expected number of times this should occur by chance is only 6.2. Obviously the real number of alignments exceeds the number expected by chance by a significant amount. Therefore, we find that for just the class of galaxies that has the most significant relationship to the QSR's in terms of mean distance, this same class of galaxies independently shows a significant tendency for alignment between these same close QSR's.

In the NGH we observe a strong line of intermediate-brightness QSR's originating from the Virgo Cluster. This line does not seem to be associated with any fainter galaxies in the Shapley–Ames catalogue, but as we have just seen, fainter QSR's and non-Virgo galaxies running roughly along the supergalactic equator (de Vaucouleur 1959) are strongly associated with each other, and bright galaxies and faint QSR's to the west of this are also associated. In the SGH we have nothing comparable to the Virgo Cluster, but some of the intermediate-brightness QSR's possibly form a similar line through NGC 1068, which is one of the brightest galaxies in the SGH **(Fig. 1(b))**. There is a

sparse cluster (or group) of galaxies associated with NGC 1068 and fainter galaxies running along the suggested line of intermediate-brightness QSR's to the northwest and southeast, Fig. 5(b). We therefore test first the combined intermediate and faint QSR's against the galaxies in the SGH. (Since most of the SGH is below Dec. = +27°, we can accept QSR's down to the fainter limits of the Parkes catalogue, about $S(408) = 1.5$ in Table III.) These 28 QSR's are associated with the bright galaxies in the SGH, as Fig. 6(a) demonstrates. From about $10 \leq m_{pg} \leq 12.5$ mag, the galaxies in the SGH average 2°–3° closer to these QSR's than they do to randomly placed QSR's. Fainter galaxies in the SGH do not show any very strong association with this group of QSR's. On the other hand, Fig. 6(b) shows that the faint ($V > 17.0$ mag) QSR's alone are not

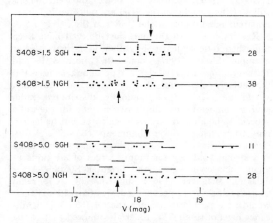

FIG. 7. Frequency distribution of QSR magnitudes is shown for the two hemispheres: *Lower* histograms compare QSR's with $S(408) > 5.0$ f.u., and *upper* histograms with QSR's of radio strength $S(408) \geq 1.5$ f.u. Arrows indicate estimated apparent magnitude of peak frequencies.

strongly correlated with the brightest galaxies, but that they are strongly associated with the galaxies slightly fainter, from about $11.8 \leq m_{pg} \leq 12.6$ mag. The angular distribution of these two groups of QSR's also shows alignment tendencies for the classes of galaxies that have the smallest mean distances to the QSR's, but the numbers involved are too small to assign a statistical measure of significance.

To summarize, we have found that in both the NGH and the SGH, the intermediate-brightness QSR's are associated with very bright galaxies. In both hemispheres the faintest QSR's are associated with slightly fainter but still bright galaxies. One fact is quite evident, however, and that is that there are fewer bright galaxies in the SGH than there are in the NGH. If the mean apparent magnitude of the bright galaxies in the SGH is appreciably fainter than in the NGH, however, we would then expect the apparent magnitudes of the QSR's in the SGH to also be fainter than the QSR's in the NGH. Figure 7 demonstrates that this is exactly what is observed. If we restrict ourselves to QSR's with flux strengths $S(408) > 5.0$ f.u., the lower histograms show the most frequent apparent magnitude in the NGH is a little greater than $V = 17.5$ mag. In the SGH the most frequent apparent magnitude is something greater than $V = 18.0$ mag. Accepting QSR's down to $S(408) > 1.5$ f.u. (upper histograms), the same result is shown with greater numbers. [In both hemispheres, small systematic magnitude corrections for those few QSR estimates near the southern declination limit are needed. This is discussed by Ekers (1967) but would appear not to appreciably affect the comparisons made here.]

Therefore, we see that not only are the bright galaxies associated with the faint QSR's in both the NGH and SGH, but that the fainter apparent magnitudes of the galaxies in the SGH are reflected in fainter apparent magnitudes of the QSR's in the SGH.

Results similar to the ones just derived have been published independently by other investigators. For example, Sharp (1967) found that both 3CR and Parkes radio sources fell closer to peculiar galaxies—and particularly closer to Shapely–Ames galaxies—than would be expected of randomly distributed radio sources. Sharp did not find any tendency for the radio sources to be aligned across the galaxies, but he looked only for much narrower alignments ($\Delta\theta = 5°$). Subsequent analysis (Arp, unpublished) shows that there is a strong tendency for radio sources of all kinds to align themselves across the brighter Shapley–Ames galaxies. Wagoner (1967) showed that 3CR radio sources fell closer to peculiar galaxies (Arp 1966a) than would be expected by chance (better than 1% significance level of result). Wagoner also showed that of the three categories of radio sources (QSR's, radio galaxies, and blank field) the category of QSR's fell closer on the average to the peculiar galaxies than the other categories of radio sources. Finally, Strittmatter et al.

(1966) pointed out the puzzling result that the distribution on the sky of QSR's with redshifts greater than $z = 1.5$ concentrated toward the Galactic poles in a way which could not be accounted for on the assumption of cosmological distances for the QSR's. The results in the present paper establish the association of faint QSR's and bright galaxies. Since the high-redshift QSR's come primarily from QSR's with $V > 17.0$ mag, the high-redshift QSR's would be expected to be distributed preferentially with the bright galaxies. Inspection of Figs. 3(b) and 5(b) shows that the explanation of the Strittmatter effect is simply that the distribution of the bright galaxies favors the Galactic polar regions.

This paper has so far established, completely independently of previous work that QSR's tend to occur in pairs, that they tend to be associated with bright galaxies, and that the QSR's tend to align themselves across bright galaxies. It is possible to strengthen these results by comparing the sizes of the separations of pairs and associations found here with specific associations reported previously. The brightest galaxies previously identified in radio source associations were NGC 908 ($m_{pg} = 11.0$ mag), NGC 5055 ($m_{pg} = 10.5$ mag), and NGC 4472 ($m_{pg} = 10.1$ mag), where radio separations were: $11°–20°$, $5°$, and $10°$, respectively. If we take twice the distance from the galaxy to the nearest associated QSR as an indicative separation of the QSR's around the bright Shapley–Ames galaxies found here, Figs. 4 and 6 show that the separations turn out to be $12°–13°$ in the NGH and $8°–9°$ in the SGH (where the average apparent magnitude of galaxies is fainter, going to as low as $6°$ for the fainter SGH galaxies). The separations found in the present paper then agree roughly with the separations of radio sources across comparably bright galaxies found in previous identifications. (In previous identifications, fainter galaxies than those discussed in the present paper had radio pairs separated by smaller amounts, as would be expected if they were viewed at a greater distance.) Table V lists faint QSR's that form spatial pairs and groups. Some of these have very similar properties. The separation of these latter pairs, as well as the pairs of intermediate-brightness QSR's which were demonstrated in Table IV to be associated, run from about $4°–13°$, which again confirms the kinds of separations found in the present and previous papers.

The purpose of the present paper is to furnish additional statistical proof of the association of QSR's and galaxies. Except for the two cases in the following section, it will be left to future analyses to find further specific identifications, as in the early papers, of individual faint QSR's with individual galaxies.

A few cautionary remarks might be made at this point, however. One is that we do not know at the start in the case of any individual pair of QSR's whether they might have been ejected as an opposite pair from a central galaxy or together in the same direction from

DISTRIBUTION OF QSR's   9

TABLE V. Faint QSR's ($V > 17.0$ mag) which fall near each other or in isolated groups (no computation of similarity of properties).

Separation	Object	V	z	$S(408)$	$S(750)$	α	Separation	Object	V	z	$S(408)$	$S(750)$	α
	NGH $S(408) \gtrsim 5.0$ f.u.*							SGH $S(408) \gtrsim 1.5$ f.u.					
3°1	3C207	18.15	0.68	7.3	⋯	−0.7	0°7	PHL 5200	(18.2)	1.98	$S(178) =$	3.2	⋯
	3C208	17.42	1.11	(10.0)	⋯	−1.0		3C446	18.39	1.40	10.3	⋯	−0.5
5.4	3C287	17.67	1.06	⋯	9.7	−0.52	QSR's in the NGC 520 group shown in Fig. 7 and discussed in Sec. V						
	3C286	17.30	0.85	⋯	19.2	−0.38	4.3	0214+10	(17)	0.41	2.3	⋯	−0.7
7.3	3C205	17.8	1.53	⋯	4.0	−0.81		0229+13	17.71	2.07	(2.3)	⋯	+0.2
	3C204	18.21	1.11	⋯	2.4	−0.95	5.5	3C47	18.10	0.43	⋯	7.0	−0.99
7.5	3C277.1	17.93	0.32	⋯	3.6	−0.57		0123+25	(18)	2.38	1.4	⋯	−0.8
	3C288.1	18.12	0.96	⋯	2.9	−1.08	5.6	3C9	18.21	2.01	⋯	3.9	−1.05
7.3 ⎫ 10°5′ 7.7 ⎭	13−011	17.68	0.63	10.1	⋯	−0.8		2354+14	18.18	1.81	3.7	⋯	−1.1
	1317−00	17.32	0.89	5.4	⋯	−0.7	6.1	2209+08	(18.5)	0.49	4.1	⋯	−0.6
	3C279	17.75	0.54	13.5	⋯	C		CTA 102	17.32	1.04	7.1	⋯	C
8.65 ⎫ 8°86 8.82 ⎭	3C245	17.25	1.03	7.6	⋯	−0.7	6.6	3C454	18.40	1.76	⋯	3.5	−0.80
	1055+20	17.07	1.11	(7.1)	⋯	−0.6		2223+21	(18)	1.96	$S(635) =$ (2.0)		⋯
	1116+12	19.25	2.12	5.5	⋯	−0.6	8.5	PHL 1078	18.25	0.31	$S(178) =$	2.2	⋯
0.15 ⎫ 10°2 0.17 ⎭	3C280.1	19.44	1.66	13.5	⋯	C		0159−11	(17.5)	0.67	6.5	⋯	−0.5
	3C268.4	18.42	1.40	⋯	3.6	−0.73	10.9 ⎫ 13.1 12.6 ⎭	0336−01	18.41	0.6	3.5	⋯	C
	3C270.1	18.61	1.52	⋯	5.0	−1.09		0420−01	(18)	1.7	1.5	⋯	+0.2
								0403−13	17.09	0.57	8.7	⋯	−0.5

* Note that fainter than $S(408) \gtrsim 5.0$ f.u. are the pairs 0932+02, 0957+00 and 1222+21, 1225+20, which have closely similar properties as well as falling spatially close together.

the parent galaxy, or whether they might have been ejected as a single body and subsequently fissioned in a secondary explosion. If they did originate in a secondary ejection, there might be a conspicuous luminous body at that secondary point of origin, or that point might have been explosively dispersed, fragmented, or otherwise diminished. Another factor that is important to keep in mind is that occasionally an ejection will be oriented more or less in our line of sight to the galaxy of origin. In that case, the projected separation of the radio sources on the sky would be much smaller than normal for that magnitude (distance) galaxy. But, since the ejections normally deviate from a straight line by somewhere between 0° and 30°, as shown in previous work (Arp 1967), or between 0° and 40°, as indicated here, the galaxy of origin in that case could project at a considerable relative distance from the center of the pair, and some difficulty might be experienced in identifying the origin of the pair. The point is that the ejection hypothesis by no means requires a galaxy to be found between every pair of radio sources. We simply establish the reality of the phenomenon in those associations that can be demonstrated to be physically connected, and then use further associations to confirm the predictions of that hypothesis and extend our understanding of the nature of the process.

V. CONCENTRATIONS OF OPTICALLY FAINT QSR'S

A. Association with NGC 520 (Atlas 157)

It has been shown that the distribution characteristics exhibited by QSR's of different apparent

brightness are marked more by concentrations in fairly large areas rather than very close groupings. Nevertheless, there is a very conspicuous grouping of QSR's that appears in the SGH. It is shown in Fig. 5(a), at about R.A. = 1^h10^m and Dec. = +2°. There is also a concentration of galaxies at this same position, four of which are slightly fainter than plotted in Fig. 5(b). Among those fainter galaxies is NGC 520 at $m_{pg} = 12.4$ mag. NGC 520 is an outstanding example of a violently disrupted galaxy. This galaxy was one of the prime examples of an M82-type irregular galaxy in the Hubble Atlas (Sandage 1961). M82 later became a prototype of galaxies in which the shattered aspect was hypothesized to have been caused by a "violent event" (see, for example, Lynds 1961; Lynds and Sandage 1963; Burbidge et al. 1963). In the Atlas of Peculiar Galaxies (Arp 1966a), NGC 520 was one of the prime examples of a "disturbed galaxy with interior absorption" and was placed in the same category with the very active NGC 5128 (Cen A). The latter galaxy was numbered 153 in the Atlas, and NGC 520 was numbered 157. In the analysis of "Peculiar Galaxies and Radio Sources" (Arp 1967), it was shown (p. 343) that radio sources extended on a line on either side of NGC 520. It was shown that a particularly strong line of radio sources extended southwest from the galaxy, and in that paper it was concluded that the radio sources lay in a narrow "ejection cone" originating in NGC 520. The point of this recapitulation of previous work on this galaxy is that up to the date of the writing of the present paper, one of the best candidates anywhere in the sky for

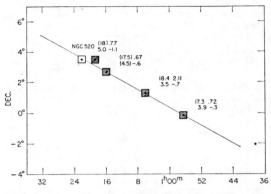

FIG. 8. NGC 520 (*Atlas of Peculiar Galaxies*, No. 157) is plotted as an open square. All the faint QSR's (*V* > 17.0 mag) in an area 11°×15° southwest of NGC 520 are plotted as filled squares. To the upper right of each QSR is written its optical magnitude *V*, redshift *z*, then, on the next line, its radio flux, *S*(408), and its radio index α. The point on the lower right represents a QSR from a new faint strip survey by Bolton (unpublished).

explosive ejection of excited material was just this disturbed galaxy, NGC 520.

Figure 8 shows NGC 520 and an 11° × 15° region of the sky to the southwest. All the faint QSR's which are now known from current work as summarized in Table III are plotted in this region. The extraordinary result shown in Fig. 8 is that the concentration of four QSR's in this region falls on a line proceeding from NGC 520 to the southwest. In fact, three of these four QSR's fall so accurately on this line that there is less than 10′ deviation from the line over its entire length of about 7°.

The importance of this configuration cannot be overemphasized. It would seem that there could be only two possibilities—either it is accidental or it is not. If it is not accidental, there does not seem to be any other conclusion other than that these four QSR's have been ejected from the exploding galaxy NGC 520. At the same time, the concentration, accuracy of alignment, and arrangement with respect to NGC 520 shown in Fig. 8 make it clear that the chances are negligible that this configuration could be accidental.

Since we do not know what intrinsic factors cause the variation in optical and radio properties of QSR's, it might be, *a priori*, that these four QSR's would not show any similarity or pattern to their properties. On the other hand, any similarities they do show, other than those to be expected from a random group of QSR's, would be simply additional confirmation of the physical reality of the configuration.

But there are, in fact, striking similarities of the properties of the QSR's shown in Fig. 8. We will come, in a few paragraphs, to consider the very important member of the line whose redshift is *z* = 2.11. For the

moment, however, let us consider only the three whose redshifts are *z* = 0.77, 0.67, and 0.72. This is a remarkable set of coincidences in redshift. To see just how improbable such a chance agreement in redshifts would be, we can consider the 28 QSR's in the SGH with *V* > 17.0 mag. Table III shows that only four out of these 28 redshifts occur in the range *z* = 0.67–0.77. But three of these four fall in the NGC 520 line!

There is one final consideration that fulfills a prediction of previous work on the ejection of QSR's from galaxies in such a way as to interconnect and confirm both the present results and the earlier work. One of the major results reported in Arp (1968b) was: of the radio sources which were ejected from galaxies (including QSR's), the more distant had "a preferentially flatter and weaker radio spectrum." At the time that *Astrofizika* paper was being prepared, Jon Sharp (1967), in response to those results, included the radio strength and radio index in his computer analysis of radio-source relationships to peculiar and Shapley–Ames galaxies. Sharp found, in agreement with the *Astrofizika* results, that both 3CR and Parkes radio sources showed a clear tendency to become weaker in flux strength and flatter in radio spectrum as their distance from both peculiar galaxies or Shapley–Ames galaxies became greater.

This result has been established from two previous, independent investigations, therefore, and it is extremely significant that this same behavior is now demonstrated again in the NGC 520 chain of QSR's. Proceeding from the QSR nearest NGC 520 to the one furthest, we see the flux strengths get generally weaker in the sequence: *S*(408) = 5.0, (4.5), 3.5, and 3.9. The spectral indices get flatter in the sequence: α = −1.1, −0.6, −0.7, and −0.3. Finally, there is a fifth QSR, indicated by a point in Fig. 8, that falls close to an extension of the QSR line. This is the only QSR in this region so far discovered in a strip from about Dec. = +2° to −4°, which has been the subject of a recent faint flux-level, high-frequency survey with the Parkes radio telescope (information kindly communicated in advance of publication by John Bolton). Since this QSR does not show up in previous radio surveys, it is undoubtedly fainter in radio flux than the four QSR's closer to NGC 520. It therefore fulfills the relationship of fainter radio fluxes falling farther from the ejecting galaxy. Since the new strip was surveyed at higher frequency, it may well be of flatter radio flux, also.

We now come to consideration of that particular QSR in the NGC 520 line with the high redshift of *z* = 2.11. The unusual property of this QSR is that its optical magnitude is almost 1 mag fainter than the average of the other three. Since all these QSR's must be at essentially the same distance from us, this result implies that the high-redshift one is of a relatively fainter intrinsic luminosity. We will comment on another substantiating example of this effect presently, but let us first consider a well-established but nevertheless puzzling property of QSR's which this result now explains.

It is well known that the number of QSR's per unit area on the sky increases as we consider higher redshifts. If we accept their redshifts as measures of their distance from us, the increase is much greater than would result from a uniform distribution of QSR's in space. But, at a redshift slightly greater than about $z=2$, the inferred space density stops increasing and drops abruptly to essentially zero (Schmidt 1968; Bolton 1969). The explanation advanced by Schmidt is that conditions in the recent cosmic past were more favorable for the formation of QSR's, but that in the far cosmic past they were highly unfavorable for the formation of QSR's. This sudden transition, as a function of past time, for conditions in large regions of the universe would seem to me to be not a very plausible model. The results here indicate, I believe, a much more satisfactory explanation for these accepted observational facts. If, as we have concluded, the intrinsic luminosity of QSR's drops rapidly for redshifts higher than about $z=2$, then we see abruptly smaller volumes of space at increasingly higher redshifts, and a cutoff in number density on the sky at this redshift is a natural consequence.

B. Other Concentrations

Aside from the close pairs discussed in Sec. III, the concentration of QSR's around NGC 520 is really the only strong grouping shown in any of the diagrams of QSR sky distributions. As we have seen, even the faint QSR's are spread out around their galaxies of origin to an order of magnitude of 10°. This causes considerable overlapping of distributions and prevents high density concentrations of QSR's from building up. There are two triplets of QSR's, however, that stand out. One is shown in Fig. 3(a) at R.A.=11^h, Dec.=15°. It consists of the following QSR's: 3C245, 1055+20, and 1116+12, whose properties are listed in Table V.

The flux strengths and particularly the radio indices are similar enough to confirm that, in fact, this group of radio sources is physically associated. The first two, in addition, have exceedingly similar V magnitudes and redshifts. Just as in the NGC 520 group, the redshift greater than $z=2$ is associated with a sharply fainter apparent magnitude, indicating a fainter intrinsic brightness. This triplet of QSR's coincides with a cluster of bright galaxies which includes NGC 3368–3379 and NGC 3623–3627.

There is another fairly isolated triplet on the eastern side of the southern extension of the Virgo Cluster. The QSR's involved are 3C279, 1317−00, and 1335−06. The properties of these QSR's resemble each other only moderately, however, and it is therefore not certain whether they are physically associated.

VI. ASSIGNMENT OF THE REMAINING, OPTICALLY BRIGHT QSR'S

The preceding sections have established that the faintest QSR's ($V > 17.0$ mag) are generally associated with galaxies brighter than about $m_{pg} = 12.5$ mag. We would expect the intermediate-brightness QSR's to be associated with even brighter galaxies, and, indeed, we saw evidence of their association with objects brighter than $m_{pg} = 10.0$ mag, such as the Virgo Cluster, M81, and NGC 1068. We would then expect the brightest QSR's to be associated with the brightest galaxies we have in our sky.

When the brightest QSR's are examined, the following properties emerge:

1. These brightest QSR's tend to appear in that half of the sky nearest M31.

2. Twelve of the 17 QSR's brighter than $V \leq 16.4$ mag show pairings in redshift values, and the lowest redshift pairs have almost exactly matching redshifts.

3. One of these 12 pairs appears to be the likeliest candidate for ejection from our own Galaxy.

4. In the remaining ten pairs, one pair member falls roughly 40°–80° on side of M31, and the other pair member falls an equal distance on the other side of M31.

It is considered that these brightest (in apparent magnitude) QSR's have been ejected on either side of M31 to distances of the order of from $\frac{1}{2}$–2 Mpc and have absolute magnitudes in the $M_v = -11$ to -12 mag range.

Details of the association of QSR's with local group members are being prepared for later publication.

ACKNOWLEDGMENTS

I would like to acknowledge the aid of J. F. Bartlett who designed the computer program that statistically analyzed some of the relationships discussed in the present paper. It is also a pleasure to thank J. G. Bolton who helped me to check the basic data used for the QSR's in this paper against his own extensive records.

REFERENCES

Arp, H. 1966a, *Atlas of Peculiar Galaxies* (California Institute of Technology, Pasadena); *Astrophys. J. Suppl.* 14, 1.
——. 1966b, *Science* 151, 1214.
——. 1967, *Astrophys. J.* 148, 321.
——. 1968a, *ibid.* 152, 633.
——. 1968b, *Astrofizika* 4, 59.
Barbieri, C., Battistini, P., and Nasi, E. 1967, *Publ. Oss. Astron. Padova*, No. 141.
Bennett, A. S. 1962, *Mem. Roy. Astron. Soc.* 68, 163.
Bolton, J. G. 1969, *Astron. J.* 74, 131.
Bolton, J. G., Shimmins, A. J., Ekers, J., Kinman, T. D., Lamla, E., and Wirtanen, C. A. 1966, *Astrophys. J.* 144, 1229.
Burbidge, E. M. 1967, *Ann. Astron. Astrophys.* 5.
Burbidge, E. M., Burbidge, G. R., and Sandage, A. R. 1963, *Rev. Mod. Phys.* 35, 947.

12   HALTON ARP

Day, G. A., Shimmins, A. J., Ekers, R. D., and Cole, D. J. 1966, *Australian J. Phys.* **19**, 35.
Edge, D. O., Shakeshaft, J. R., McAdam, W. B., Baldwin, J. E., and Archer, S. 1959, *Mem. Roy. Astron. Soc.* **68**, 37.
Ekers, R. D. 1967, Ph.D. thesis, Australian National University.
Lynds, C. R. 1961, *ibid.* **134**, 659.
Lynds, C. R., and Sandage, A. R. 1963, *ibid.* **137**, 1005.
Moffet, A. T. 1965, *ibid.* **141**, 1580.
Sandage, A. R. 1961, *Hubble Atlas of Galaxies* (Carnegie Institution of Washington, Washington, D. C.).
Schmidt, M. 1966, *Astrophys. J.* **146**, 7.
——. 1968, *ibid.* **151**, 393.

Shapley, H., and Ames, A. 1932, *Harvard Ann.* **88**, No. 2.
Sharp, J. R. 1967, M.S. thesis, San Diego State College.
Shimmins, A. J., and Day, G. A. 1968, *Australian J. Phys.* **21**, 377.
Shimmins, A. J., Clarke, M. E., and Ekers, R. D. 1966, *ibid.* **19**, 649.
Strittmatter, P., Faulkner, J., and Walmsley, M. 1966, *Nature* **212**, 1441.
Vaucouleurs, G. de 1959, "The Local Super Cluster of Galaxies," *Astron. Zh.* **36**, 977 [*Sov. Astron.—AJ* **3**, 897 (1960)].
Vaucouleurs, G. de, and Vaucouleurs, A. de 1964, *Reference Catalogue of Bright Galaxies* (University of Texas Press, Austin, Texas).
Wagoner, R. V. 1967, *Nature* **214**, 766.

communication among interested statisticians, scientific workers and philosophers and historians of science.

ALLAN BIRNBAUM

New York University,
Courant Institute of Mathematical Sciences,
251 Mercer Street,
New York, NY 10012.

Received November 28, 1969; revised January 19, 1970.

[1] Edwards, A. W. F., *Nature*, **222**, 1233 (1969).
[2] Birnbaum, A., *J. Amer. Stat. Assoc.*, **57**, 269 (1962).
[3] Birnbaum, A., in *Philosophy, Science and Method: Essays in Honor of Ernest Nagel* (edited by Morgenbesser, S., Suppes, P., and White, M.) (St Martin's Press, NY, 1969).
[4] *Likelihood* in *International Encyclopaedia of the Social Sciences* (Crowell-Collier, NY, 1968).

LETTERS TO THE EDITOR

GENERAL

Statistical Methods in Scientific Inference

IT is regrettable that Edwards's interesting article[1], supporting the likelihood and prior likelihood concepts, did not point out the specific criticisms of likelihood (and Bayesian) concepts that seem to dissuade most theoretical and applied statisticians from adopting them. As one whom Edwards particularly credits with having "analysed in depth . . . some attractive properties" of the likelihood concept, I must point out that I am not now among the "modern exponents" of the likelihood concept. Further, after suggesting that the notion of prior likelihood was plausible as an extension or analogue of the usual likelihood concept (ref. 2, p. 200), I have pursued the matter through further consideration and rejection of both the likelihood concept and various proposed formalizations of prior information and opinion (including prior likelihood). I regret not having expressed my developing views in any formal publication between 1962 and late 1969 (just after ref. 1 appeared). My present views have now, however, been published in an expository but critical article (ref. 3, see also ref. 4), and so my comments here will be restricted to several specific points that Edwards raised.

If there has been "one rock in a shifting scene" or general statistical thinking and practice in recent decades, it has not been the likelihood concept, as Edwards suggests, but rather the concept by which confidence limits and hypothesis tests are usually interpreted, which we may call the confidence concept of statistical evidence. This concept is not part of the Neyman–Pearson theory of tests and confidence region estimation, which denies any role to concepts of statistical evidence, as Neyman consistently insists. The confidence concept takes from the Neyman–Pearson approach techniques for systematically appraising and bounding the probabilities (under respective hypotheses) of seriously misleading interpretations of data. (The absence of a comparable property in the likelihood and Bayesian approaches is widely regarded as a decisive inadequacy.) The confidence concept also incorporates important but limited aspects of the likelihood concept: the sufficiency concept, expressed in the general refusal to use randomized tests and confidence limits when they are recommended by the Neyman–Pearson approach; and some applications of the conditionality concept. It is remarkable that this concept, an incompletely formalized synthesis of ingredients borrowed from mutually incompatible theoretical approaches, is evidently useful continuously in much critically informed statistical thinking and practice.

While inferences of many sorts are evident everywhere in scientific work, the existence of precise, general and accurate schemas of scientific inference remains a problem. Mendelian examples like those of Edwards and my 1969 paper seem particularly appropriate as case-study material for clarifying issues and facilitating effective

PHYSICAL SCIENCES

Redshifts of Companion Galaxies

SINCE 1966 (ref. 1) there has been observational evidence[2-4] that quasi-stellar radio sources (QSRs) and compact radio galaxies are ejected from galaxies. Whatever distance scale one uses, one of the characteristics of QSRs and compact radio galaxies is that a great deal of energy is compressed into very small dimensions. Unless very special conditions prevail, such concentrated sources would be expected to expand with time. Indeed, the short lifetimes which are inferred from their rapid rate of energy expenditure, the associated jets of material and the prevalence of ejected radio pairs across the sources all attest to expansion and eruption.

We would therefore expect that when a compact object which is initially not optically resolvable (that is, a QSR or a QSO) expands sufficiently it eventually must become resolvable (a compact galaxy). As it continues to expand we would expect the surface brightness to diminish and eventually the object should look only slightly more compact than a galaxy of normal surface brightness. I have recently reported results[5] which lead to the conclusion that compact galaxies and companion galaxies in general have also been ejected from large, central galaxies. The observations also indicate that the ejections are initially very compact and rapidly expand. The model which suggests itself is one where matter is ejected in a compact, QSR-like form, which evolves into a compact radio galaxy and finally into a less and less compact companion galaxy. (This is, of course, an evolutionary interpretation. As far as the observational evidence goes, one may simply wish to consider that smaller bodies, with a continuity of properties that reaches to the extremes of QSRs, are associated with large, normal galaxies.)

It is usually assumed in astronomy that galaxy redshifts are only due to Doppler, velocity shifts. Objects which differ greatly in redshift are always presumed to be at different distances so as not to violate the (expansion) velocity–distance relation. The only way such an assumption could be shown to be incorrect would be to demonstrate that objects of greatly different redshifts cluster together on the sky. The observational evidence on the QSRs and radio compact galaxies referred to simply argues that they do in fact cluster on the sky around lower redshift galaxies and tend to be aligned and paired across these galaxies.

The proposed continuity of characteristics of the ejected compact bodies, however, suggests another kind of test on the nature of their redshifts. We can investigate smaller companions around galaxies, companions which astronomers have long accepted as belonging to the central galaxy and being at essentially the same distance as the larger galaxy. If the intrinsic redshift evolves from high to low as ejected objects evolve toward brighter and less compact companions, then we would expect even the

NATURE VOL. 225 MARCH 14 1970

normally accepted companions to have at least some small residual, excess redshift left. Even if there is no evolution involved and it is simply the case that objects of various redshifts are ejected, we would still expect the companion galaxies to have somewhat higher redshifts, on the average, than the parent galaxy.

In order to test this prediction I have listed in Table 1 all the galaxies having companions which are unquestionably accepted as belonging to the larger galaxies. The Andromeda Nebula and the four well known companions that fall within 15 arc degrees are listed, followed by the M81 and NGC 5128 groups with the members as defined by de Vaucouleurs[12]. In eleven out of these thirteen cases the companion galaxies have higher redshifts than the central galaxies.

The data for the first three groups are taken from redshifts, corrected for galactic rotation, listed by de Vaucouleurs and de Vaucouleurs[6]. These redshifts represent, for the most part, means of two or more optical determinations. The independent determinations usually agree within about 20 km s⁻¹. In addition, there are redshifts from radio (21 cm) observations for five of the objects which confirm the optical redshifts to within about this same accuracy. Mean errors of the redshifts in the first part of Table 1 may be obtained from the weights listed in column 23 of the de Vaucouleurs Reference Catalogue. Of course, the errors of measurements, like orbital or ejection velocities around the central galaxy, will simply spread the distribution of redshifts around their mean value. The important point, as Fig. 1 indicates, is whether the mean residual redshift is significantly different from zero in terms of the dispersion regardless of the cause of that dispersion.

Also listed in Table 1 are six spiral galaxies with companions on the ends of arms[5]. These can be included because the connexion to the companion galaxy ensures that the companion is at the same distance as the main galaxy. The listed redshifts represent all that I know of in this class of galaxies.

Altogether the redshift of the companion galaxy is higher than that of the main galaxy in sixteen out of nineteen cases in Table 1. This could occur by accident in less than two out of a thousand chances. If we were to continue to interpret the entire redshift of each companion as a velocity shift, we would have an absurd situation where the companions were systematically receding from the main galaxy along our line of sight to that main galaxy.

It should be emphasized that Table 1 represents a conservative assignment of companion galaxies. For

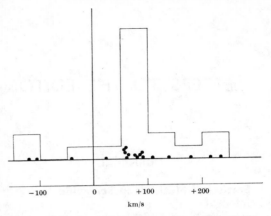

Fig. 1. Residual redshifts of nineteen galaxies known to be companions of larger galaxies.

example, NGC 404 (not listed in Table 1) is situated on a line between M31 and M33, and is believed to be a companion of M31 (refs. 4 and 12). Its redshift is + 237 km s⁻¹ greater than M31.

Of the three negative redshifts, the two most negative are questionable for the following reasons. (1) The companion to Atlas 49 registers only emission lines which may well come from material in the disk of the parent galaxy excited by the object which appears to be in the process of ejection. There is therefore no assurance that the measured redshift really represents the intrinsic redshift of the compact companion. (2) NGC 3077 forms a pair with M82 on either side of M81. If we consider them to be an ejected pair, the mean of their redshifts is + 65 km s⁻¹. This is just about the expected redshift of these kind of companions as judged from Fig. 1. The remaining redshift would then represent a separation of about 170 km s⁻¹ in mutually opposite directions from M81.

Table 1 essentially exhausts those cases where redshifts are known in groups where we are certain that the smaller companions belong to the large central galaxies. If we try to extend this list we encounter groups where there are a number of galaxies of comparable size and therefore it is not certain which are the companions and which are the main galaxies. We also encounter looser groups in which membership is less certain. Of the five nearest groups which de Vaucouleurs[12] lists, three are represented in Table 1. Of the remaining two, the Sculptor Cloud is a loose group of comparable Sc to Sm nebulae. (Nevertheless the faintest member of the group, NGC 45, has a redshift about 400 km s⁻¹ greater than the group mean.) The other group, the Canes Venatici I Cloud, is very heterogeneous and overlaps with various, presumably more distant, groups in that general direction in the sky.

Although, as I have mentioned, clusters of galaxies represent a somewhat different situation than dominant galaxies with companions, if we investigate some of the fainter clusters we find further evidence for fainter galaxies in the clusters tending to have higher redshifts. Holmberg[7] demonstrated that there is an increasing redshift with increasing faintness among galaxies belonging to the Virgo cluster. It is well known that late type (young population) spiral galaxies in Virgo tend to have considerably higher redshifts than the mean of the cluster[13]. Even when we exclude Sc spirals, however, the remaining Virgo cluster galaxies still show a correlation of redshift with apparent magnitude (unpublished results).

The same situation is true with respect to the four largest groups listed in Humason, Mayall and Sandage[8]

Table 1. GALAXIES KNOWN TO BE COMPANIONS OF LARGER GALAXIES

Dominant galaxy	Companion galaxy	Differential redshift (km s⁻¹)
M31	M32	+ 85
	NGC 205	+ 62
	NGC 185	+ 58
	M33	+ 57
M81	M82	+ 234
	NGC 2976	+ 81
	NGC 3077	− 104
	IC 2574	+ 91
	HO II	+ 215
NGC 5128	NGC 5102	+ 77
	NGC 5236	+ 64
	NGC 5253	− 42
	NGC 5068	+ 139
M51	NGC 5195	+ 109
Atlas 48	Companion	− 120
Atlas 58	Companion	+ 60
Atlas 82	Companion	+ 90
Atlas 86	Companion	+ 23
Atlas 87	Companion	+ 180

(G383, Leo, G5846 and G7619). The first of these groups in particular, the galaxies involved in the Markarian chain centred on NGC 383, shows a marked correlation of higher redshifts with fainter magnitudes. Recently Sargent[9] reported one of the fainter galaxies in another chain to have an excess redshift of considerable magnitude, and earlier I argued[10] that IC4296 was a member of the NGC 5128 chain despite a considerably higher redshift.

The slope of the excess redshift–magnitude relation looks steeper (in km s^{-1} per magnitude) for clusters than it does for the dominant-galaxy-with-companions case treated in Table 1. It might have something to do with the possibility of larger galaxies fragmenting[11,14], which would make each original, low redshift fragment fainter and thus steepen the relation. It may also be related to the much higher dispersion of redshifts which is encountered in the more distant clusters. Much higher redshift objects are therefore included as members of clusters and the slope can thus be much steeper. The higher dispersion of velocities, of course, makes the cluster membership criterion generally more dependent on areal projection on the cluster area. The generally small differences in redshift between the main galaxies and companions in Table 1, on the other hand, are added confirmation of their accepted group membership.

In conclusion, the systematically higher redshifts for the companion galaxies supports the conclusion that compact, intrinsically higher redshift objects are ejected from nearby galaxies. At the same time it contradicts the current assumption that galaxy redshifts are only due to Doppler velocities. That is, it observationally contradicts exactly that principle which has been invoked to reject the previous observational evidence for association of high and low redshift objects.

Halton Arp

Mount Wilson and Palomar Observatories,
Carnegie Institution of Washington,
California Institute of Technology,
Pasadena, California 91106.

Received December 11, 1969.

[1] Arp, H., Science, 151, 1214 (1966).
[2] Arp, H., Astrophys. J., 148, 321 (1967).
[3] Arp, H., Astrophyzika, 4, 59 (1968).
[4] Arp, H., Astron. J. (in the press).
[5] Arp, H., Astron. Astrophys. (in the press).
[6] de Vaucouleurs, G., and de Vaucouleurs, A., Reference Catalogue of Bright Galaxies (Univ. Texas Press, Austin, 1964).
[7] Holmberg, E., Astron. J., 66, 120 (1961); Uppsala Astr. Obs. Medd. No. 138.
[8] Humason, M. L., Mayall, N. U., and Sandage, A. R., Astron. J., 61, 97 (1956).
[9] Sargent, W. L. W., Astrophys. Lett., 153, 135 (1968).
[10] Arp, H., Pub. Astron. Soc. Pacific, 80, 129 (1968).
[11] Sersic, J. L., Astron. Soc. Pacific Leaflet No. 453 (1967); Astrophyzika, 4, 105 (1968).
[12] de Vaucouleurs, G., Stars and Stellar Systems (edit. by Kuiper, G., and Middlehurst, B.), 9 (1969).
[13] de Vaucouleurs, G., and de Vaucouleurs, A., Astron. J., 68, 96 (1963).
[14] Vorontsov-Velyaminov, B., Astron. Soc. Pacific Leaflet No. 458 (1969).

Low-frequency Spectrum of the Crab Nebula

THE compact low-frequency radio source[1] in the Crab nebula is exceptional in most of its properties. Its spectral index[2] of ~1·75 between 26·5 and 1,407 MHz is the highest known for a galactic continuum source, and its brightness temperature of ~10^{14} K at 38 MHz implies an unusual emission mechanism with a high volume emissivity. The high spectral index suggests that only a small fraction of the total radio luminosity of the source may have been observed in the frequency range studied so far. The error field of its radio position[3] contains the pulsar NP 0532, and these objects may therefore be related; if the relationship were substantiated, it would be the only continuum source of small angular size known to coincide with a pulsar. This report presents evidence that the high spectral index is maintained to 10 MHz.

Observations of the flux density of radiation from the Crab nebula at frequencies below ~40 MHz must be corrected for absorption by free-free transitions of electrons in the intervening interstellar medium to determine the spectrum of the emitted radiation. The interstellar absorption along the 2,020 pc line of sight[4] to the nebula ($l^{II} = 184°·6$, $b^{II} = -5°·8$) can be estimated from low-frequency studies of other continuum radio sources at low galactic latitudes, and also from 21 cm line observations of the interstellar medium on this line of sight.

Observations[5] of discrete radio sources at 10 MHz have recently been used to define models of the low frequency absorption by the interstellar medium. The 10 MHz optical depth of the line of sight to the Crab nebula predicted by a uniform-disk model[6] of the interstellar electron gas is $\tau(10) = 0·68 \pm 0·17$. A non-uniform model[7] which allows for differences in the absorption coefficient $\varkappa(10)$ between the local spiral arm and the inter-arm region predicts $\tau(10) = 0·62 \pm 0·17$ on this line of sight. These predictions are derived from the same 10 MHz data and are not independent; the agreement between them indicates that the expected value of $\tau(10)$ on this line of sight is only weakly dependent on the model used to interpret the 10 MHz data.

The optical depth $\tau(10)$ of the line of sight to IC443, which is 1,500 pc (ref. 8) from the Sun in the direction $l^{II} = 189°·0$, $b^{II} = +3°·2$, should be similar to that of the line of sight to the Crab nebula and the value observed[7], $\tau(10) = 0·65 \pm 0·25$, agrees with the other estimates.

Hjellming et al.[9] have estimated the electron density and temperature within interstellar clouds and the inter-cloud medium on the line of sight to the Crab nebula from the 21 cm absorption of the nebular continuum. The optical depth $\tau(10)$ calculated from their data on this line of sight is $0·42 \pm 0·10$, with approximately equal contributions from clouds at ~50 K and from the inter-cloud medium at ~1,000 K.

Table 1. CORRECTIONS FOR INTERSTELLAR ABSORPTION AT LOW FREQUENCIES

Frequency (MHz)	Measured intensities multiplied by
10	1·82
12·5	1·46
14·7	1·31
16·7	1·23
20	1·15
22·25	1·12
25	1·09
26·3	1·08
35	1·04
38	1·04
81·5	1·01

Observations of the integrated flux density of the Crab nebula at low frequencies[5,10-15] have been corrected for interstellar free-free absorption by assuming that $\tau(10) = 0·6$ on the line of sight, and calculating the corresponding optical depths at higher frequencies from the relation $\tau(\nu) = k\nu^{-2}[17·7 + \ln(T^{3/2} \nu^{-1})]$. The corrections at each frequency are given in Table 1. Fig. 1 shows the estimated spectrum of the radiation leaving the nebula between 10 and 15,500 MHz. The spectrum shows an enhancement of the total radiation at low frequencies which is consistent with a power-law extrapolation of the spectrum of the compact source to 10 MHz; a marginally significant enhancement is noticeable in the uncorrected data.

(Reprinted from Nature, Vol. 233, No. 5314, pp. 41–44, September 3, 1971)

On the Nature of Mass

F. HOYLE & J. V. NARLIKAR

Institute of Theoretical Astronomy, University of Cambridge

The increasing number of observations of discrepant redshifts means that no longer can these be passed off as chance juxtapositions. A possible explanation of the data is given here in terms of a theory that incorporates a gravitational "constant" that is decreasing with time.

EVIDENCE concerning "discrepant redshifts" has accumulated fairly rapidly in recent months. The case where it is hardest to deny the evidence was reported at the beginning of the present year by Arp[1]. The galaxy NGC 7603 was found to be connected to an appendage by two arms which intersect at the appendage. The redshift of the galaxy is $z = 0.029$ and that of the appendage $z = 0.056$. In the customary Doppler interpretation of the redshift this difference exceeds 8,000 km s^{-1}. A redshift difference of the same order was found many years ago[2] by Minkowski, Humason and Zwicky in "Zwicky's triple system". The galaxies in this system are also connected by a bridge.

If it is accepted that a discrepancy of ~8,000 km s^{-1} exists for NGC 7603, then there seems no good reason for denying the two cases with discrepancies ~20,000 km s^{-1} obtained by Sargent (ref. 3, and W. L. W. Sargent's contribution to the Vatican Conference, 1970). Both the latter cases are compact groups of galaxies in which one member has the discrepant redshift. Less than ten such groups have been examined. The probability of a more distant galaxy being projected against a nearer cluster by chance is small. Arp (private communication) has obtained what appears to be a jet connecting the quasi-stellar object Makarian 205, $z \simeq 0.07$, to the galaxy NGC 4319. The discrepancy is again of order 20,000 km s^{-1}.

Three cases have been observed of a quasi-stellar object apparently associated with a small compact cluster of faint galaxies. In one case Gunn has reported[4] the same redshift for the QSO as for one of the galaxies of the cluster. In the other two cases there appear to be large redshift differences (Hazard, Jauncey, Sargent and Gunn, private communication).

In a recent preprint Burbidge, Burbidge, Solomon and Strittmatter have drawn attention to a bridge which appears on the Palomar sky prints to connect the quasi-stellar object PHL 1226 to the bright galaxy IC 1746. The redshift of PHL 1226 is $z = 0.404$, whereas that of the galaxy has not yet been measured, but must be small because of its brightness.

Because all objects are projected on the sky there must be some apparently peculiar juxtapositions of objects that are really at very different distances. This has hitherto led to a situation in which all but a very few astronomers have dismissed apparent redshift discrepancies as simply unusual projection effects. Yet there has to be a point of balance in one's judgment as to the extent of the chance juxtapositions one is willing to accept. One of us was aware at an early stage of the case of Zwicky's system (Zwicky, private communication) but it seemed that, in spite of the bridge, there might be a very peculiar chance juxtaposition. But the case of VV 172[3] formed the point of balance and NGC 7603 has turned the balance.

If a highly convincing theory of discrepant redshifts were available then we think there is little doubt that the data would today be considered reasonably clear-cut. It has been the absence of such a theory that has caused most astronomers to prefer to believe in unusual projection effects. If, as seems very possible, the accumulation of data forces us over a watershed (not only in our thinking but in the history of astronomy) it will clearly become necessary to arrive at a theory of discrepant redshifts. We wish to emphasize the need for a thoroughly radical assessment of the problem, considering it unlikely that a satisfactory theory will be achieved by a small change in our concepts. Explicitly, we do not think discrepant redshifts will be explained adequately either as simple Doppler peculiar motions or as excess reddening due to gravitation.

In the following discussion we shall present the usual Friedmann cosmological models and the Einstein de Sitter case in particular, from an unusual point of view. Mathematically, this point of view is at first entirely equivalent to the usual formulation. Then at a later stage we shall arrive at a possible major shift from the usual theory. The balance between the usual theory and an entirely different view both of physics and astrophysics hinges on a question raised by Dirac[5] in 1937.

It is well known that when c and $\hbar$ are set equal to unity, only a single dimensionality is needed for the whole of physics. We take this to be length and we denote the unit by L. Every quantity has a dimensional form L^n, for example, mass $\sim L^{-1}$, frequency $\sim L^{-1}$, charge $\sim L^0$, magnetic field $\sim L^{-2}$ and the gravitational 'constant' has dimensionality L^2.

Every observation is concerned with a dimensionless number, so that every observation is concerned with a product of quantities such that the dimensional dependencies on L cancel to zero. So far as experimental physics and engineering are concerned it is possible to convert a quantity of dimensionality L^n into a quantity of dimensionality L^m by means of a linear device provided $n = m$. Non-linear devices are needed if $n \neq m$. This property makes it comparatively easy to see what kind of physical device is needed to relate one quantity to another. There is no doubt that physics and engineering are made quite unnecessarily complicated by the current practice of using multiunit systems.

Because observed quantities are dimensionless they are unaffected by a scale change in the length unit. But what if we elect to change the length unit in accordance with a well behaved function $\Omega(X)$ (no zeros of Ω as well as no infinities) which varies with the position of the space-time point X? Nothing should be changed provided that first, we always combine physical quantities at the same X, and, second, every physical quantity is scaled according to its dimensionality —a quantity of dimensionality L^n being scaled by Ω^n.

Physical theories with these properties are said to be conformally invariant. This is a different kind of invariance from the coordinate invariance of relativity. Not all physical theories, however, are conformally invariant. Maxwell's theory is conformally invariant. The derivative terms of Dirac's equation are conformally invariant, but the mass term is not. Einstein's theory is not conformally invariant and physical theories cease to be conformally invariant whenever "mass" is involved. This is because the second property is then not satisfied. Clearly, we cannot expect classical dynamics to be conformally invariant unless the classical action

$$S = -\int m \, da \qquad (1)$$

of particle a is invariant. Since the element da of the path of the particle becomes multiplied by Ω we would require the mass m to be multiplied by Ω^{-1}, which indeed would accord with the dimensionality of mass, $\sim L^{-1}$. But Ω can now be different at different space–time points, so that in general the mass of the particle will not be the same for all points on its path. Plainly then, we cannot expect to arrive at a conformally invariant system of dynamics so long as the mass of a particle is considered a fixed quantity belonging autonomously to the particle.

We have been concerned[6,7] since 1964 in developing a conformally invariant theory of dynamics and of gravitation. The central idea of this work is that the mass of a particle m say, is given by

$$m(X) = \lambda \, M(X) \qquad (2)$$

where λ is a coupling constant belonging to the particle itself and $M(X)$ is a "mass field" generated by all the particles in the universe. In a very general way we can write

$$M(X) = \sum_b \int P(X, B) \, db \qquad (3)$$

the summation being over all particles b, the line integral being over the world line of b, and $P(X, B)$ being a propagation function from point B at db to the field point X. To avoid self-action problems in the classical theory it is convenient to add the refinement that in determining the mass of any particular particle a this particle is omitted in the summation with respect to b. A comparable situation occurs in classical electrodynamics.

Mathematically we require that when all lengths on the right hand side of (3) are changed by the function Ω the resulting mass field is multiplied by $\Omega^{-1}(X)$. This requirement is met by putting

$$P(X, B) = \lambda \, \tilde{G}(X, B) \qquad (4)$$

where $\tilde{G}(X, B)$ is the time-symmetric elementary solution of

$$\Box_x \tilde{G}(X, B) + 1/6 \, R(X)\tilde{G}(X, B) = [-g(B)]^{-1} \, \delta_4(X, B) \qquad (5)$$

in which $\Box_x$ is the d'Alembertian operator, R the scalar Riemannian curvature and δ_4 the 4-dimensional Dirac delta-function. So far as we are aware the wave-equation (5) was first considered with respect to its conformal properties by Penrose.

The action of particle a is now

$$-\lambda^2 \int da \cdot \sum_{b \neq a} \int \tilde{G}(A, B) \, db \qquad (6)$$

and the total action of all particles is given by

$$S = -\tfrac{1}{2} \lambda^2 \sum_a \sum_b \iint \tilde{G}(A, B) \, da \, db \qquad (7)$$

Subtleties arise for a universe with infinitely many particles concerning the finiteness of (7), but the situation in this respect is not really different from the usual theory.

Enough has already been done to make the Dirac equation conformally invariant, including the mass term. To obtain conformally invariant gravitational equations we work with the usual condition that S be stationary for small changes of the metric tensor, $\delta S = 0$ for $g_{ik} \to g_{ik} + \delta g_{ik}$. The resulting equations are different and more complicated than those of Einstein. For cosmology however, and for all problems concerned with weak gravitational fields, the equations can be reduced to the Einstein form with an exceedingly high degree of approximation. The situation in regard to the classical tests of general relativity—the bending of light, the perihelion of Mercury—is therefore exactly the same as with Einstein. To effect the reduction to Einstein's equations one chooses the conformal frame so that $M(X)$ becomes independent of X. If, to begin with, $M(X)$ varies with X we simply choose $\Omega \propto M$. Particle masses are independent of X in this particular conformal frame, just as we normally take them to be.

The gravitational 'constant' comes out to be given by

$$G = \frac{3}{4\pi \, M^2} \qquad (8)$$

Again in the special conformal frame in which M is constant, G is indeed a constant. It is interesting that G is necessarily positive, so that weak gravitational fields are necessarily attractive.

It should be added that the gravitational equations do not reduce to the Einstein form in certain cases where the field is strong. The discussion of "black holes" is greatly changed, for example.

Because the gravitational theory is conformally invariant we can now effect a major simplification in discussing the isotropic homogeneous models usually employed in cosmology. These models are usually discussed with respect to the Robertson–Walker form of the line element

$$ds^2 = d\tau^2 - Q^2(\tau) \left[\frac{d\rho^2}{1 - k\rho^2} + \rho^2 (d\theta^2 + \sin^2\theta \, d\varphi^2) \right] \qquad (9)$$

Here τ is cosmic time and $Q(\tau)$ is the expansion function. The coordinates ρ, θ, φ are spherical polars for an observer at the origin. The constant k can be zero or ± 1. $k = 0$ gives the Einstein de Sitter model, $k = +1$ is closed, $k = -1$ hyperbolic. It is well known that a conformal transformation function Ω can be found that changes the line element (9) into flat Minkowski space.

It is not usually possible to take advantage of this geometrical simplification because the physical theory is not conformally invariant. Here, however, the physics is conformally invariant and we can transform to Minkowski space. When we have done so the theory is as good as it was in the Robertson–Walker representation. It turns out that while the $k = 0$ case is still homogeneous in the Minkowski space

$$dt^2 - dr^2 - r^2 (d\theta^2 + \sin^2\theta \, d\varphi^2) \qquad (10)$$

the cases $k = \pm 1$ are simply localized clouds of no special interest.

The full Minkowski space, $t < 0$ as well as $t > 0$, can be used. There is, however, nothing corresponding to both halves of Minkowski space in the Robertson–Walker representation. The reason for this emerges when we make the inverse transformation from (10) to (9). The inverse transformation has a singularity at $t = 0$, and this introduces a singularity at $\tau = 0$ in (9). This singularity, which now emerges as a mathematical construct, is usually interpreted as the 'origin' of the universe. There need be no 'origin' in the Minkowski representation.

Taking the particle masses to be constant in the Robertson–Walker frame, and applying Ω^{-1} to the mass, Ω being chosen for transformation from (9) with $k = 0$ to (10), we obtain

$$m = (\text{constant}) \cdot t^2 \qquad (11)$$

The particle masses are a quadratic function of cosmic time t in the Minkowski representation. Even though space is now flat this leads to a redshift effect. Remembering that atomic sources emit frequencies proportional to m, we see that such a source at distance r from an observer at $r = 0$ would be judged to have a redshift given by

$$1 + z = \left(\frac{t}{t - r} \right)^2 \qquad (12)$$

the time of observation being t and the time of emission $t - r$.

A standard galaxy may be considered to have a specified number of stars each containing a specified number of particles arranged in the same way with respect to chemical composition. The luminosity of a star has dimensionality L^{-2}. This is proportional to m^2, and therefore by (11) proportional to the fourth power of the cosmic time. The bolometric flux from such a standard galaxy observed at time t is therefore

$$\frac{L(t - r)}{4\pi r^2} \propto \frac{(t - r)^4}{r^2} \qquad (13)$$

where $L(t - r)$ is the luminosity of the galaxy at the time $t - r$ of emission of the observed radiation. Eliminating r from (13) with the aid of (12) we find that the bolometric flux depends on z according to

$$\frac{1}{1 + z} \cdot \frac{1}{[\sqrt{1 + z} - 1]^2} \, . \qquad (14)$$

which is the classic Hubble redshift-magnitude relation for the Einstein de Sitter model. The reader will note how much more easily we have arrived at (14) than in the usual treatment.

The present treatment is still deficient in one important respect. The form of equation (11) for the particle mass was obtained for the Minkowski representation by assuming constant mass in the Robertson–Walker representation. To make the argument complete it is essential to show that (11) follows from an explicit evaluation of (2), (3), (4), (5). This evaluation turns out to involve interesting problems which will not be considered here, since details have been given elsewhere[8]. It is convenient to choose the length unit L so that the particle density is everywhere unity. We then obtain from (2), (3),(4), (5),

$$m = \tfrac{1}{2}\lambda^2 t^2 L^{-3} \qquad (15)$$

which has the required quadratic dependence on t.

Expanding (12) for $r/t \ll 1$ we obtain

$$z \simeq 2r/t \qquad (16)$$

Since the Hubble constant H is defined by $z = Hr$ for small r, we have

$$t = 2H^{-1} \qquad (17)$$

for the present epoch. Empirically $H^{-1} \simeq 2 \times 10^{28}$ cm. Empirically also, if we consider the particles to be hydrogen atoms, the average particle density n is given by

$$n = L^{-3} \simeq 2.5 \times 10^{-6} \text{ cm}^{-3} \qquad (18)$$

From (17), (18), and $H^{-1} = 2 \times 10^{28}$ cm,

$$nt^3 \simeq 10^{80} \qquad (19)$$

This cosmological number is dimensionless, and is the number of particles contributing to the mass field $M(X)$ at the present epoch.

The dimensionless number λ^2 has no effect on gravitation, since it multiplies the action (7), and therefore disappears entirely from $\delta S = 0$, $g_{ik} \rightarrow g_{ik} + \delta g_{ik}$. But the value of λ^2 has a profound effect on the behaviour of individual particles when electromagnetic terms are added to the action. The value of

λ^2 can be obtained empirically in the following way. Applying (15) to protons,

$$m_p = \tfrac{1}{2}\lambda^2 t^2 L^{-3} = 2\lambda^2 H^{-2} L^{-3} \qquad (20)$$

for the present epoch. The empirical values (18) for L^{-3} and $H^{-1} \simeq 2 \times 10^{28}$ cm then lead to

$$m_p \simeq 2 \times 10^{51} \; \lambda^2 \text{ cm}^{-1} \qquad (21)$$

The laboratory value of the Compton wavelength of the proton m_p^{-1} is $\sim 10^{-14}$ cm, so that empirically $m_p \simeq 10^{14}$ cm^{-1}, and

$$\lambda^2 \simeq 5 \times 10^{-38} \qquad (22)$$

is therefore required by atomicity.

The remarkable fact now emerges that

$$\lambda^2 (nt^3)^{\frac{1}{3}} = O(1) \qquad (23)$$

On the left λ^2 apparently belongs autonomously to the particle, whereas $(nt^3)^{\frac{1}{3}}$ is a cosmological number. Apparently also λ^2 and n are independent of t, so that (23) would seem only a coincidence of the present epoch. This is the usual view.

At this point we reach our watershed. How if we require that (23) shall apply at all t? Then λ^2 or n, or both, must depend on t. This was the proposal of Dirac[5]. It is clear that (23) can be maintained by $n\lambda^4 \propto t^{-3}$, which allows many possible combinations of n, λ^2. Here we discuss a possibility which involves no change in the mass field $M(X)$. This possibility is similar in some respects to the case discussed by Dirac but is different in other respects. According to (2), (3), (4),

$$M(X) = \lambda \sum_b \int \tilde{G}(X, B) \, \mathrm{d}b \qquad (24)$$

from which it is not hard to see that provided the product λn is independent of t there will be no change in $M(X)$. From

$$\lambda \propto n^{-1}, \; n\lambda^4 \propto t^{-3} \qquad (25)$$

we then get

$$\lambda \propto t^{-1}, \; n \propto t \qquad (26)$$

Moreover, the total action (7) is the same as before. It is as if more particles are being created by arranging that the masses of individual particles, $m = \lambda M$, shall increase linearly with t, instead of as t^2. In so far as we discuss the gravitational aspects of cosmology in terms of a smooth fluid, instead of in terms of discrete particles, nothing is changed. The interaction of each particle with the mass field M weakens as t^{-1}, but there are more particles because of $n \propto t$, and the total interaction of all particles remains the same.

An important by-product of these considerations is that the universe now becomes a perfect absorber in the sense of Wheeler and Feynman. The original Einstein de Sitter model did not meet the perfect absorber condition, which in our view is a fatal flaw of this model. The present model is not the steady state model, but by moving some way towards the steady state model, by permitting $n \propto t$, we have managed to meet an essential electrodynamic requirement.

It is possible to make a conformal transformation $\Omega \propto t$ from the Minkowski representation to a representation in which individual particle masses are constant, as we consider them to be in all terrestrial and astrophysical problems. The line

$$\mathrm{d}s^2 = \mathrm{d}T^2 - 2T[\mathrm{d}r^2 + r^2(\mathrm{d}\theta^2 + \sin^2\theta \; \mathrm{d}\varphi^2)] \qquad (27)$$

which is the same as the line element of a radiation-dominated Friedmann model. The present model, however, is not radiation-dominated and it has a gravitational 'constant' that varies with time. In this representation the mass field M is proportional to $T^{\frac{1}{2}}$ and G, given by (8), is proportional to T^{-1}. Such a variation would have a profound effect on astrophysics and geophysics.

Stars would be much brighter at early T than we usually suppose them to be. There are advantages to be gained from such behaviour, but $G \propto T^{-1}$ may be too drastic[9], for example, in its effect on the past luminosity of the Sun. On the other hand, there are aspects of geophysics that seem as if they would be greatly helped by this kind of dependence. Steadily weakening gravity would gradually release the interior of the Earth from compression. It can be calculated[8] that the radius of the Earth would be required to increase at about 10 km per 10^8 yr. There is no possibility of this expansion being resisted by the crust, which must be cracked open repeatedly to make way for new surface material. At all times there would be an excess upward force on the crust at the limit of its strength. The possibility of large horizontal pressure differences, of order 10^9 dyne cm^{-2}, also exists, provided in particular regions that excess pressure is conveyed to the immediate subsurface by fluid material.

We are reminded in this connexion of the old controversy concerning continental drift. Our impression is that, while modern evidence shows unequivocally that drift actually takes place, the early calculations demonstrating the need for exceedingly large forces, really remain valid. If this is so, we would be inclined to think that some such behaviour of G as is given by our model becomes essential for an understanding of the geophysical evidence.

It may be added that Shapiro, Smith, Ash, Ingalls and Pettengill[10] have recently placed an observational upper limit of 4×10^{-10} yr^{-1} on G/G. The variation expected here, with a Hubble constant of 5×10^{-11} yr^{-1}, would be 10^{-10} yr^{-1}.

We turn finally to the problem of how we should interpret $\lambda \propto t^{-1}$ in relation to the redshift problems discussed at the outset. The proportionalities $n \propto t$, $\lambda \propto t^{-1}$ maintain cosmological homogeneity. We must contemplate, however, that these proportionalities represent the smoothed-out effect of changes that could possess spatial irregularities. There could be spatially adjacent variations of λ. This we feel to be the kind of concept needed to come to grips with the problem of redshift anomalies. The smoothed-out homogeneous behaviour of the particle masses gives the normal cosmological redshift, obeying the usual Hubble relation. Local variations give anomalies. A more rapid decrease locally of λ is required in order that the anomalies be in the sense of an increased redshift. One can speculate that such variations occur as a consequence of the physical conditions in regions of strong gravitational fields. These are characterized by the condition that local particles make a contribution to the total mass field M that is comparable with the contribution of particles at cosmological distances. Such localities may be able to produce their own environmental conditions, leading to the local variations of λ.

Received July 27, 1971.

[1] Arp, H. C., *Astrophys. Lett.*, **7**, 221 (1971).
[2] Zwicky, F., *Naturwissenshaften*, **29**, 344 (1956).
[3] Sargent, W. L. W., *Astrophys. J. Lett.*, **153**, L135 (1968).
[4] Gunn, J., *Astrophys. J. Lett.*, **164**, L113 (1971).
[5] Dirac, P. A. M., *Nature*, **139**, 323 (1937).
[6] Hoyle, F., and Narlikar, J. V., *Proc. Roy. Soc.*, A, **294**, 138 (1966).
[7] Hoyle, F., and Narlikar, J. V., *Ann. Phys.*, **62**, 44 (1971).
[8] Hoyle, F., and Narlikar, J. V., *Mon. Not. Roy. Astron. Soc.* (in the press).
[9] Pochoda, P., and Schwarzschild, M., *Astr. J.*, **139**, 587 (1964).
[10] Shapiro, I. I., Smith, W. B., Ash, M. B., Ingalls, R. P., and Pettengill, G. H., *Phys. Rev. Lett.*, **26**, 27 (1971).

THE ASTROPHYSICAL JOURNAL, **170**:233–240, 1971 December 1

APPARENT ASSOCIATIONS BETWEEN BRIGHT GALAXIES AND QUASI-STELLAR OBJECTS

E. M. BURBIDGE, G. R. BURBIDGE, P. M. SOLOMON, AND P. A. STRITTMATTER
Department of Physics, University of California, San Diego
Received 1971 June 14; revised 1971 August 30

ABSTRACT

A comparison of the spatial distribution of the 47 identified QSOs in the 3C and 3CR catalogs with the small-redshift galaxies contained in the *Reference Catalog of Bright Galaxies* shows that four QSOs with redshifts in the range 0.5–1.4 are much closer to bright galaxies than would be expected if the 47 QSOs were distributed randomly. An extensive analysis of the distributions shows that the probability that this is a chance occurrence is less than 5×10^{-3}. It is also shown that the radio-quiet QSO PHL 1226 with a redshift of 0.4 lies very close to the comparatively nearby galaxy IC 1746 to which it appears to have a physical connection.

I. INTRODUCTION

Two methods are currently available to determine the distances to faint extragalactic sources of unknown intrinsic brightness. The first is to measure a redshift and, assuming that it is due to the expansion of the Universe, obtain a distance by using the Hubble constant H_0 (and a chosen value of the deceleration parameter q_0 if the redshift is large). The second is to establish that the source is physically associated with a second object whose distance is known. It has recently been shown that some objects with highly disparate redshifts appear to be physically associated. Examples are (*a*) NGC 7603 and its companion (Arp 1971*a*) and (*b*) Markarian 205 and NGC 4319 (Weedman 1970; Arp 1971*b*). At the same time Gunn (1971) has shown that the quasi-stellar object PKS 2251+11 has the same redshift as a somewhat peculiar galaxy in a faint cluster in its neighborhood. These results suggest (*a*) that there appear to be some extragalactic objects with substantial redshift components which are not due to the expansion of the Universe and (*b*) that there are probably some QSOs whose redshifts are largely cosmological in origin. In this paper we discuss further evidence bearing on (*a*).

II. 3C QUASI-STELLAR OBJECTS AND BRIGHT GALAXIES

Among the radio sources in the 3C revised catalog (Bennett 1962) there are only 40 QSOs; these are listed by Schmidt (1968). Here we are dealing with a small but virtually complete sample of objects[1] selected because they are radio sources with flux densities greater than 9 flux units and are optically brighter than ~18.5 mag. If we consider the original 3C catalog, however, the number of QSOs with measured redshifts is increased to 47 (Burbidge and Burbidge 1969). While the original radio positions allowed a considerable degree of uncertainty with regard to identifications (e.g., 3C 275.1 was originally identified with NGC 4651[2] by Longair 1965), further observations have in general reduced the discrepancy between radio and optical coordinates of QSOs to such an extent that for the objects cited here the identifications are beyond dispute.

---

[1] A study of the 3CR identification lists reveals at most four more blue stellar objects brighter than 19.25 mag which may turn out to be QSOs when spectroscopic evidence is obtained. They may, however, already have been excluded by Sandage and Schmidt (unpublished work quoted by Schmidt 1968). The inclusion of these in our analysis makes essentially no difference to our statistical results. It has been suggested to us that 3C 275.1 is biased in the sense that spectroscopic evidence may have been obtained precisely because of its previous misidentification with NGC 4651. This possibility, even if true, is irrelevant since our sample is essentially complete to 19 mag.

[2] See n. 1.

234   E. M. BURBIDGE *ET AL.*   Vol. 170

Comparison of the positions of these sources with those of bright galaxies in the revised Shapley-Ames catalog (de Vaucouleurs and de Vaucouleurs 1964) shows that four of these QSOs lie comparatively near to bright galaxies. They are shown in Figure 1 (Plate 1) on reproductions from the *Palomar Sky Survey* O prints. Data on the positions, magnitudes, redshifts, and separations of these pairs are listed in Table 1. Among the galaxies, NGC 3067 and NGC 4138 have redshifts which have previously been published. We have obtained more spectra of NGC 3067 and have also measured the redshifts of NGC 4651 and NGC 5832 at the Lick Observatory.

Since there are no easily detectable luminous features connecting these QSOs to the galaxies, we must rely on probability arguments to estimate whether these pairs are likely to be physically associated or whether their apparent proximity may be ascribed to chance. A preliminary estimate may be made as follows.

The de Vaucouleurs' catalog contains about 2600 bright galaxies, so that the average space density, allowing for the regions of obscuration in our own Galaxy, is $\langle\lambda\rangle \approx 0.1$ per square degree. Thus, if the galaxies were randomly distributed, the probability of finding one or more galaxies within r minutes of a given point is

$$p \approx \tfrac{1}{3600}\langle\lambda\rangle\pi r^2 = 8.7 \times 10^{-5}r^2 , \tag{1}$$

provided that p is small. The expected number of cases for a sample of N points is thus

$$n = Np = 8.7 \times 10^{-5}r^2 N . \tag{2}$$

If we restrict ourselves to 3CR QSOs and consider the number expected within a radius $r \sim 7'$ of each QSO, we obtain $n = 0.17$. This is to be compared with the three objects (3C 268.4, 3C 275.1 and 3C 309.1) which are observed. Since the variance of the predicted mean is $\sigma_p \sim 0.41$, this formally represents significance at the 7 σ level. It is not necessary, however, that the sample of QSOs be complete, but merely that it be well defined and contain no selection bias with respect to bright galaxies. The original 3C QSOs provide such a sample, and with $N = 47$ and $r = 7'$ we find $n \approx 0.20$. Since four

TABLE 1

POSITIONS, MAGNITUDES, AND REDSHIFTS FOR CLOSE QSO-GALAXY PAIRS

| OBJECT | POSITION | | m_v | z | $\Delta\theta$ |
	α(1950)	δ(1950)			
3C 232:					
Optical	$09^h55^m25^s44$	$32°38'23''2$	15.8	0.534 ⎫	
Radio	09 55 25.39	32 38 23.0	...	... ⎬	1'9
NGC 3067 (Sb)	09 55 26	32 36 36	12.8	0.0050 ⎭	
3C 268.4:					
Optical	12 06 42	43 56	18.4	1.400 ⎫	
Radio	...	...	...	... ⎬	2'9
NGC 4138 (Sa)	12 07 00	43 58	...	0.0036 ⎭	
3C 275.1:					
Optical	12 41 27.68	16 39 18.7	19.0	0.557 ⎫	
Radio	12 41 27	16 39 22	...	... ⎬	3'5
NGC 4651 (Sc)	12 41 12	16 40	11.8	0.0025 ⎭	
3C 309.1:					
Optical	14 58 57.6	71 52 19	16.8	0.904 ⎫	
Radio	14 58 57.45	71 52 10	...	... ⎬	6'2
NGC 5832 (SBb)	14 57 54	71 53	~13	0.0020 ⎭	
PHL 1226	01 51 48	04 34	18.2	0.404 ⎫	
IC 1746 (Sb)	01 51 48	04 34	~14	0.026 ⎬	0'8

cases are observed (3C 232 is contained in the 3C catalog, but not in the 3CR), and $\sigma_p \approx 0.45$, the result is formally significant at the 8.5 σ level.

With the present small sample, these results must, of course, be viewed with caution. In particular, the quoted significance levels certainly cannot be interpreted as having their usual meaning. A second objection is that the significance levels are clearly affected by our choice of r which in turn has been influenced by the data. A third objection to the above analysis is that it assumes that the galaxies are randomly distributed over the sky when in fact they are known to be clustered. Each of these objections will therefore be considered in an attempt to assess its importance.

To allow for the clustering of galaxies we have estimated the local density of galaxies λ_i ($i = 1, 2, \ldots, 47$) in the neighborhood of all 47 QSOs. This was done by first calculating λ_i (1) within 9 square degrees of the source, (2) within 25 square degrees, (3) within 180 square degrees, and (4) within 180 square degrees, treating local clusters as one point if the density in the cluster exceeded by a factor 10 or more the average density computed in calculation (3) above. The highest value thus computed was set equal to λ_i^W and may be considered the least favorable value to adopt. A second set of local densities λ_i^B ($1 \leq i \leq 47$) was produced to give what, in our opinion, was the best estimate (*not* the most favorable) of galaxy density. Values of λ^W and λ^B for the four cases in question are given in Table 2, together with probabilities of finding a galaxy within the actual distance $\Delta\theta$ to their respective nearest neighbors. We note that while $\lambda^B > \langle\lambda\rangle$ for 3C 268.4 and 3C 275.1, $\lambda^B < \langle\lambda\rangle$ for 3C 309.1 and 3C 232. The distribution of $\lambda^{W,B}$ for all 47 QSOs is illustrated in Figure 2.

The expected number of QSOs with galaxies within distance r, based now only on the assumption of *local randomness*, is

$$n^{W,B} = \tfrac{1}{3600}\Sigma_i \lambda_i^{W,B}\pi r^2 = \tfrac{1}{3600}N\langle\lambda^{W,B}\rangle\pi r^2 , \qquad (3)$$

where $\langle\lambda^{W,B}\rangle$ represents the average value of $\lambda_i^{W,B}$. We obtain $\langle\lambda^W\rangle = 0.14$ and $\langle\lambda^B\rangle = 0.076$ per square degree, values which straddle our previous estimate of $\langle\lambda\rangle$. The corresponding expectation numbers within a 7' radius are $n^W = 0.28$, $\sigma^W = 0.53$, and $n^B = 0.15$, $\sigma^B = 0.39$. With four cases observed, this formally gives significance at the 7 σ and 10 σ levels, respectively. We may therefore conclude that clustering in the galaxy distribution does not account for the high number of very close pairs. Indeed, the significance for $\langle\lambda^B\rangle$ is increased over that for $\langle\lambda\rangle$ essentially because the QSOs do *not* sample preferentially regions occupied by large clusters which nonetheless contribute strongly to $\langle\lambda\rangle$.

TABLE 2

SEPARATIONS, LOCAL GALAXY DENSITY ESTIMATES AND CORRESPONDING PROBABILITIES FOR FOUR CLOSE PAIRS

Object	$\Delta\theta$	λ^W	λ^B	p^W	p^B
3C 232 } NGC 3067 }	1'.9	0.11	0.08	3.6×10^{-4}	2.5×10^{-4}
3C 268.4 } NGC 4138 }	2'.9	0.80	0.40	6.5×10^{-3}	3.3×10^{-3}
3C 275.1 } NGC 4651 }	3'.5	0.50	0.25	5.3×10^{-3}	2.7×10^{-3}
3C 309.1 } NGC 5832 }	6'.2	0.02	0.02	6.6×10^{-4}	6.6×10^{-4}

236   E. M. BURBIDGE *ET AL.*   Vol. 170

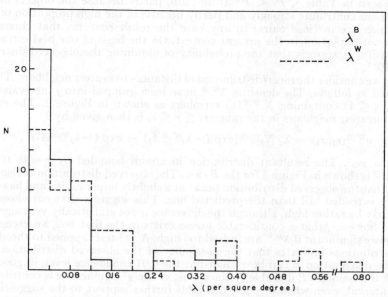

Fig. 2.—Distribution of local galaxy densities near 47 3C QSOs based on two methods (*W*, *B*) of estimation described in the text.

To estimate the true significance of these results and to demonstrate the effects of our choice of *r*, we proceed as follows. The probability of finding a galaxy within a given radius *r* is given by equation (1) with λ taken as the local density. If we rely for the moment on the mean densities $\langle \lambda^{W,B} \rangle$, the probability of finding *m* or more galaxy–QSO pairs within radius *r* is then given by

$$P = \sum_{i=0}^{N-m} {}^{N}C_{m+i} \, p^{m+i}(1-p)^{N-m-i} . \tag{4}$$

This probability is evaluated for the *W* and *B* cases for a number of values of *r*, and the results are listed, together with the formal expectation numbers, in Table 3. We note here that the source 3C 351 is within 18′ of a group of three galaxies, NGC 6306, NGC 6307, and NGC 6310; it is included where appropriate. $P^{W,B}$ is the probability that the observed number (or more) of coincidences could occur by chance and thus provides a good measure of the significance of our results. At the larger radii the significance declines as expected when *r* begins to approach the mean separation of the galaxies. At lower values of the radius, however, our results do not depend crucially on the choice of *r* and would normally be considered highly significant.

To estimate the effects of clustering on the significance test, we have considered the *B* distribution of local densities shown in Figure 2. Taking the higher densities ($\lambda^B > 0.2$) as a single group with $\lambda = 0.30$, we may consider each interval separately. The probability P_i of *m* close pairs within radius *r* for the N_i QSOs with local density λ_i is given by equation (4) with *N* replaced by N_i. Since the probabilities of close pairings in each group are independent of the other groups, the total probability P^C of obtaining the observed distribution is thus

$$P^C = \Pi_i P_i . \tag{5}$$

As can be seen in Table 3, $P^C < P^B$ at all radii, partly because the objects in lower-density regions contribute strongly and partly because of the high proportion of high-λ objects which are in close pairs. In any case the result confirms that clustering of galaxies is not important in the present context. On the basis of our best estimates of galaxy density, it appears that the probability of obtaining the observed distribution is $<5 \times 10^{-3}$.

Finally, we consider the mean distribution of distances to nearest neighbors. This may be predicted as follows. The densities $\lambda^{W,B}$ have been grouped into j intervals of 0.04 in λ ($0 \leq \lambda \leq 1$) containing $N_j^{W,B}(\lambda)$ members as shown in Figure 2. The expected number of nearest neighbors in the range $r_1 \leq r \leq r_2$ is then given by

$$n^{W,B}(r_1, r_2) = \Sigma_j \, N_j^{W,B} \left[\exp \left(-\lambda_j^{W,B} A_1 \right) - \exp \left(-\lambda_j^{W,B} A_2 \right) \right], \qquad (6)$$

where $A_i = \pi r_i^2$. The resultant distribution in annuli bounded by $r = 0$, $0°.33$, $1°$, $2°, \ldots, 10°$ is shown in Figure 3 for the B case. The observed distribution is also shown. It is clear that the observed distribution peaks at a slightly larger radius and has a somewhat more extended tail than the predicted one. This suggests that our chosen local densities may be rather high, although the difference is not statistically very significant. It is clear, however, that a considerable excess exists in the first box, an excess which becomes *more* significant if $\lambda^{B,W}$ are indeed too high. A χ^{-2} test[2] applied to the observed and B distributions indicates that the possibility of the observed distribution arising from a true distribution B can be excluded at the 0.01 confidence level, in good agreement with the estimate P in Table 3. Since most of the χ^2 significance is contributed by the first interval, even when $r = 20'$, this adds further support to the suggestion that the close proximity of these galaxies and QSOs does not arise by chance.

The above arguments, while not conclusive, suggest that a physical association does exist between close galaxy–QSO pairs described here.

III. OPTICAL DATA ON SOME OF THE PAIRS LISTED IN TABLE 1

a) NGC 3067 and 3C 232

This pair was first pointed out to one of us (G. B.) by Dr. Campbell Wade who also supplied us with a very accurate radio position for 3C 232. He pointed out that there are two blue stellar objects, Ton 469 (= 3C 232) and Ton 470, very close to the galaxy. In Figure 1, Ton 470 is the object lying between 3C 232 and NGC 3067. Two plates of NGC 3067 were kindly taken for us by Mr. E. Harlan at the Lick Crossley 36-inch

TABLE 3

OBSERVED (n) AND EXPECTED ($n^{W,B}$) NUMBERS OF GALAXIES WITHIN r MINUTES OF 3C QSOs; ALSO SHOWN ARE LIKELIHOODS $P^{W,B,C}$ OF THE OBSERVED DISTRIBUTION

r	n	n^W	P^W	n^B	P^B	P^C
3.....	2	0.05	1.3×10^{-3}	0.012	3.8×10^{-4}	7.7×10^{-5}
5.....	3	0.14	3.6×10^{-4}	0.077	7.4×10^{-5}	9.4×10^{-6}
7.....	4	0.28	1.8×10^{-4}	0.15	1.6×10^{-5}	1.0×10^{-6}
10.....	4	0.56	2.2×10^{-3}	0.30	2.6×10^{-4}	1.8×10^{-5}
15.....	4	1.28	3.0×10^{-2}	0.69	4.6×10^{-3}	3.5×10^{-4}
20.....	5	2.28	5.4×10^{-2}	1.22	6.5×10^{-3}	5.5×10^{-4}

[2] For the χ^2 test to be strictly valid it is necessary that at least five cases be expected in each subgroup, a condition that is not satisfied in this case. However, application of the χ^2 test is generally considered to be justified (cf. Clarke, Coladarci, and Coffrey 1965) if no more than one group has a predicted content less than 5 and if the predicted number in that group exceeds 1. Our sample satisfies these requirements if we consider groups at larger radius corresponding to intervals $2° \leq r \leq 4°$, $4° \leq r \leq 6°$, and $r \geq 6°$.

238   E. M. BURBIDGE *ET AL.*   Vol. 170

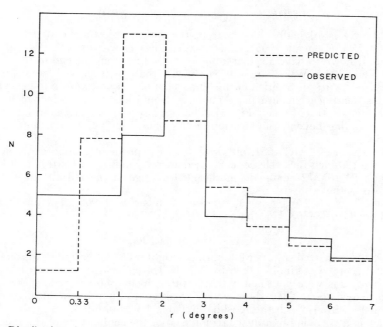

Fig. 3.—Distribution of angular separations between 3C QSOs and their nearest neighboring bright galaxy. Four cases for which *r* exceeds 7° are not shown. The predicted distribution (based on λ^B) is also shown.

reflector with exposures of 30 and 5 minutes; they are shown in Figure 4 (Plate 2). On them one can see a bright knot north of the center and dusty structure throughout the galaxy; there is no well-defined nucleus. Three spectrograms were taken with the photographic prime-focus spectrograph on the Lick 120-inch telescope; for two, the slit was in P.A. 12°.8, aligned through Ton 470, through the bright knot, and through the central region of NGC 3067. The third was taken with the slit in P.A. 112°, about 10° to the direction of the major axis of the outer isophotes, through the brightest areas of the galaxy. Lines measured were [O II] λ3727 and Ca II λ3933. The spectrum near the major axis showed [O II]λ3727 to be inclined by rotation of the galaxy. The bright knot, lying near the minor axis, has the same velocity as the central region. There are indications of noncircular motions, but they are not in any way exceptional, based on the spectra obtained so far. The object Ton 470 turns out to be an early-type star with strong Balmer absorption lines.

The redshift of NGC 3067, given in Table 1, was obtained from our spectra and from Humason, Mayall, and Sandage (1956). The agreement between the two values is satisfactory. The redshift of 3C 232 was measured by Schmidt (1968) and by Lynds and Wills (1968) and is based on Hγ, [Ne III]λ3869, [O II]λ3727, He II λ3203, Mg II λ2798, and C II]λ2326. In addition, there is a line at λ4508 which could possibly be Mg II λ2798 at *z* = 0.61 according to Schmidt, or [Mg V]λ2931 at a redshift *z* = 0.53 according to Lynds and Wills.

b) NGC 4138 and 3C 268.4

No new optical spectra of this galaxy have been obtained. The redshift in Table 1 is taken from Humason *et al.* (1956). The redshift of 3C 268.4 was measured by Schmidt (1968) and is based on C III]1909 and C IV λ1549; He II λ1640 is also probably present.

c) NGC 4651 and 3C 275.1

This galaxy was originally thought to be the optical object associated with the radio source 3C 275.1 (Longair 1965), and on this basis it was investigated optically by Sandage and shown to have a remarkable jet extending from the end of the major axis. Sandage, Veron, and Wyndham (1965) showed that an improved radio position for 3C 275.1 agreed with the position of the blue stellar object now known to be a QSO; they showed this identification in a plate made with multiple printing in which the jet is beautifully demonstrated (their Fig. 3). There is also a counterjet, and faint luminous material extending beyond the location of the QSO; the QSO is not, however, in line with the jet.

A spectrogram of NGC 4651, obtained with the prime-focus photographic spectrograph on the 120-inch Lick telescope, showed absorption lines of Ca II H and K and the G-band and [O II]λ3727 emission in the nuclear region; the redshift in Table 1 was measured from these lines.

The redshift of 3C 275.1 was measured by Lynds et al. (1966). The lines which they identified are [O III]λ4363, Hγ, [Ne V]λ3426, He II λ3203, O III λ3133, and Mg II λ2798.

d) NGC 5832 and 3C 309.1

NGC 5832 is a barred spiral of low surface brightness; a 30-minute exposure with the Crossley telescope was kindly taken by Mr. E. Harlan and is reproduced in Figure 4. Two spectrograms were obtained with the prime-focus photographic spectrograph on the 120-inch telescope. They show a very weak continuum and weak extended [O II] λ3727 emission, from which the redshift in Table 1 was obtained. A stellar object can be seen in Figure 4, lying in a dusty region east of the center. One of the spectrograms was taken with the slit through this object, and it appears to be a foreground star in our Galaxy, with Ca II H and K absorptions at a slightly negative radial velocity.

The redshift of 3C 309.1 was measured by Burbidge and Kinman (1966) and by Lynds (1967). The lines which have been identified are Mg II λ2798, C III]λ1909, and C II]λ2326.

e) IC 1746 and PHL 1226

The most direct way of demonstrating that some QSOs are nearby is to find evidence of a physical connection between a QSO with a large redshift and a comparatively nearby galaxy. We have recently come upon what appears to be a good example of such an object. The object PHL 1226 was first picked out as a radio-quiet QSO by Sandage and Luyten (1967), and its redshift of $z = 0.404$ was measured by one of us (Burbidge 1968). Reproductions of the *Palomar Sky Atlas* (both red and blue plates) are shown in Figure 5 (Plate 3). It is immediately obvious that there appears to be a bridge of luminous material joining the galaxy to the QSO which is much more plainly visible on the red print. The separation between the two is 50″. The galaxy is IC 1746 and may be roughly classified Sb. It is not listed by de Vaucouleurs and de Vaucouleurs (1964) because it is too faint, but it is listed by Vorontsov-Velyaminov (1963) (MCG 1-5-43) and by Zwicky, Karpowicz, and Kowal (1965). Optical studies of the galaxy and the luminous bridge are under way. We estimate the magnitude of the galaxy to be about 14.5 mag, while Zwicky et al. (1965) have estimated it to be 15.1 mag (see Note Added in Proof).

IV. CONCLUSIONS

To summarize, we believe that the fact that four 3C or 3CR QSOs lie very close to bright galaxies and that the probability that this should occur by chance is very small ($< 5 \times 10^{-3}$) lends further credibility to the idea that at least some QSOs are comparatively local objects, which are genetically related to galaxies, and that their redshifts are not of cosmological origin.

240   E. M. BURBIDGE *ET AL.*

In addition, the discovery of a QSO with a redshift of 0.4 which appears to have a luminous connection to a galaxy that apparently is comparatively nearby is further evidence of the same effect.

If this conclusion is correct, it may well be asked where the galaxies are which are associated with the rest of the QSOs in this sample which have not been found to lie very close to a parent (or daughter) galaxy. There are several possible answers to this question. It may be that the majority of the QSOs have moved so far from the bright galaxies (several degrees) that their relationship is masked because their separations have become comparable with (or greater than) the average galaxy separation. The method discussed by Abell, Neyman, and Scott (1964) in their study of subclustering of galaxies should, with modifications, prove useful in analyzing this question. Another possibility is that they are associated with galaxies which are not found in the catalog we have used, because they are very much fainter than 13 mag, either because they are farther away than the majority of galaxies studied here, or because they are intrinsically fainter.

Note Added in Proof.—After this paper was submitted, a redshift for IC 1746 was measured (see Table 1). More important, Dr. H. C. Arp has obtained an excellent plate of IC 1746 with the Palomar Hale reflector, which he has kindly allowed us to reproduce in Figure 6 (Plate 4). This plate shows no obvious luminous bridge, such as appeared in the red print in Figure 5. Instead, a compact nonstellar object is clearly visible in Figure 6, between the QSO and the galaxy; this did not show on the *Sky Survey* prints. Further studies of this remarkable situation are under way. In any event, a chain of objects consisting of a QSO, a compact object, and a bright galaxy is just as remarkable as the original luminous bridge that was apparent in Figure 5.

We would like to acknowledge the help of Dr. C. Wade, Mr. E. Harlan, and Mrs. Janet Strittmatter at different stages in this investigation. Extragalactic research at UCSD is supported in part by the National Science Foundation and by NASA under grant NGL 05-005-004.

REFERENCES

Abell, G. O., Neyman, J., and Scott, E. L. 1964, *Ap. J.*, **69**, 5.
Arp, H. C. 1971*a*, *Ap. Letters*, **7**, 221.
———. 1971*b*, preprint.
Bennett, A. S. 1962, *Mem. R.A.S.*, **68**, 163.
Burbidge, E. M. 1968, *Ap. J. (Letters)*, **154**, L109.
Burbidge, G. R., and Burbidge, E. M. 1969, *Nature*, **222**, 735.
Burbidge, E. M., and Kinman, T. D. 1966, *Ap. J.*, **145**, 654.
Clarke, R. B., Coladarci, A. P., and Coffrey, J. 1965, *Statistical Procedures* (New York: Charles Merrill).
Gunn, J. 1971, *Ap. J. (Letters)*, **164**, L113.
Humason, M. L., Mayall, N. U., and Sandage, A. R. 1956, *A.J.*, **61**, 97.
Longair, M. 1965, *M.N.R.A.S.*, **129**, 419.
Lynds, C. R. 1967, *Ap. J.*, **147**, 837.
Lynds, C. R., Hill, S. J., Heere, K., and Stockton, A. N. 1966, *Ap. J.*, **144**, 1244.
Lynds, C. R., and Wills, D. 1968, *Ap. J. (Letters)*, **153**, L23.
Sandage, A. R., and Luyten, W. J. 1967, *Ap. J.*, **148**, 767.
Sandage, A. R., Veron, P., and Wyndham, J. 1965, *Ap. J.*, **142**, 1307.
Schmidt, M. 1968, *Ap. J.*, **151**, 393.
Vaucouleurs, G. de, and Vaucouleurs, A. de. 1964, *Reference Catalog of Bright Galaxies* (Austin: University of Texas Press).
Vorontsov-Velyaminov, B. A. 1963, *Morphological Catalog of Galaxies* (Moscow).
Weedman, D. 1970, *Ap. J. (Letters)*, **161**, L113.
Zwicky, F., Karpowicz, M., and Kowal, C. T. 1965, *Catalog of Galaxies and Clusters of Galaxies* (Pasadena: California Institute of Technology).

14.05.10 <u>The Correlation of Redshift with Magnitude
in the Coma Cluster</u>. W.G.Tifft, Steward Obs., Univ.
of Ariz., Tucson, Ariz.-Redshifts for a large sample
of faint galaxies in the core of the Coma Cluster have
now been obtained. The sample of about 70 galaxies is
nearly complete to seventeenth magnitude. Nuclear
region magnitudes show a good correlation with redshift.
<u>Galaxies fall in bands sloping fainter with increasing
redshift. A magnitude decrease of 0.6 magnitudes per
1000 km/sec redshift increase is observed.</u> The deviation from the expected random correlation of nuclear
region magnitude and redshift is significant at a
probability much less than one per cent by x^2 testing.
After allowance for the redshift-magnitude pattern and
data uncertainty, a residual true internal Doppler
motion dispersion in the Coma Cluster appears not to
exceed 100 km/sec. Nearly all the highest redshift
galaxies in the Coma Cluster are nonellipticals. This
is not expected from a normal dynamical model and is
remarkably similar to the pattern in the Virgo Cluster
where the spiral galaxies show a distinctly greater
redshift than the ellipticals. No known physical
mechanisms can explain the observed patterns, and it
is suggested that an "intrinsic" redshift may be
required and that the redshift-magnitude states of
galaxies may be quantized. The observed redshift-
magnitude pattern is tested by application to more
outlying concentrations of galaxies in the Coma Cluster
which show identical patterns slightly shifted in redshift and by comparison with the m-logz plot for nearby
galaxies for which the Coma pattern predicts a steeper
than expected slope as is actually observed.

20.03.10   Redshift, Morphology, and Integrated
Magnitude Relationships in the Coma Cluster. W. G.
Tifft, Steward Obs. - Previous results on the
relationship of redshift, morphology, and nuclear
magnitudes in the Coma Cluster (Tifft, W. G., 1971,
Bull. Am. Astron. Soc. 3, 391) indicated that galaxies
lay in narrow bands in the redshift magnitude diagram
sloping to fainter magnitudes at higher redshifts.
It is now further shown that the morphological
dependence on redshift and magnitude is a maximum
along the direction of the bands.
        In order to extend the sample of galaxies
to verify the redshift-magnitude correlation, redshifts, morphological types, and integrated magnitudes have been adopted for a large sample of bright
galaxies within 4° of the Coma center based upon
redshifts by Kintner (1971, Astron. J. 76, 409).
This sample of galaxies is shown to conform with the
band structure and to show strong morphological
sorting along the band directions. The dependence
of morphology on redshift is not a simple function
of redshift as recently discussed in the literature
(1971, Nature 234, 505).
        The three parameters, galaxy luminosity,
morphological type, and differential redshift can
apparently be described by two variables--a quantized
variable specifying band membership and a continuous
"morphological" variable specifying position within a
band. A third variable, distance from the cluster
center, may also slightly influence the correlations.

534

NATURE VOL. 234 DECEMBER 31 1971

LETTERS TO NATURE

PHYSICAL SCIENCES

On the Redshifts of Galaxies

THERE have been several investigations of possible parts in the redshifts of galaxies which are apparently not due to their radial systemic velocities[1-7]. In most of them the correlation between redshift and magnitude has been studied. To obtain more information about the nature of the possible redshift effect, other parameters should be studied. I have made a preliminary study of the redshifts of galaxies in clusters, groups and pairs in respect to type, magnitude and peculiarities of the galaxies. Systematic differences between the redshifts of different types of galaxies were found. I wish to report here some further data on the redshifts of galaxies with respect to their types, magnitudes, diameters, colours and inclinations.

Five systems of galaxies have been considered: (a) the Virgo cluster, (b) five other clusters with more than fifteen measured redshifts in each, (c) four distant clusters and one nearer cluster, each with two to eight measured redshifts, (d) thirty-five groups with three to nineteen redshifts, and (e) eighteen pairs of galaxies. The Virgo cluster was considered separately because it may consist of two overlapping clusters[8]. Groups and pairs were taken from I. D. Karachentsev's lists[9,10], and only systems with both spirals and SO and/or E galaxies with measured redshifts were included in the study. To avoid including the same galaxies in more than one

system when all data are considered, groups and pairs which are subsystems of greater systems have been excluded. The numbers of the galaxies in (a)–(e) are 113, 175, 16, 210 and 36, respectively. The total number of galaxies studied, with known redshifts and types, is 550. The amount of data for other parameters is smaller (Table 1). Most data were taken from the reference catalogue of G. and A. de Vaucouleurs[11] and were complemented using the morphological catalogue of Vorontsov–Velyaminov, Krasnogorskaya and Arhipova[12] and an article by Zwicky and Humason[13].

So that systems with different sizes and velocity dispersions could be studied together, the dimensionless quantity $u = \Delta V / \sigma_V$ was used. $\Delta V = V_{galaxy} - V_{system}$ is the residual redshift of the galaxy (in units of velocity) and σ_V the redshift (velocity) dispersion of the system. The mean dispersions 194 km/s and 89 km/s, calculated for the SE and SSO pairs, respectively, were used for σ_V in the case of pairs. Fig. 1 gives the distributions of E, SO, S and Sbc-Sc galaxies to different u values for the five kinds of systems.

In almost all the kinds of systems, ellipticals and lenticulars show an excess of negative and spirals an excess of positive residual redshifts. Ten out of eighteen spirals belonging to mixed field pairs have positive residuals. But eight of twelve spirals in pairs which are members of larger systems have positive residuals, and thus the effect is also seen in pairs. A more accurate analysis revealed that, in particular, there was an excessive number of Sbc and Sc galaxies with positive residual redshifts. For earlier type spirals the situation is the reverse.

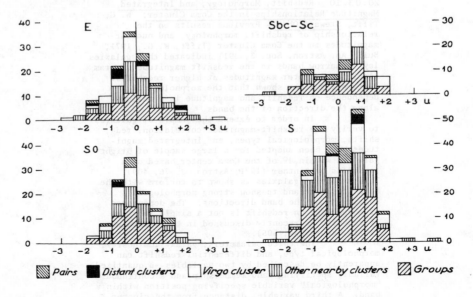

Fig. 1 Distributions of E, SO, S and Sbc-Sc galaxies to different u values.

NATURE VOL. 234 DECEMBER 31 1971   535

The total numbers of different type galaxies n_V and the corresponding frequencies of negative residual redshifts p^- are given in Table 1, lines 1 and 2. P^- (line 3) represents the probability that a frequency appears by chance. The chance probability, that in all the five different kinds of systems a nearly similar situation is valid simultaneously, is much smaller than p^-. The next lines give the mean redshift differences $\overline{\Delta V}$, the corresponding chance probabilities $P(\overline{\Delta V})$, the redshift dispersions σ_V, the mean values of u and the dispersions of u, respectively. Field pairs are not included in the calculations.

Most of the elliptical and spiral galaxies whose types are incompletely given or not at all in the reference catalogue have relatively large positive residuals. For 16 elliptical galaxies of this kind $\overline{\Delta V} = +177$ km/s, and for 24 spirals $\overline{\Delta V} = +289$ km/s.

An analysis has been made to test the assumptions[1,5] of a systematic redshift effect depending on the magnitude of the galaxy. According to these assumptions, fainter galaxies have larger redshifts. I have calculated the linear regressions between the quantity u and the magnitude for each morphological type. This has been done separately for the absolute magnitudes M and for ΔM, the difference between M and the average absolute magnitude of the galaxies with measured redshift in the system. The parameters were normalized by dividing them by σ_M, the standard deviation of the absolute magnitudes determined for each type. The quantity b_M (line 10) is the mean of the coefficients of linear regressions calculated by these two methods. This procedure diminishes random fluctuations involved in the methods.

The problem appears to be more complicated than had been supposed. First, faint elliptical galaxies have an excess of positive residual redshifts. Six out of nine galaxies with $\Delta M > 1^m.5$ have $\Delta V > 0$ and their mean residual redshift corresponds to $+155$ km/s. (See also the third column, lines 9–11, of Table 1.) Similarly, in the case of groups of galaxies, the spirals have a weak trend in the same direction ($b_M = +0.07 \pm 0.10$ for all 102 galaxies and $\overline{\Delta V} = +194$ km/s for five faint spiral galaxies with $M > 18^m.0$). These points give some support to the assumptions made particularly by Arp[5]. Of course, part of this effect can be caused by background galaxies. Another possible explanation is that these galaxies

are related to peculiar, compact galaxies and to QSOs, for which non-Doppler redshifts have been proposed.

The second point of the redshift-magnitude problem is that Sb and Sc galaxies in clusters seem to behave conversely to that discussed above. As can be seen from the lines 12–14, bright late-type spirals have the largest positive residual redshifts. If real, this kind of effect cannot be explained by means of foreground or background galaxies. It follows that there is no such redshift effect, in the direction assumed previously[1,5], which could have influence within the whole magnitude region.

When galaxies with negative velocities are excluded, the spirals in the Virgo cluster have $b_M = -0.08 \pm 0.14$. This is against the assumption that the Virgo cluster could be a superimposition of two separate clusters.

The principal results for the other physical parameters are given in the lines 15–23. The parameters in question are shown as subscripts. D means the linear diameter (Hubble constant $H = 75$ km/s/Mpc), C_0 the colour index, log R the logarithm of the ratio of major and minor axes of the galaxy. The regression coefficients have been calculated by a method similar to that used for the magnitudes.

As expected, because of the mutual dependence between magnitude and diameter, the coefficients of regression for D roughly follow those for M with opposite sign. The lines of regression for C_0 are in most cases in the direction which is expected when the redshift differences and the colour variation are taken into account. This suggests a continuous change of the amount of the additional redshift as a function of type and thus gives a further proof of the reality of the redshift effect.

If the redshift effect is connected to physical conditions in the nucleus of the galaxy, one might not be able to observe it in very inclined galaxies. Indeed, according to lines 21–23 of Table 1 this seems to be the case. Most galaxies with known rotation curves have quite large inclinations. Possibly for this reason distinct redshift differences between the nucleus and other parts of the galaxy have not been observed. These differences would be expected if the effect were due to the nucleus.

It is shown that E, SO and Sa galaxies have excessive negative and Sb and Sc ones excessive positive residual redshifts. At the same time, for the latter there are several statistically significant correlations between the redshift and other parameters. The result means that in some part even for normal galaxies, of late Hubble types, with small colour indices and small inclinations, part of the redshift cannot be explained by systematic velocity. Because the average redshifts have been considered here, the explanation proposed by Lewis[6], that outward gas motion from nuclei of central galaxies would be the reason for systematic differences in redshifts, is not valid.

Toivo Jaakkola

Observatory and Astrophysics Laboratory,
University of Helsinki

Received October 18, 1971.

Table 1 Mean Residual Redshifts and Coefficients of Linear Regression between the Redshifts and Various Parameters for Different Type Galaxies

Line	Quantity	E	SO	S	Sa	Sb	Sc
1	n_V	131	114	253	73	84	51
2	p^- (%)	57.3	57.9	47.4	58.9	44.0	41.2
3	P^-		0.05	0.05	0.21	0.06	0.14	0.11
4	$\overline{\Delta V}$ (km/s)	-39	-28	$+30$	-124	$+56$	$+131$
5	$P(\overline{\Delta V})$	0.20	0.28	0.18	0.02	0.17	0.04
6	σ_V (km/s)	645	433	487	463	406	517
7	$\overline{u}$		-0.07	-0.05	$+0.06$	-0.18	$+0.13$	$+0.20$
8	σ_u		0.98	0.80	1.01	0.95	0.96	1.16
9	n_M	120	109	242	70	83	51
10	b_M	$+0.09$	$+0.02$	-0.01	$+0.09$	-0.12	-0.02
11	D_M	±0.10	±0.07	±0.06	±0.11	±0.11	±0.15
12	n_M (clust)	75	59	139	39	40	30
13	b_M (clust)	$+0.11$	-0.02	-0.07	$+0.19$	-0.24	-0.30
14	D_M (clust)	±0.12	±0.11	±0.09	±0.15	±0.16	±0.22
15	n_D	121	109	243	68	84	51
16	b_D	-0.05	-0.02	$+0.02$	-0.15	$+0.11$	$+0.10$
17	D_D	±0.09	±0.08	±0.07	±0.12	±0.11	±0.16
18	n_{C_0}	65	75	153	44	58	31
19	b_{C_0}	-0.09	$+0.05$	-0.10	-0.06	-0.05	-0.36
20	D_{C_0}	±0.12	±0.10	±0.08	±0.15	±0.13	±0.18
21	$n_{\log R}$	121	109	243	68	84	51
22	$b_{\log R}$	-0.04	$+0.08$	-0.14	-0.21	-0.13	-0.18
23	$D_{\log R}$	±0.09	±0.08	±0.06	±0.12	±0.10	±0.17

[1] Holmberg, E., *Uppsala Obs. Medd.*, **138** (1961).
[2] de Vaucouleurs, G., and A., *Astron. J.*, **68**, 96 (1963).
[3] Jaakkola, T., *Vestnik KGU, Ser. Astron.*, **13** (in the press).
[4] Arp, H., *Pub. Astron. Soc. Pacific*, **80**, 129 (1968).
[5] Arp, H., *Nature*, **225**, 1033 (1970).
[6] Lewis, B. M., *Nature Physical Science*, **230**, 13 (1971).
[7] Arp, H., *Nature Physical Science*, **231**, 103 (1971).
[8] de Vaucouleurs, G., *Astrophys. J.*, Suppl. 6, No. 56, 213 (1961).
[9] Karachentsev, I. D., *Vestnik KGU Ser. Astron.*, **12** (1970).
[10] Karachentsev, I. D., *Sbornik Problemy Kosmicheskoi Fiziki*, vypusk 5 (1970).
[11] de Vaucouleurs, G., and A., *Reference Catalogue of Bright Galaxies* (University of Texas Press, Austin, 1964).
[12] Vorontsov-Velyaminov, B. A., Krasnogorskaya, A. A., and Arhipova, V. P., *Morfologischeskii katalog galaktik, I-IV*, MGU (1962–1968).
[13] Zwicky, F., and Humason, M. L., *Astrophys. J.*, **139**, 269 (1964).

NATURE VOL. 236 MARCH 24 1972

compositions at different distances along the same line of sight. There are, however, equally good arguments supporting the concept of a simple cluster at a well defined mean distance[4]. If so, non-velocity shifts must be considered a real possibility.

We wish to caution, however, that small systematic differences (some tens of km s[-1], although not hundreds as discussed above) can easily arise from errors in the rest wavelengths adopted for many of the broad lines (H, K) or blends (G band, and so on) used to measure redshifts of early-type galaxies (E, L) on low dispersion spectrograms and which may not be strictly consistent with the system defined by the emission lines (Hα, N_1, N_2, λ3727) available in most spirals[5,6].

Finally, we wish to point out that the initial discovery of large redshift differences in interconnected galaxies was made jointly by Zwicky, Minkowski and Humason nearly 20 years ago in their study of the famous triple system including IC 3481–3483 (see ref. 7).

G. DE VAUCOULEURS
A. DE VAUCOULEURS

Department of Astronomy,
University of Texas at Austin,
Austin, Texas 78712

Received January 31, 1972.

[1] Nature, 234, 505 (1971).
[2] de Vaucouleurs, G., and de Vaucouleurs, A., Astron. J., 68, 96 (1963).
[3] de Vaucouleurs, G., Astrophys. J. (Suppl.), 6, No. 56, 213 (1961).
[4] Kowal, C. T., Pub. Astron. Soc. Pacific, 81, 608 (1969).
[5] Mayall, N. U., and de Vaucouleurs, A., Astron. J., 67, 363 (1962).
[6] de Vaucouleurs, G., and de Vaucouleurs, A., Astron. J., 72, 730 (1967).
[7] Zwicky, F., Ergeb. Exakt. Naturwiss., 39, 354 (1956).

Non-Velocity Redshifts in Galaxies

A RECENT editorial in Nature[1] on non-velocity redshifts in galaxies overlooks an important early reference to the subject[2]. In that work we demonstrated from an analysis of all galaxy redshift data available at that time (1963) that (a) the Holmberg effect (a supposed systematic error depending on galaxy type, exposure time and nuclear magnitude) does not exist—that is, the effects noted by Holmberg were accidental and disappeared when a larger sample was considered; and (b) the mean redshift of the Virgo cluster galaxies increases systematically and monotonically from $\langle V \rangle \simeq +900$ km s[-1] at type E to $\langle V \rangle \simeq +1,800$ km s[-1] at type Sc, and more precisely as follows:

Ellipticals	(E, $n=17$):	$+900 \pm 120$ (km s[-1])
Lenticulars	(L, $n=20$):	$+1,080 \pm 117$ (km s[-1])
Spirals	(Sa–Sb, $n=10$):	$+1,398 \pm 194$ (km s[-1])
Spirals	(Sbc, Sc, $n=7$):	$+1,789 \pm 231$ (km s[-1])

The difference $\langle V(S) \rangle - \langle V(E, \quad L) \rangle = (1,559 \pm 148) - (990 \pm 85) = +569 \pm 170$ is significant at the 3.3 σ level.

This systematic difference was first reported by G. V.[3] in a study of the structure of the Virgo cluster in which it was suggested that the gross differences between the velocity distributions, as well as in the apparent distributions on the sphere of the E and L galaxies, on the one hand, and S and I galaxies on the other, could be most simply interpreted by the assumption that the traditional Virgo "cluster" is actually the optical superposition of two clusters of different types, shapes and

NATURE VOL. 236 MARCH 17 1972

Are Quasars Local or Cosmological?

THE debate about the distances of the quasars resembles in many ways the nineteenth century debate on the nebulae. One of the chief defining characteristics used, that of being stellar in appearance, is apparent rather than intrinsic. And rather compelling arguments have been advanced to show both that quasars are at the large distances implied by a cosmological interpretation of their redshifts (continuity with radio-galaxies[1,2], association with cluster of same redshift[3]) and that they are comparatively weak, nearby objects (continuity with Seyfert nuclei[4,5], associations with foreground galaxies[6], the anomalous 1.95 redshift[7], rapid lateral motions[8] in 3C 273 and 279, anisotropic distribution on the sky[9]).

NATURE VOL. 236 MARCH 17 1972

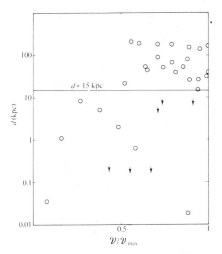

Fig. 1   Largest radio dimension, d(kpc), against $\mathcal{V}/\mathcal{V}_{max}$, for 35 3CR quasars brighter than $V = 18.5$, Einstein-de Sitter model.

The solution I propose also resembles that reached for the nebulae. There are two totally different types of object masquerading under the title of quasar. In one class the origin of the redshift is cosmological, the optical luminosities range from several to several hundred times that of a normal galaxy, and the radio luminosities and dimensions are in the range occupied by the strongest of the radio galaxies. In the second class the cosmological component of the redshift is close to zero and the objects are comparatively local, with some intrinsic redshift mechanism, perhaps gravitational, at work. The optical and radio luminosities are similar to those of radio-emitting Seyfert nuclei like NGC 1275 or NGC 1068.

A separation of quasars into two distinct classes suggests itself when the distribution of the quasars in depth is examined as a function of radio dimensions. The distribution in depth of a set of objects brighter than given optical and radio limiting flux levels can be examined by considering, for sources of a particular class (that is, with particular optical and radio luminosities and spectra), their distribution within the "observable volume" for this class of source[10,11]. If we imagine any particular quasar, which is visually brighter than 18.5 magnitudes and radio-wise brighter than the limit of the 3CR Survey[12], say, moved to greater and greater distances (and redshifts), there will come a point, corresponding to $z = z_{max}$, at which it would become fainter than either the optical or radio limiting flux levels. The corresponding (comoving) volume which is observable using this class of quasar is denoted by $\mathcal{V}_{max}$. If there are no evolutionary effects, that is, neither the luminosity nor the (comoving) number-density of the sources change with epoch (the proper number-density changes, of course, due to the expansion of the universe), then the number of objects of a particular class with redshift $\leq z$ would, for $0 \leq z \leq z_{max}$, simply be proportional to the comoving volume $\mathcal{V}(z)$. Equivalently, if the quantity $\mathcal{V}/\mathcal{V}_{max}$ is calculated for each object, it should then be uniformly distributed between 0 and 1. This is true for each class of quasar, so it is true for the whole sample.

It has been found that the 3CR quasars are not uniformly distributed, in fact the mean value of $\mathcal{V}/\mathcal{V}_{max}$ is about 0.7 (ref. 10) (depending slightly on the cosmological model chosen for calculating $\mathcal{V}$), significantly different from the expected value for a uniform distribution of 0.5. This is interpreted as being due to evolutionary effects, either the frequency, power or

lifetime of the events giving rise to the quasar phenomenon, or the properties of some external medium with which the quasars interact, changing with epoch. It is naturally of interest to ask whether this evolution is related to the radio sizes of the quasars, since it has been suggested that different radio sizes correspond to different phases in the life of a quasar, compact quasars later evolving into quasars with extended radio sources[13].

Fig. 1 shows $\mathcal{V}/\mathcal{V}_{max}$ against intrinsic radio size, d(kpc), for thirty-five quasars brighter than visual magnitude 18.5 identified with sources from the revised 3C catalogue[13]. Where the radio dimensions differ at different frequencies, the dimensions characteristic of the higher frequencies (where most of the energy is emitted) have been used, for example, for 3C 273, those of component B rather than component A (ref. 14).

There seems to be a rather abrupt change in the distribution of $\mathcal{V}/\mathcal{V}_{max}$ at around $d = 15$ kpc. For the larger sources the distribution is strongly non-uniform: none of the objects have $\mathcal{V}/\mathcal{V}_{max} < 0.5$, and the mean value is 0.80. The smaller sources, on the other hand, have values of $\mathcal{V}/\mathcal{V}_{max}$ fairly uniformly distributed between 0 and 1, mean value 0.56.

At first sight this effect could be explained within the picture that the smaller quasars radio-wise evolve into the larger, by saying that the evolutionary effect implied by the non-uniform distribution of $\mathcal{V}/\mathcal{V}_{max}$ for the larger sources is due to the interaction of the radio-emitting components with some hypothetical intergalactic medium. But this type of explanation is inconsistent with the fact that the radio-quiet quasars also show a strong evolutionary effect[15,16], which must therefore be due to some change with epoch in the population of the central (optical) objects. A similar result is obtained for radio-galaxies[17].

An equally sharp division into two groups is obtained if apparent, rather than intrinsic, radio size is used. In this case the dividing line occurs at around $\theta = 3''$.

The most surprising fact to emerge from this division into two classes is that all 7 3CR quasars belonging to Arp's proposed associations of radio-sources with peculiar galaxies[6] fall in the group with smaller radio dimensions. The statistical significance of Arp's associations has been challenged by several authors[18-21]: Arp has replied in ref. 22. The situation would seem to be that the association of 3CR sources with a particular section of Arp's catalogue of peculiar galaxies is significant if regarded in isolation, but may be thought of as a statistical fluctuation with respect to the catalogue as a whole. The probability that seven out of sixteen 3CR quasars with $\theta < 3$ arc s (or $d < 15$ kpc) should be in Arp associations (if the latter are pure chance), while none of the twenty-four with $\theta > 3$ arc s are, is rather low $(16/40)^7 \sim 0.0016$.

The group of 3CR quasars with smaller radio dimensions contains eleven out of twelve objects with flat radio spectra, all three objects which are known to have varied violently in the optical[23] (3C 2, 345, 454.3), and all four which are known to have varied at radio frequencies[24] (3C 273, 345, 380, 454.3). When the same classification is applied to all quasars with known redshift and radio structure[25], all six quasars with more than one absorption line, all four of the "1.95" objects[7], and all eleven which have shown radio variations[24] belong to this group. Twelve out of thirteen quasars with $z > 1.9$ belong to this group (and it is the large redshift quasars which are allegedly anisotropically distributed on the sky[9]).

I propose that this class of quasars consists primarily of comparatively local objects (10–100 Mpc); that the associations with Arp's peculiar galaxies are real, and that several other of these quasars may be associated with "near" Zwicky clusters[26]. A list of possible associations and estimated cosmological redshifts is given in Table 1. The radio and optical luminosities, as well as the radio dimensions and spectral indices, would then be in the same range as those of the more active Seyfert nuclei.

The group of quasars with larger radio dimensions, on the other hand, show a complete continuity with the stronger radio.

114   NATURE VOL. 236 MARCH 17 1972

Table 1 Properties of "Local" Quasars

3C	In association with Arp peculiar galaxy No.	3C number of other member of association	Inside or just outside (*) boundary of Zwicky cluster No.	Adopted cosmological redshift	$\text{Log}_{10}\, P_V$ (W ster^{-1} Hz^{-1})	$\text{Log}_{10}\, P_{178}$ (W ster^{-1} Hz^{-1})
2				?		
48	35,201	15, 17 (G)	0107.5+3212 *	0.014	19.4	24.0
138						
147			0544.4+5036	0.03	19.45	24.8
186	143	194		0.013	18.75	23.35
191			0800.0+0946	0.03	19.25	24.05
273	134	274 (M87)	Virgo	0.0038	19.7	23.0
277.1	160	266 (G ?)		0.024	18.25	22.95
286			1319.6+3.35 *	0.02	19.0	23.65
287	139,196	277.3 (G)		0.038	19.7	24.3
298			1416.0+0752 *	0.025	19.55	24.5
309.1	109	330		?		
345	125	337 (G)	1625.5+4006 *	0.029	20.2	24.0
380			1916.8+4855	0.006	18.45	23.35
454			2256.5+1933 *	0.03	19.25	24.0
454.3			2255.8+1350	0.025	20.0	24.0

galaxies if a cosmological interpretation of their redshifts is assumed. Moreover, Gunn's object, PKS 2251+11 (ref. 3), for which the redshift has been demonstrated to be the same as that of the nearby cluster, falls into this group. Thus I propose that these quasars are indeed "cosmological".

The proportion of 3CR sources which are "local" quasars would be only 5–10%. It should decline at first with decreasing flux density at 178 MHz, though it may become dominant at the very lowest flux levels. At high radio frequencies (above ~2,700 MHz), on the other hand, where a high proportion of the sources have flat radio spectra[26], the contribution of "local" quasars may be dominant even at high flux densities, so that the flatter source-count slope[28] can easily be understood. On the other hand, the steep optical counts of radio-quiet quasi-stellar objects[15] suggest that the majority of these objects are not "local".

A detailed discussion of the arguments presented here, of some of the implications of the proposal that some quasars are "local" and some "cosmological", and of possible explanations of the intrinsic redshift component in the "local" quasars, will be given later.

This work was started while I was a visitor at the Institute of Theoretical Astronomy, Cambridge, and completed at the Radio Astronomy Group, Bologna. I thank Professors Hoyle and Setti for the hospitality of their respective establishments. I also thank Professor G. Burbidge for stimulating this work and Professor G. Setti and Drs A. Braccesi, C. Lari, J. Sutton and C. Mackay for discussions.

M. ROWAN-ROBINSON

Department of Applied Mathematics,
Queen Mary College,
*Mile End Road, London E*1

Received December 15, 1971; revised January 3, 1972.

[1] Sandage, A. R., *Astrophys. J. Lett.*, **150**, L9 (1967).
[2] Miley, G. K., *Mon. Not. Roy. Astron. Soc.*, **152**, 477 (1971).
[3] Gunn, J., *Astrophys. J. Lett.*, **164**, L113 (1971).
[4] Demoulin, M. H., and Burbidge, G. R., *Astrophys. J.*, **154**, 3 (1968).
[5] Burbidge, G. R., *Ann. Rev. Astron. Astrophys.*, **8**, 369 (1970).
[6] Arp, H., *Astrophys. J.*, **148**, 321 (1967).
[7] Burbidge, G. R., *Astrophys. J. Lett.*, **154**, L41 (1968).
[8] Shapiro, I. I., *Science*, **172**, 52 (1971).
[9] Strittmatter, P. A., Faulkner, J., and Walmsley, M., *Nature*, **212**, 1441 (1966).
[10] Schmidt, M., *Astrophys. J.*, **151**, 393 (1968).
[11] Rowan-Robinson, M., *Mon. Not. Roy. Astron. Soc.*, **138**, 445 (1968).
[12] Bennet, A. S., *Mem. Roy. Astron. Soc.*, **68**, 163 (1962).
[13] Ryle, M., and Longair, M. S., *Mon. Not. Roy. Astron. Soc.*, **136**, 123 (1967).
[14] Kellermann, K., and Pauliny-Toth, I. I. K., *Ann. Rev. Astron. Astrophys.*, **6**, 417 (1968).
[15] Braccesi, A., and Forminggini, L., *Astron. and Astrophys.*, **3**, 364 (1969).
[16] Braccesi, A., *Proc. Intern. Astron. Union Symposium*, No. 44 (edit. by Evans, D. S.) (Reidel, Dordrecht, 1970).
[17] Rowan-Robinson, M., *Nature*, **229**, 388 (1971).
[18] Holden, D. J., *Observatory*, **86**, 229 (1966).
[19] Lynden-Bell, D., Cannon, R. D., Penston, M. V., and Rothman, V. C. A., *Nature*, **211**, 838 (1966).
[20] Wagoner, R. V., *Nature*, **214**, 766 (1967).
[21] Van der Laan, H., and Bash, F. N., *Astrophys. J.*, **152**, 621 (1968).
[22] Arp, H., *Astrophys. J.*, **152**, 633 (1968).
[23] Penston, M. V., and Cannon, R. D., *Roy. Obs. Bulletins*, **159**, 85 (1970).
[24] Kellermann, K., and Pauliny-Toth, I. I. K., *Ann. Rev. Astron. Astrophys.*, **6**, 417 (1968).
[25] Miley, G. K., *Mon. Not. Roy. Astron. Soc.*, **152**, 477 (1971).
[26] Zwicky, F., *et al.*, *Catalogue of Galaxies and Clusters of Galaxies* (California Institute of Technology, 1960–68).
[27] Wall, J. V., and Merkelijn, J. K., *Austral. J. P. Astron. Suppl.*, **19**, 1 (1971).
[28] Shimmins, A. J., Bolton, J. G., and Wall, J. V., *Nature*, **218**, 154 (1968).

Reprinted from:

THE ASTRONOMICAL JOURNAL   VOLUME 77, NUMBER 7   SEPTEMBER 1972

Radio Galaxies, Quasars, and Cosmology*

K. I. KELLERMANN

National Radio Astronomy Observatory,† Green Bank, West Virginia

(Received 28 June 1972)

This paper discusses the role of radio observations of galaxies and quasars in observational cosmology. Particular emphasis is given to the interpretation of radio-source counts, and it is concluded that when properly analyzed, the counts are not a function of wavelength, and that the claims for a large observed excess of weak sources have been greatly exaggerated. Evidence for a nonuniform spatial distribution of radio sources is inconclusive, and rests on (a) the significance of a small deficiency of relatively strong sources, (b) the nature of the unidentified sources, and (c) whether or not quasar red shifts are in fact of cosmological origin. The rather pronounced change in number density observed between 5 and 10 sources sr^{-1}, the apparent nonuniform angular distribution, and the absence of any observed spectral index–flux density relation suggest that the data may not be cosmologically relevant. In particular, the radio-source counts do not provide evidence for cosmic evolution independent of the distribution of quasar red shifts. The dependence of observed angular size of radio sources on flux density and red shift is also examined, and it appears to reflect an apparent change in linear dimensions with red shift, rather than the effect of space curvature, and appears to be of little value in distinguishing between various world models. Finally, the apparent "super-light" motions of quasar components is discussed and it is concluded that the simplest explanation of the data suggests that the sources are nearer than indicated by their red shift, and that either the motions are nonrelativistic or there are no real motions at all and only variations in the intensity of one or more stationary components.

EVER since the realization about 20 years ago that most of the discrete radio sources are extragalactic, radio galaxies and later quasars have for two reasons become one of the most exiting subjects of modern astronomy. First, there are the well known perplexing problems connected with their source of energy and its conversion to radio emission; and second, there is the potentially important role of extragalactic radio sources in observational cosmology, which is the topic of this paper. Although there has been considerable progress in recent years in the first of these problems and we are obtaining some understanding of the physical process going on in radio galaxies and quasars, in contrast, the contribution to cosmology has been disappointing, in spite of widespread early optimism.

Why, we may ask, have radio astronomers always been so enthusiastic about the cosmological implications of studying the discrete extragalactic radio sources? There are, in fact, a number of simple reasons.

(1) The radio emission from at least some galaxies and quasars is very intense, and may be detected from a very great distance, even with radio telescopes of modest dimensions. The first hint of this potential power of radio astronomy in exploring the distant parts of the universe came with the identification by Baade and Minkowski in 1954 of the most intense extragalactic source, Cygnus A, with a galaxy having a red shift of $z = 0.056$. Today, modern radio telescopes can observe sources more than one million times weaker than Cygnus. If these have the same absolute luminosity as Cygnus, then they are expected to be 100 to 1000 times more distant, and far beyond the limits of even the largest optical telescopes. In 1960 3C 295, which is about the 10th strongest radio source in the sky, was identified by Baade and Minkowski with a faint galaxy whose red shift of 0.46 provided dramatic confirmation of the ability of radio telescopes to detect very distant radio sources.

(2) The sky at radio frequencies is very cold, particularly at short wavelengths. The typical brightness temperature at decimeter and centimeter wavelengths is about 20°K due to radiation from the atmosphere, the galactic background, and the 3°K cosmic background. The brightness temperature of radio sources is much larger than this, typically between 10^3 and 10^{12} °K, so they are easily distinguished from the background. In contrast, at optical wavelengths the sky is relatively bright and it is difficult to make quantitative measure-

* This paper is based on the Helen B. Warner lecture given at the 136th meeting of the American Astronomical Society, 6 Dec. 1971, in San Juan, Puerto Rico.

† Operated by Associated Universities, Inc., under contract with the National Science Foundation.

532   K. I. KELLERMANN

ments, especially for faint objects. In particular, there are important systematic errors in trying to determine the magnitude or angular extent of the fainter galaxies.

(3) The radio spectra of galaxies and quasars can be determined over a wavelength range of 1000 or more, so that K corrections to observed flux densities may be easily calculated even for very large red shifts. Optical spectra, on the other hand, are for the most part restricted to the familiar narrow window, so that the K correction is uncertain for objects with cosmologically interesting red shifts.

(4) The effects of evolution in intrinsic properties of galaxies and quasars is probably less important for radio emission than for optical emission. This is because the typical age of even the oldest radio sources is probably not more than 10^8 or 10^9 years, much less than the characteristic Hubble time. Thus the age distribution is not expected to be a function of epoch, and even for $z \gtrsim 1$, we may still observe sources with the same distribution as locally. At optical wavelengths, since the typical age of a galaxy is comparable with the Hubble time, galaxies observed at large z are expected to be young and thus have intrinsically different properties in any evolutionary cosmology. Of course radio sources may evolve statistically in the sense that a sample of objects selected at large z or early epochs *may* differ from those selected from a local sample. But, unlike optical measurements, the properties do not *necessarily* differ with cosmological epoch.

(5) Galactic absorption at the shorter radio wavelengths is negligible and the correction for uncertain and varying amounts of extinction necessary at optical wavelengths is avoided.

With all of these apparent advantages, why then have radio sources made so little contribution to observational cosmology? There are, in fact, two important and fundamental reasons. First, since there are no sharp features in the radio spectra of galaxies and quasars, there is no way to determine their distance from radio measurements alone. Second, the origin and evolution of radio sources is not well understood; their source of energy, its conversion to relativistic particles, and the subsequent formation of the multiple extended complex clouds of radio emission are still unsolved problems. So, while we radio astronomers may claim to be able to see further out in the universe than anyone, unfortunately we don't know *where* we are looking, and we don't know *what* we are looking at.

For many years, therefore, the emphasis in extragalactic radio astronomy was on the determination of precise positions and optical identifications, not only to obtain red shifts and distances, but also in the hope that detailed optical studies of identified galaxies and quasars would shed some light on the problem of the nature of radio galaxies and quasars.

Since the early 1960's when it first became possible to routinely measure radio-source positions to an accuracy of about 10″ of arc or better, optical identifications have been obtained in increasing numbers, first for radio galaxies, and later for quasars. Today, red shifts are available for about 100 radio galaxies and 250 quasars. It is interesting that 10 years ago just before the discovery of quasars, there were about 75 radio-galaxy red shifts compared against none for quasars. The relatively small number of radio-galaxy red shifts obtained during the past 10 years is no doubt due in part to the much greater effort required to measure the red shifts of faint galaxies compared with that required to observe quasars; but, also, unfortunately, it reflects an apparent lessening interest in the study of galaxies compared with the more exotic but perhaps cosmologically less meaningful quasars.

In spite of the great effort which has gone into the determination of red shifts, the lack of progress in using these data to distinguish between the different cosmological models has been disappointing, particularly in view of the large amount of observing time which has been used on the largest optical telescopes.

In the case of the identified radio galaxies, the dispersion in absolute magnitude is quite small and the red shift–magnitude diagram reflects a well defined Hubble law. The observed red shifts, however, appear to be too small to definitively distinguish among different world models, especially considering the distance-dependent systematic observational errors discussed earlier.

For the quasars, the red shifts are of course sufficiently large for cosmological effects to be important, but the large apparent dispersion in absolute magnitude appears to be too great for the effect of space curvature to be detected from the quasar red shift–magnitude diagram. When the studies are confined to a complete sample of quasars, the results are dominated by an apparent evolution of source density or luminosity with distance (or epoch) (e.g., Schmidt 1968), and little is revealed about the geometry of the universe. Moreover, the number of quasars used in these studies is small since the identification and measurement of complete samples are necessary, and the required observing time is discouragingly long. Also there is, of course, still considerable doubt among some about the cosmological origin of the quasar red shifts.

The continued dependence on optical identifications appears to me to be unsatisfactory for several reasons.

(1) Many sources have a complex distribution of brightness extending up to several minutes of arc. Since there may be more than one quasar or galaxy in the region defined by the radio emission, an unambiguous identification is difficult, particularly if the optical search includes faint objects. Identification with the brightest object in the field is also not necessarily correct, and a number of radio-source identifications based on this assumption have later turned out to be wrong. In the early days it was fashionable to look for

RADIO GALAXIES, QUASARS, COSMOLOGY   533

some optical peculiarity when searching for optical identifications, because it happens that four of the earliest identifications, M87, NGC 5128, NGC 1275, and Cygnus A had some apparent abnormality. However, most of the recent identifications are with more normal looking galaxies, and the consideration of "peculiarities" appears to be mostly wishful thinking when one is faced with the choice of choosing the proper identification.

(2) Distant radio galaxies which are beyond the plate limit of the largest optical telescopes may be included in radio surveys of even modest sensitivity. The identification of a complete sample of radio sources is then impossible. Since optical identifications are easiest for those objects with a large ratio of optical to radio luminosity, this may introduce undesirable selection effects in statistical studies.

(3) If we want to investigate the origin and evolution of the intense *radio* emission, then clearly we should observe at *radio* not *optical* wavelengths.

(4) Finally, as long as radio astronomy has to depend on optical identifications, the full potential of the radio telescope to explore distant parts of the universe will not be fully realized.

It is interesting, therefore, to see what can be done by using the radio data alone without making direct use of individual optical identifications.

I. NUMBER COUNTS

The most widely discussed procedure used by radio astronomers to study cosmology has been the notorious number counts. These radio counts are analogous to the galaxy counts made by Hubble (e.g., 1936a, b) and others which have been inconclusive due to the limited range of distance sampled by the galaxies and to the systematic errors introduced in measuring the magnitude of faint galaxies, and to variations in extinction. As I have already described, these problems are not expected to be important at radio wavelengths. Unfortunately, though, the early optimism generated by the apparent ability to do number counts which might include objects at very large red shifts has not proved justified, and the results have so far been as inconclusive as the optical-galaxy counts.

The interpretation of the radio-source counts may be summarized as follows:

In a universe assumed to be Euclidean, static, and uniformly filled with radio sources of luminosity L, the following realtions hold.

$$S \propto L/r^2, \quad N \propto r^3$$

where S is the observed flux density, and N the number of sources in a sphere of radius, r. The number of sources, N, with flux density greater than S is then given by

$$N \propto S^{-\frac{3}{2}} \quad \text{or} \quad \log N = -1.5 \log S + \text{const.}$$

This is the so-called $\log N$–$\log S$ relation, which in this example is independent of the radio luminosity function, since it is valid for any luminosity, and thus for any combination of luminosities.

In fact there are no grounds for any of the above assumptions. In particular, the universe is generally accepted to be not static, but expanding. The effect of the expansion is that the flux densities of distant sources decrease faster than $1/z^2$, due to the red shift and the Hubble "number-effect" and "energy effect" (Hubble 1936b). The slope of the $\log N$–$\log S$ relation is thus expected to be smaller than -1.5 unless $z \ll 1$. The specific value of the slope at any flux level depends on the particular cosmology and the radio luminosity function (to relate flux density to distance), but for any expanding universe with zero cosmological constant and uniform density of sources, the slope must be smaller than -1.5. It is this effect which Hubble attempted, with only partial success, to observe by means of optical-galaxy counts, and which in fact led him to question the interpretation of the red shifts as a recession velocity, 25 years before the discovery of quasars (Hubble 1936a)!

For many years the experimental value of the slope for radio-source counts was a subject of great controversy among radio astronomers, with the observed values ranging from -3 for the early Cambridge 2C survey to -2 for the improved 3C survey to near -1.5 for the southern-hemisphere Mills Cross survey. We know now that in some of these early surveys, particularly the Cambridge 2C survey, and to a lesser extent the 3C survey, the experimental errors were grossly underestimated, leading to a systematic overestimate of the slope. Today, as a result of the vastly increased sensitivity and resolution of modern radio telescopes, the experimental situation is much clearer and there is good general agreement among radio-source surveys made between wavelengths of 6 cm and 3.7 m. However, the interpretation of the significance of the number counts is still being debated.

The most extensive studies of source counts have come from the series of surveys made by the Cambridge group using radio telescopes of steadily increasing sensitivity and resolution. Following the early surveys which were limited to the strongest sources, the later surveys of weaker sources appeared to show a huge apparent excess of weak sources compared to the number expected from the observed number of strong sources and an assumed slope of -1.5.

Figure 1(a) shows the result of the Cambridge source counts made at 75-cm wavelength. The data shows a steep slope of near -1.8 for the strong sources which gradually becomes flatter for the weaker sources, reaching a slope of about -0.8 near source densities

534   K. I. KELLERMANN

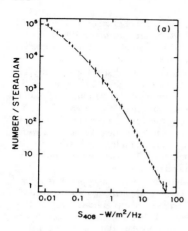

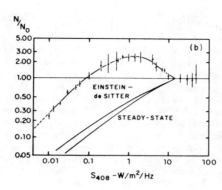

Fig. 1(a). The Cambridge source counts at 75 cm taken from Ryle (1968). (b) The same data as in (a) normalized to an ideal static Euclidean universe. The lower curves represent an Einstein–de Sitter model with a uniform density of sources, and the steady-state model.

of 10^5 sr^{-1}. In Fig. 1(b), the same data is shown normalized to what is expected in the ideal uniform, static, Euclidean universe, and typical expanding universes with steady-state and Einstein–de Sitter cosmologies. The usual interpretation is that at a level of about 1 f.u., corresponding to a source density of about 1000 sr^{-1}, there is an apparent excess in the observed number of sources by a factor of about 2.5 over the static universe and a factor of 10 over the typical expanding universes. Over the whole sky this represents a total number of about 10^4 observed sources, compared with a predicted 10^3, or an apparent excess of about 10^4 sources (10^4 minus 10^3) over a uniformly filled universe. This is generally accepted as evidence that at large red shifts, corresponding to early epochs, the density of radio sources is much larger than it is locally, a result which is naturally expected from any of the evolving cosmological models.

Attempts to be quantitative about the form of the evolution indicate that the density or luminosity of radio sources must increase approximately as $(1+z)^6$ and then must cut off near 1000 sources sr^{-1} corresponding to red shifts of 2 or 3 in order to explain the observed convergence of the counts at low flux-density levels. Moreover, it is required that the density of only the strongest sources evolve in this way so that the luminosity function near the cutoff can be sufficiently narrow to account for the observed sharp cutoff in source density (e.g., Longair 1966).

But the details are unclear since it is first necessary to have a specific world model and radio luminosity function in order to relate the source density (or flux density) to distance (or epoch). Thus the number-count data may be fitted by many world models, luminosity functions, and forms of evolution. It is particularly difficult to find a unique solution because even for the strongest sources the red shifts may be large, so that it is hard to determine even the "local luminosity function" free of evolution, or independent of world model.

Although the original goal of specifying the correct world model has not been achieved, most workers have, nevertheless, agreed that the huge apparent excess of distant sources is direct evidence against the steady-state model which requires that except for possible statistical fluctuations, the density of radio sources (or anything else) be constant throughout the universe. On the other hand, it has been argued by Mills (1959), Hoyle (1968), and others that the observed steep slope can equally well be interpreted as a small deficiency of strong sources as by a large excess of weak ones.

As emphasized by Jauncy (1967) and others, more insight into the interpretation of the source counts can be obtained by looking at the differential counts of sources, rather than the integral or cumulative counts, which suffer from two serious defects. First, the data are not independent so the effect of statistical errors is hidden and the apparent significance of the difference between the observed and "expected" counts can be vastly exaggerated. Second, the result of any small anomaly in the number of sources at any particular flux level will propagate back toward lower flux-density levels in the cumulative presentation.

When the best available data are examined in the form of differential counts, it is found that the uncertainties are dominated by statistical errors and not by the uncertainty in the measurement of individual flux densities. This is especially true for counts based on recent data at short wavelengths, where the flux densities are determined with considerably greater accuracy than at the longer wavelengths where all of the earlier source counts were made. Within the limits of the statistical error, it appears to me that contrary to some published reports there is no convincing evidence that the counts vary with the wavelength of observation. In particular, the 11-cm counts of Shimmins, Bolton, and Wall (1968), which were claimed to differ from the long wavelength counts, in fact are in satisfactory agreement with the best data available at 75 cm

RADIO GALAXIES, QUASARS, COSMOLOGY   535

(Pooley and Ryle 1968), 20 cm (Bridle *et al.* 1972), and 6 cm (Kellermann, Davis, and Pauliny-Toth 1971), when the results are compared at the same number density (Pooley 1968) or when proper account is taken of the dispersion in spectral indices (Kellermann, Pauliny-Toth, and Davis 1968).

Figure 2 shows the differential counts at wavelengths of 75, 20, 11, and 6 cm normalized to a slope of -1.5. The most obvious thing seen here is that the steep slope which appeared to extend to 1000 sources sr^{-1} in the integral counts is actually confined to the region between 5 and 10 sources sr^{-1}. Between $n = 10$ and $n = 10^5$ sources sr^{-1}, and for $n < 5$ sr^{-1}, the integral slope is indeed -1.5 or less as is expected from a nonevolving expanding universe. Between 10 and 10^3 sources sr^{-1} the data do not differ significantly from that expected in a static universe, *and nearly all of the presently catalogued radio sources fall in this interval.* (While the number densities discussed here are large, the data is based on only a few hundred observed sources for $n > 10^3$ sr^{-1} and obviously only 100 sources for $n < 10$ sr^{-1}. For $10 < n < 1000$ sr^{-1}, about 10^4 sources have been catalogued.)

What is the significance of the apparent deficiency of strong sources below densities of 10 sr^{-1}? First, is it statistically significant? The answer appears to be yes at the 2 or 3 sigma level (Kellermann, Davis, and Pauliny-Toth 1971; Bridle *et al.* 1972). But is it cosmologically significant? This is a more important and more difficult question. The steep slope represents a deficiency of only 30 to 50 strong sources over the whole sky. Of the strongest sources, about 40% are identified with galaxies of low red shift, 25% with QSO's, and 35% are unidentified. Do we want to base a cosmology on 50 strong sources, 40% which are known to have small red shifts, and the others whose nature is unknown

or in question; or on the absence of 50 of these sources? Perhaps this may be compared with the 19th century astronomer trying to do cosmology with the equally inhomogeneous Messier catalogue!

Also, we note the apparently sharp discontinuity in the source counts between 5 and 10 sources sr^{-1}. Considering that the luminosity distribution of radio galaxies covers an apparent range of more than 10^4 to 1, it is difficult to see how this can be a cosmological effect since even over such a small flux-density range, a wide variety of red shifts are sampled and any spatial discontinuity is smoothed by the wide luminosity function of the extragalactic sources.

It is argued by some that if the steep slope is interpreted as a local deficiency, then the apparent isotropy requires that the Earth be in a preferred place in the universe (e.g., Pooley and Ryle 1968). And of course we categorically reject this explanation with the same confidence that pre-Copernican astronomers rejected putting the Earth anywhere but at the center of the universe. But in fact the so-called isotropy refers only to the weaker sources where the observed slope is -1.5 or less and the interpretation is not in question. For the strong sources, where the slope is apparently steep, the data are less clear, particularly at the shorter wavelengths where the data are more accurate, but where as yet only small areas of the sky have been observed. For example, in a recent 6-cm survey made at NRAO (Kellermann, Davis, and Pauliny-Toth 1971), the steep slope observed for $n < 130$ sr^{-1} ($S_6 \sim 0.6$ f.u.) is confined to the region north of the galactic equator which includes the nearby Virgo and Coma clusters, and where the slope is -2.1 ± 0.2. South of the galactic equator, by contrast, the slope is only -1.6 ± 0.1 (Pauliny-Toth and Kellermann 1972; Yahil 1972). The difference appears to be significant.

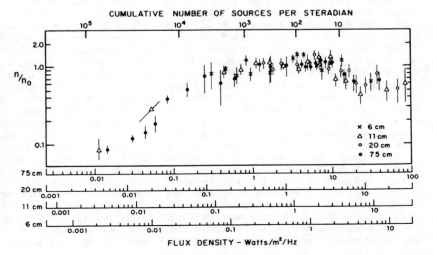

FIG. 2. Combined differential source counts, (n), at 75, 20, 15, and 6 cm normalized to a static, homogeneous, Euclidean universe, (n_0). The source of data is as follows. 6 cm: Kellermann, Davis, and Pauliny-Toth 1971; 11 cm: $S < 0.1$, Fomalont, private communication; $0.1 < S < 10$, Wall, Shimmins, and Merkelijn 1972, $S > 10$ Parkes catalogue; 20 cm: Bridle *et al.* 1972; 75 cm: $S > 6$, Parkes Catalogue; $S < 6$ Pooley and Ryle 1968. The individual flux-density scales are given at the bottom for each wavelength, and the *common* cumulative counts at the top.

536   K. I. KELLERMANN

TABLE I. Paradoxes.

Either	Or
The density of radio sources depends on red shift in just such a way to cancel the geometrical effects and the effect of the red shift.	The effect of the red shift on the observed fluxes for most of the observed sources is negligibly small.
The intrinsic spectral-index distribution depends on red shift in just such a way to cancel the effect of the red shift.	The effect of the red shift on the spectra for most of the observed sources is negligibly small.
The linear dimensions of radio sources depend on red shift in just such a way to cancel the geometrical effects of the red shift.	The geometric effect of the red shift on apparent size is negligibly small.
The intrinsic intensity of the two components of 3C 279 differ by just the right amount to cancel the effect of the Doppler shifts.	The distance of 3C 279 is much closer than indicated by the red shift and the motion is actually nonrelativistic.

But perhaps more important, the significance of the difference between the two regions is comparable with that of the apparent deficit of strong sources taken from the whole sample. It is tempting to interpret this apparent anisotropy as a "local" inhomogeieity. However, when the analysis is confined to sources with $n < 50$ sr^{-1} ($S_6 \sim 1$ f.u.), both regions show a similar steep slope of about -1.9.

Even among the weak sources, the counts appear to depend on galactic latitude. At 6 cm the observed slope for $300 < n < 3000$ is only -1.16 ± 0.15 for $5 < |b_{II}| < 20$, and $-1.72 \pm .21$ for $|b| > 20°$ (Davis 1971), although very few, if any, of the sources in the range $5 < |b| < 20$ are thought to be galactic. And even for the weakest observed sources near 10^5 sr^{-1}, the apparent difference observed between the 5C1 and 5C2 areas (Maslowski 1972; Pauliny-Toth et al. 1972) suggests clustering on a scale of one degree. Clearly it is difficult to reconcile the presence of these apparent nonuniformities in the angular distribution of radio sources on scales ranging from a few square degrees to 2π sr, with the source counts being cosmologically significant.

It is often argued that in a uniformly filled expanding universe the slope must be flatter than -1.5 if the red shifts are sufficiently large. So even an observed slope of -1.5 is then evidence for evolution. *But then we have the curious situation listed as Paradox 1 in Table I.* Similar conclusions were reached by Hubble (1936a) on the basis of optical-galaxy counts, and more recently from the quasar v/v_m test (Lynds and Wills 1972).

If most radio sources are at distances comparable to those identified with classical galaxies, whose red shifts are mostly small [the mean red shift of all identified radio galaxies (excluding N type) listed by Burbidge (1970) is 0.058 and out of 77 galaxies only one (3C 295) has $z > 0.25$] and have not been questioned (at least by

most astronomers), then the observed radio number counts alone do not present a convincing case for a nonuniform distribution of radio sources. However, if indeed the red shifts of the compact objects such as N galaxies and quasars are cosmological, then even a -1.5 slope requires evolution. The evidence for evolution thus depends entirely on the acceptance of the quasar red shifts as cosmological. Without this the data agree with both evolutionary and nonevolutionary models. Thus the source counts, which were devised as a technique to eliminate the need for identifications, are of little value by themselves, and in particular do not provide *independent* evidence of cosmological evolution.

It is not the intention of this paper to enter the debate as to the origin of the quasar red shifts. However, I do want to comment that perhaps the main objection today to the noncosmological interpretation of the quasar red shifts is simply the lack of any other plausible expanation. But the growing acceptance of the large difference between emission- and absorption-line red shifts as being intrinsic to the quasars (e.g., Lynds 1972) appears to me to greatly reduce the weight of this argument.

Those who firmly believe in an evolutionary cosmology find some support in the steep slope observed for the 50 or so brightest sources. By balancing the effect of the red shift by changes in source density, they are also able to reconcile the form of the number-flux relation for the next million or so sources where the slope is 1.5 or flatter, with their evolutionary model. On the other hand, it is clear if you start out without any special preferences, and postulate until proven otherwise a simple model where the density of radio sources is uniform throughout the universe, then it appears to be difficult to conclude from the source counts alone that the density of sources depends strongly on epoch.

In any event, aside from the cosmological relevance of the source counts, it is now clear that the widely accepted excess of weak distant sources has been greatly exaggerated, and that the apparent excess of 10^4 weak sources is merely due to a deficiency of about 50 strong sources. The value of extending the counts to weaker sources is thus not clear, since the dominant "cosmological effect", if any, is confined to the strongest sources. Also, because of the rapid convergence in the counts near 10^5 sr^{-1}, extension to weaker sources will mostly include an increasing fraction of nearby low luminosity sources. Moreover, since the experimental uncertainties for the important strong sources are negligible compared with the statistical errors, no improvement is expected from future investigations made with more powerful instruments.

It is perhaps an historical accident that for so many years the data have been interpreted in terms of a large excess of weak sources, rather than a small deficiency of strong ones. Most of the early surveys reached source

RADIO GALAXIES, QUASARS, COSMOLOGY   537

densities of only about 10 sr⁻¹, and so the later surveys which reached much weaker sources showed an apparent excess over what was expected from a few observed strong sources and the extrapolation of the −1.5 law. If, on the other hand, the much more numerous weak sources had been surveyed first, then perhaps no one would have gotten excited about the lack of a few strong sources. This also explains in part the different interpretation of the Cambridge 75-cm and Parkes 11-cm counts (Shimmins, Bolton, and Wall 1968) for which the data are in fact in good agreement.

In all of the above discussion we have implicitly assumed that the weaker radio sources are in general more distant. In fact there is no direct evidence that this is true. It is well known that identified radio galaxies and quasars show no relation between flux density and red shift. It has been argued (e.g., Burbidge and Burbidge 1967) that this, therefore, means that either there is no flux density–distance relation (in which case the source counts do not tell us anything about the universe) or that there is no distance–red shift relation (in which case, as we have just seen, the interpretation of the counts is ambiguous). In either case one concludes that the counts are dominated by the form of the radio luminosity function, and are meaningless, as far as cosmology is concerned.

However, this argument is based on incomplete identifications which may obscure any real flux density–red shift relation. An improved understanding will be possible if the counts can be limited to only the intrinsically strong sources, such as Cygnus A and 3C 295, by means of radio observations only. Two weak relations are known between radio luminosity and a distance-independent observable property. One is the tendency for the most luminous sources to have steeper radio spectra (Heeschen 1960); the other is the relation between high radio luminosity and high surface brightness (Heeschen 1966). In fact as expected from the latter relation, there is recent evidence that only the high brightness, presumably more luminous and thus more distant sources contribute to the steep log N log S slope (Bridle *et al.* 1972).

But, still in order to assess the cosmological significance of the counts, clearly we need to find some way of estimating distances from the radio data alone, without recourse to optical identifications.

II. SPECTRAL INDEX–FLUX DENSITY RELATION

I have already mentioned the problems one encounters when trying to use optical identifications to determine the distance.

The absence of any strong radio-frequency spectral lines in radio galaxies and quasars precludes any direct radio determination of the red shift in the usual way. But, the wide dispersion in the spectral shape of extragalactic sources can be used to estimate the mean red shift of a group of sources by comparing the observed

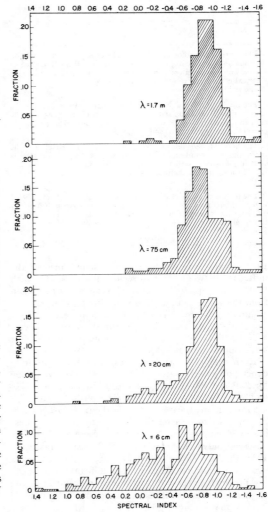

Fig. 3. Spectral-index distribution for sources selected from surveys at (a) 1.7 m, 3CR survey, (b) 75 cm, Parkes survey, (c) 20 cm, Bridle *et al.* 1972, (d) 6 cm, Kellermann, Davis, and Pauliny-Toth 1971. The spectra are determined from flux densities measured at the survey frequency and a nearby frequency, and so approximately represent the index at the survey frequency.

spectral indices with those of nearby sources. The spectral dependence of the red-shift effect on apparent luminosity, which is analogous to the K correction for the measurement of optical magnitudes, causes sources with steep spectra to become weaker with increasing red shift more rapidly than sources with flat spectra. Thus, *if there is a flux density–distance relation*, then we expect the weak sources to show an increase in the relative fraction of flat-spectra sources. Or in other

538   K. I. KELLERMANN

words the log N–log S relation for flat-spectra sources should show a relative excess of weak sources. In this way a comparison of the number counts for sources with different spectra, or comparison of the spectral-index distribution at different levels of flux densities, provides an estimate of the characteristic red shift as a function of flux density.

At long wavelengths this effect is not important since nearly all sources have a spectral index near −0.8. But, as shown in Fig. 3, as we go toward shorter wavelengths the fraction of sources with flat spectra increases. The effect of the red shift is to observe sources at a wavelength longer by a factor of $(1+z)$ than the wavelength of emission, so if weaker sources are in fact more distant than the stronger ones, they should show an excess of flat spectra.

Many studies have been made of the dependence of radio spectra on flux density and so far no effect of the red shift has been found down to source densities of about 10^3 sr^{-1}. The conventional analysis of the source counts (e.g., Longair 1966) indicates that the maximum excess which occurs at 10^3 sr^{-1} is due to sources at a red shift of about 2 or 3. However, the absence of any measurable spectral index–flux density relation suggests that the red shifts of the weaker sources are not significantly greater than for the stronger sources, and so the correct interpretation of the source counts is not clear. Of course, it is possible that the expected spectral index–flux density effect is masked by an intrinsic spectral-index distribution which depends on red shift, *but then we are faced with the second paradox shown in Table I.*

III. ANGULAR SIZE–DISTANCE RELATION

Another potentially effective radio measure of distance is the angular size. This can be used as a particularly sensitive test of world models, since as pointed out by Hoyle (1959), in exploding cosmologies with $q_0 > 0$, the observed angular size has a minimum near $z \sim 1$, and then *increases* for larger red shifts.

A purely radio test is the angular size–flux density relation, since both angular size and flux density vary with red shift in a way which depends on the specific world model. Similar tests have been tried optically, but since these measurements refer to *isophotal* diameters, rather than *metric* diameters, as obtained from the radio data, and as a result of the systematic errors in determining the magnitude and angular size of faint galaxies, no definitive results have yet been obtained (e.g., Sandage 1961).

The limiting factor in the radio case is the wide dispersion of absolute luminosity which extends from 10^{37} to 10^{45} erg/sec, and in linear dimensions which range from less than 1 pc to hundreds of kpc. These wide dispersions in intrinsic properties will, of course, hide any relation between observed angular size and flux density due to cosmological effects. However, during the past few years we have begun to better understand and to classify the radio sources, so that it is possible to restrict the intrinsic scatter of a selected sample directly from observable properties. For example, the intrinsically small sources are optically thick, so that their radio spectra show a low frequency cutoff. By restricting the analysis to sources that do not show this cutoff, the very compact sources are

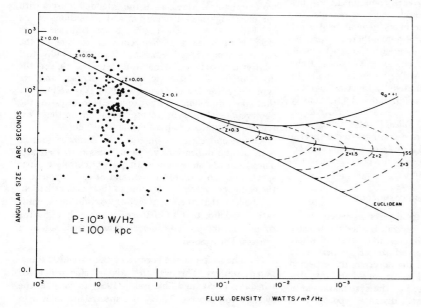

FIG. 4. Angular size vs flux density at 75 cm for double radio sources which do not have a low frequency spectral cutoff. The lines show the expected relation for several cosmological models.

eliminated. Similarly, the intrinsically weak sources generally have a simple radio structure while those with double or more complex structure usually are more luminous.

Data for all known double radio sources which do not show a low frequency spectral cutoff are plotted in Fig. 4 together with three commonly discussed cosmological models. The scatter of the data is, however, still large, and although there does appear to be a weak relation between the observed flux density and angular size, it is not of the form expected from any of the simple world models.

A more direct test is to compare the measured angular size and red shift, which eliminates the scatter due to the large spread in radio luminosity. Since the linear dimensions of most identified double sources is between 30 and 300 kpc, the scatter of the θ–z diagram may be expected to be much less than that of the θ–S diagram. The disadvantage is, however, that the measurement of the red shift, and thus optical identification, is required for each individual source. Figure 5 shows the θ–z diagram for identified galaxies and QSO's taken from a paper by Miley (1971) with additional data added by Wardle (private communication). The models shown are not exactly the same as in Fig. 4 but are similar. The fairly well defined upper envelope suggests that the scatter may be due mostly to projection effects. As expected, there is a clear decrease in angular size with increasing red shifts in apparent support of the cosmological interpretation of the quasar red shifts. However, there is no evidence for any change in the slope at large red shifts as expected from any of the models. [Von Hoerner (private communication) has shown that the steady-state model represents the minimum limiting diameter in all general relativistic world models.] At $z \sim 2.2$ for example there is nearly an order of magnitude discrepancy with the $q_0 = 1$ model.

Now, of course, this can be easily explained if the more distant sources are smaller, as in fact might be expected if the sizes are limited by inverse Compton scattering against the 3-K background, or by the density of the intergalactic medium. *However, this leads to the third coincidence shown in Table I.* Also there is no evidence in any of the short wavelength radio spectra of the sharp cutoff expected if inverse Compton scattering is important, nor is there any direct evidence of the required intergalactic medium.

IV. PHYSICS AND COSMOLOGY

In each of three cosmological tests discussed above, the number counts, the spectral index–flux density relation, and the angular size–red shift relation, where we tried to use the observed source properties to investigate the geometry of the universe, the expected geometrical effect is found to be mixed up with the possible evolutionary effect; i.e., the systematic variation in source properties with distance (epoch). It is

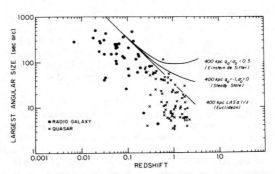

FIG. 5. Angular size vs red shift for identified double radio galaxies and quasars taken from Miley (1971) with additional data added by Wardle (private communication).

becoming increasingly clear that if any progress is to be made in separating the two we must better understand the nature of the sources themselves, such as what determines their properties and how they evolve with time.

Here the prospects are encouraging due primarily to the extraordinary improvements in sensitivity and resolution made during the past decade which now allow radio astronomers to make detailed measurements of properties such as source structure, polarization, spectra, and time variations with precision comparable to or superior to that previously enjoyed only by optical astronomers. As a result we now recognize two essentially distinct classes of radio source, both of which are commonly thought to radiate by the synchrotron mechanism. One class is the extended sources with complex brightness distribution and dimensions of the order of 100 kpc. These sources are transparent at radio frequencies and their emission is due to the superposition of the emission from the individual relativistic particles. Thus their spectra reflect the distribution of relativistic electron energies, and are typically power law or near power law. It is well known that the minimum energy required to explain the observed level of synchrotron radiation in these extended transparent sources is about 10^{60} erg.

The second class of source is the compact sources with angular dimensions very much less than 1 arc sec. These sources show a low frequency cutoff in their spectra due to self-absorption, and at low frequencies the compact sources are opaque to their own radiation.

I believe that there is now very convincing quantitative evidence that all of these sources radiate by the commonly accepted incoherent synchrotron radiation process. For example

(1) The shape of the spectra of the extended transparent sources. This was the original clue which led Ginzburg (1951) and Shklovsky (1952) to consider the synchrotron process. Today the spectra are known in

540 K. I. KELLERMANN

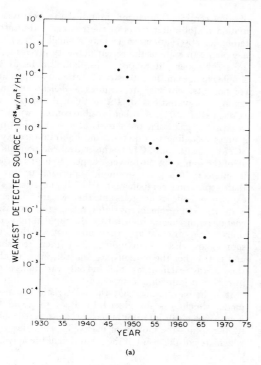

(a)

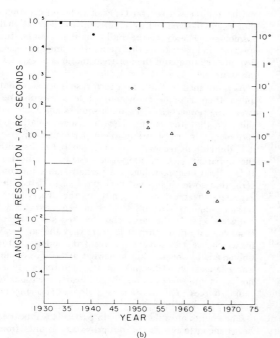

(b)

considerably greater detail, and provide quantitative support for the synchrotron mechanism.

(2) Among the compact sources, the smallest have spectral peaks at the shortest wavelengths as predicted by the synchrotron mechanism, and the measured angular sizes are in good agreement with those estimated from the observed self-absorption cutoff frequency.

(3) The maximum observed brightness temperature is $\sim 10^{12}$ °K, as is expected from an incoherent synchrotron source which is "cooled" by inverse Compton scattering.

(4) The observed pattern of intensity variations and its dependence on wavelength is in surprisingly good agreement with the simple model of synchrotron radiation from an expanding cloud of relativistic electrons.

I do not, however, put much weight on arguments supporting the synchrotron mechanism which are based on the observed existence of polarization, since almost any nonthermal emission process will produce partially polarized radiation.

Although the synchrotron mechanism appears to adequately explain the observed properties of radio galaxies and quasars, the ultimate source of energy and the way it is converted to relativistic particles has remained a basic problem. A major advance came with the discovery by Dent (1965) of variations in the intensity of several quasars on time scales of a few years or less. Subsequent observations at many observatories have shown that both radio galaxies and quasars may show frequent radio outbursts which often appear first at short wavelengths and then propagate with reduced amplitude toward longer wavelengths. The form of the observed variations can be simply explained in terms of an expanding cloud of relativistic particles which is initially opaque out to short wavelengths but which becomes transparent at successively longer wavelengths as the source expands.

The problem is that if the quasars are at cosmological distances the rapid intensity variations appear to require unacceptably large energies (e.g., Pauliny-Toth and Kellermann 1966) and a rapid exhaustion of the relativistic electrons by inverse Compton cooling (e.g., Hoyle, Burbidge, and Sargent 1966). The reason for this is that, if, as has been commonly assumed, significant fluctuations in intensity cannot occur in times less than

FIG. 6(a). Sensitivity of the "best" radio telescopes plotted as a function of time since the detection of the Sun in 1943 and Cygnus A in 1946. The weakest source which has been observed is one observed in a "deep synthesis survey" using the NRAO 3-element interferometer at 11 cm (Fomalont, private communication) which is about 2 m.f.u. (b) Best resolution of radio telescopes from the time of Jansky to the present. The symbols are as follows: (●) Filled-aperture antennas; (○) cable-linked interferometers; (△) radio-link interferometers; (▲) independent oscillator–tape recorder–interferometers. These plots show that both the sensitivity and resolution have continued to increase by about a factor of 2, each year, representing a formidable challenge to engineers designing future radio telescopes which may take 10 or more years to fund and construct.

the light travel time across the source without "blurring" the observed variation, then the observations put a limit on the linear dimensions typically of the order of one light year. If the quasars are at cosmological distances, then this corresponds to angular sizes well under 10^{-3} arc sec and magnetic fields less than 10^{-5} G. In such weak fields, the energy in the form of relativistic particles required to explain the observed radio emission is then typically greater than 10^{58} erg generated in one year or less. Even in these days when violent events in quasars and the nuclei of galaxies are commonly accepted, the thought of energy generation at the rate of 10^{51} erg/sec lasting one year or more seems too fantastic for most to accept.

For this reason, it was suggested that either the quasars were not at cosmological distances or that the radiation mechanism was by a more efficient process than incoherent synchrotron radiation, notwithstanding the success of the ordinary synchrotron theory in interpreting the spectra, the observed peak brightness temperature, and the form of the observed time variations.

A way out of this dilemma was suggested several years ago by Rees (1967) who pointed out that the cosmological interpretation could be saved if the sources were expanding at velocities close to the speed of light, in which case the differential light travel time to various parts of the source would cause an illusion of an expansion rate $v > c$. In this case since the true angular size can be greater than the size deduced from the simple assumption that $v < c$, the magnetic field is then greater, and the electron energy requirements are greatly reduced.

For a while many of us hoped that the direct measurement of expansion velocities by very long baseline interferometry would determine whether or not the "super-light" theory was indeed relevant. However, now that these observations have been made, the results are ambiguous, and the situation is still unclear. The reason for this is that if, as required by Rees' model, the component velocities in the line of sight are near the velocity of light, the differential Doppler shift will cause a large change in the relative intensity of the two components. The actual value depends on the specific model, but it is in the range from 10 to 100. However, the observed intensity ratio for at least one of the apparently rapidly expanding sources, 3C279, was equal to within a few percent for the four months when the component separation appeared to increase with a velocity $v \sim 8c$ (Whitney et al. 1971; Cohen et al. 1971). *This presents us with the fourth paradox shown in Table I.*

Now it is entirely possible that there are no real motions in these sources at all, but merely variations in the intensity of one or more stationary components. For example, in a ring or double source, with a small component at the center, a simple decrease in intensity of the central component imitates the relativistic motion of the outer parts of the source, if the interferometer observations cover only a small part of the u–v plane. In spite of the recent publicity given to the apparent superlight motions, the data can be fit by simple changes in the intensity of stationary expanding components, although it requires a delicate balance among the variations of the individual components (Cohen et al. 1971). Another possible concern with this interpretation is that on a basis of a sample of three sources including two quasars and one radio galaxy— 3C273, 3C279, and 3C120—the central components of all of the sources in the universe appear to be decreasing in intensity (an increase mimics approaching, rather than separating components). Also, if the relativistic expansion model is not relevant, then we are still faced with the old problem of interpreting the excessive energy requirements imposed by the rapid time variations. The observations therefore suggest that the sources are closer than indicated by their red shift, and that either the motions are nonrelativistic or that there are no real motions at all and only variations of the intensity of individual components.

It is perhaps ironic that the relativistic expansion model which was introduced by Rees to avoid the noncosmological interpretations of quasar red shifts itself requires the motion of matter at velocities $v \sim c$, which are not the result of the expansion of the universe.

V. SUMMARY

I have discussed several types of radio observations that might be expected to be relevant to observational cosmology. These are the radio number counts, the spectral index–distance relation, the angular size–distance relation, and the apparent motions of expanding sources.

As is outlined in Table I, in each case it appears that unless there is some close and subtle connection between the properties of the radio sources and their red shift, or that the entire class of general relativistic cosmologies is inappropriate to our particular universe, it is difficult to reconcile the observed red shifts with the apparent lack of any other distance indicator. It is indeed disappointing, and perhaps cause for concern, that so far it has not been possible to detect the expected effect of the red shift on the observed intensity, angular size, or spectra of the extragalactic radio sources. If the large observed red shifts are of cosmological origin, then it appears that the only way in which we can save our accepted models of the universe is by introducing somewhat arbitrary systematic changes with distance in the density or luminosity, in the spectra, and in the linear dimensions of radio sources. *The question is, "Are we drawing too many epicycles?"*

The counts of radio sources, in particular, appear to be as inconclusive as the optical-galaxy counts from

542   K. I. KELLERMANN

which Hubble concluded more than 30 years ago that

> "if the red shifts are *not* velocity shifts, the observa-
> tions lead to a rather simple, and thoroughly
> consistent, picture of the universe in which the
> velocity distance relation is linear and the large
> scale distribution of nebulae is uniform throughout
> the sample available for inspection."

And if the red shifts *are* interpreted as velocity shifts,
he comments,

> "the (velocity-distance) relation departs from
> linearity by just about the added corrections for
> recession . . . the necessary adjustments and
> compensations suggest that the model may be a
> forced interpretation of the data," and ". . . we
> may evidently choose between a curious small-scale
> universe and new principle of physics." (Hubble
> 1936a).

One can't help but wonder what choice Hubble
would make today, but his conclusion in 1936 that
"there is little excuse for the observer to weigh possibilities, because in due course, the 200 inch reflector
will presumably furnish definitive tests," stands as a
reminder that the more we learn about the universe,
the more confusing and complex the picture becomes.

ACKNOWLEDGMENTS

I am grateful to many colleagues, and in particular
to Drs. I. Pauliny-Toth, M. Davis, E. Fomalont, and
S. von Hoerner for numerous discussions of the problems
reviewed in this paper.

REFERENCES

Bridle, A. H., Davis, M. M., Fomalont, E. B., and Lequeux, J. 1972, *Nature Phys. Sci.* **235**, 123.
Burbidge, G. R. 1970, *Ann. Rev. Astron. Astrophys.* **8**, 369.
Burbidge, G., and Burbidge, M. 1967, *Quasi-Stellar Objects* (Freeman and Co., San Francisco).
Cohen, M. H., Cannon, W., Purcell, G. H., Shaffer, D. B., Broderick, J. J., Kellermann, K. I., and Jauncey, D. L. 1971, *Astrophys. J.* **170**, 207.
Davis, M. M. 1971, *Astron. J.* **76**, 980.
Dent, W. 1965, *Science* **148**, 1458.
Ginzburg, V. L. 1951, *Dokl. Akad. Nauk U.S.S.R.* **29**, 418.
Heeschen, D. S. 1960, *Publ. Astron. Soc. Pacific* **72**, 368.
———. 1966, *Astsophys. J.* **146**, 517.
Hoyle, F. 1968, *Proc. Roy. Soc. London*, A308, 1.
Hoyle, F., Burbidge, G., and Sargent, W. L. 1966, *Nature* **209**, 751.
Hoyle, F. 1959, *Paris Symposium on Radio Astsonomy* (Stanford U. Press, Stanford).
Hubble, E. 1936a, *Proc. Nat. Acad. Sci.* **22**, 621.
———. 1936b, *The Realm of the Nebulae* (Dover, New York).
Jauncey, D. 1967, *Nature* **217**, 877.
Kellermann, K. I., Pauliny-Toth, I. I. K., and Davis, M. M. 1968, *Astrophys. Lett.* **2**, 105.
Kellermann, K. I., Davis, M. M., and Pauliny-Toth, I. I. K. 1971, *Astrophys. J. Lett.* **170**, 21.
Longair, M. 1966, *Monthly Notices Roy. Astson. Soc.* **133**, 421.
Lynds, R., and Willis, D. 1972, *Astrophys. J.* **172**, 531.
Lynds, R. 1972, *IAU Symposium No. 44* (Springer-Verlag, Inc., New York).
Maslowski, J. 1972, *Astron. Astrophys.* **16**, 197.
Miley, G. 1971, *Monthly Notices Roy. Astson. Soc.* **152**, 477.
Mills, B. 1959, *Paris Symposium on Radio Astronomy* (Stanford U. Press, Stanford).
Pauliny-Toth, I. I. K., and Kellermann, K. I. 1966, *Astrophys. J.* **146**, 634.
Pauliny-Toth, I. I. K., Kellermann, K. I., Davis, M. M., Fomalont, E., and Shaffer, D. 1972, *Astron. J.*, **77**, 265.
Pauliny-Toth, I. I. K., and Kellermann, K. I. 1972, *Astron. J.* **77**, 560.
Pooley, G. G. 1968, *Nature* **218**, 154.
Pooley, G. G., and Ryle, M. 1968, *Monthly Notices Roy. Astron. Soc.* **139**, 515.
Rees, M. 1967, *ibid.* **135**, 345.
Ryle, M. 1968, *Ann. Rev. Astron. Astrophys.* **6**, 249.
Sandage, A. 1961, *Astrophys. J.* **133**, 355.
Schmidt, M. 1968, *ibid.* **151**, 393.
Shimmins, A. J., Bolton, J. G., and Wall, J. 1968, *Nature* **217**, 858.
Shklovsky, I. S. 1956, *Astron. Zhur. U.S.S.R.* **29**, 418.
Wall, J. V., Shimmins, A. J., and Merkelijn, J. K. 1975, *Aust. J. Phys. Astrophys. Suppl.*, No. 19.
Whitney, A. R., Shapiro, I. I., Rogers, A. E. E., Robertson, D. S., Knight, C. A., Clark, T. A., Goldstein, R. M., Marandino, G. E., and Vandenberg, N. R. 1971, *Science* **172**, 225.
Yahil, A. 1972, *Astrophys. J.*, in press.

05.01.10 <u>Morphology</u> <u>and</u> <u>Redshifts</u> <u>of</u> <u>Galaxies</u>. HALTON
ARP, <u>Hale</u> <u>Observatories</u>, <u>Carnegie</u> <u>Institution</u> <u>of</u> <u>Washington</u>
<u>California</u> <u>Institute</u> <u>of</u> <u>Technology</u>. Five galaxies that
fall close together on the sky, known as Stephan's Quintet,
exhibit peculiar forms and discordant redshifts. New
observations show that they are all at the same distance
and form a physical group. Evidence further indicates
the group is associated with the large, nearby spiral NGC
7331 and that four members have intrinsic excess redshifts
of from 4700 to 5700 km/sec.

The few galaxy groups similar to Stephan's Quintet are
examined and almost all of them are shown to fall close
to large spiral galaxies. Examples are seen near NGC 247,
NGC 3718, and NGC 908. All bright spiral galaxies
between apparent magnitude 10 and 11.2 are then studied.
It is shown that interacting double galaxies, a simpler
case of the nonequilibrium chains and groups, are ten
times more common near bright spirals (within 1°) than in
control fields of the order of 6° distant. The peculiar
companion galaxies tend to occur in long, straight chains
radiating from the large central galaxy. The redshifts
of these nonequilibrium companion galaxies are considerably
larger than the redshifts of the central, large galaxy.

Evidence for ejection of luminous matter from galaxies is
reviewed. It is shown that these nonequilibrium-form
companions fit all the previously established empirical
rules of ejection from large central galaxies. Individual
peculiar galaxies and high redshift objects are assembled
in a diagram which shows the relation between excess
redshift and morphological type - the excess redshift
diminishes as the type progresses continuously from quasars
through compact galaxies, the interacting doubles, and
finally through Sc, Sb, Sa, and E-like galaxies. It is
proposed that this is an evolutionary sequence. The
tendency for the nonequilibrium galaxies to be double is
related to the tendency for the compact galaxies, from which
they evolved, to be double also.

The final picture that emerges is one of the large galaxy
which ejects quasars and compact objects. These expand and
unfold, evolving into peculiar companions with some of the
excess redshift decaying at ages of about 10^{7} years. They
continue to unfold, to some extent by secondary and tertiary
ejection, and evolve into Sc-like spirals of about 10^{8} years
age. Possibly, depending on initial mass, after intervals
of the order of 10^{9} years, the young spirals evolve into Sb,
Sa, and E and SO-like forms. Jaakola's result on the
higher intrinsic redshift of Sc's compared to E's is used
to substantiate this picture.

The present observations are used inductively to conclude
that the compact objects originate in the nuclei of large
galaxies where the physical conditions approach singular
values and that their excess redshifts are related to their
young age as measured from this event. In my opinion, of
the kind of explanations that the current observations
require, one of the simplest is one along the lines of
Hoyle's suggestion that electrons and other atomic
constituents can be created with initially smaller mass.
Then smaller $h\nu$ emissions result from a given atomic
transition, and radiation from all objects in the new
galaxy is shifted to the red. As the galaxy ages, its
atomic parameters asymptotically approach that of older
matter.

THE DEVELOPING CRISIS IN ASTRONOMY[*]

FRED HOYLE

Institute of Theoretical Astronomy, Cambridge, England

and

California Institute of Technology, Pasadena, California

Henry Norris Russell Lecture

Presented at

The Seattle Meeting of the American Astronomical Society

April 8-12, 1972

[*]Supported in part by the National Science Foundation [GP-27304, GP-28027].

It is sometimes said that nothing is known from astronomy which goes outside the range of currently-known physics. Whatever point of view one takes about the interpretation of the present day observational situation this statement is incorrect. For example, if one accepts the existence of the microwave background as a survival from the origin of the Universe at some moment of time, $t = 0$ say, then it is this phenomenon of "origin" which lies outside currently-known physics.

On the other hand it can be argued that the background arises from a high spatial density of discrete sources of low intrinsic emission. To see this consider two kinds of source, one of intrinsic emission L_1 and spatial density n_1, the other of intrinsic emission L_2 and spatial density n_2. Taking both kinds to have the same radio spectrum, the condition for them to make equal background contributions is evidently

$$n_1 L_1 = n_2 L_2 \quad .\tag{1}$$

But the ratio of the numbers of sources of the two kinds with flux values greater than S (in Euclidean space for simplicity) is $(L_1/L_2)^{1/2}$. By taking $L_1 \rightarrow 0$, with the product $n_1 L_1$ nevertheless satisfying equation (1), sources of the first kind, while making just as significant a contribution to the background as sources of the second kind, are not detected individually in a discrete source count to any assigned S.

By postulating a high density population of suitable faint sources (say with a mean spacing of ~ 100 kpc) it would be possible to explain the microwave background without recourse to an origin for the Universe. Since other arguments against the steady-state model, for example the log N - log S argument from radiosource counts, are now seen to be much weaker than they were formerly thought to be, it would then be possible to return to the steady-state model. But this raises the physical problem of the creation of matter, which in principle is the same whether we are concerned with matter originating locally or with an origin for the whole Universe. We may expect the mathematical structure of the problem to be the same.

In this lecture I propose to consider the physical problem of the origin of the Universe from the standpoint of a theory which is equivalent to the usual theory so far as cosmology and weak gravitational fields are concerned. The isotropic homogeneous spacetime used in most cosmological discussions is conformally flat. This means that by appropriately changing the length scale from point to point the spacetime metric can be changed from the usual Robertson-Walker form to the flat Minkowski form,

$$ds^2 = dt^2 - dr^2 - r^2(d\Theta^2 + \sin^2 \Theta \, d\varphi^2) \quad , \tag{2}$$

in an obvious notation. By writing the physical theory in a way that is invariant with respect to such conformal transformations it is possible to take advantage of this geometrical simplification. The procedure is conceptually very simple. Only the single term

$$S = - \sum_{\text{Particles}} \int m(A) \, da \quad , \tag{4}$$

is needed to define the action for gravitation. Here A is a point at da, an element of proper time on the world line of particle a. The field term of the usual theory,

$$\frac{1}{16\pi G} \int R \sqrt{-g} \, d^4 x \quad , \tag{5}$$

2

is not required. In general the particle mess m is a function of position.
To determine $m(A)$ define

$$M(A) = \sum_{b \neq a} \int \tilde{G}(A,B) \, da \quad , \tag{6}$$

the summation being over all particles $b \neq a$. The scalar "mass propagator"
$\tilde{G}(A,B)$ is determined by a wave equation which is specified uniquely by the
conformal requirement and by the symmetry condition $\tilde{G}(A,B) = \tilde{G}(B,A)$. In
flat spacetime $\tilde{G}(A,B)$ behaves like the well-known electromagnetic propagator
so that M in flat space is like a Coulomb potential. We then write

$$m(A) = \lambda^2 \, M(A) \quad , \tag{7}$$

where λ is a coupling constant analogous to electric charge. The gravitational equations are given in the usual way, by requiring

$$\delta S = 0 \quad \text{for} \quad g_{ik} \rightarrow g_{ik} + \delta g_{ik} \quad , \tag{8}$$

i.e. by requiring an extremum with respect to small general changes of the
space-time metric.

The Einstein de Sitter cosmological model can be obtained very simply.
For a static uniform density L^{-3} of particles, and taking only the retarded
part of $\tilde{G}$, one finds

$$m(t) = \frac{1}{4} \lambda^2 \, t^2 \, L^{-3} \tag{9}$$

using (6), (7), where t is the time that has elapsed since the origin of
the Universe. The mass dependence (9), taken with the Minkowski metric
and with the gravitational equations (8), gives all the usual results of
the Einstein de Sitter model. As an example, we can easily obtain the

3

Hubble magnitude-redshift relation. Taking the velocity of light to be unity, light from a galaxy with coordinate r must start on its journey at time t - r to reach r = 0 at time t. Write L(t - r) for the intrinsic luminosity of the galaxy at that time. The flux for an observer at r = 0 at time t is just the Euclidean value

$$\frac{L(t - r)}{4\pi\, r^2} \; . \tag{10}$$

Now it can readily be seen that "luminosity" has the same dimensionality as $(mass)^2$, which at time t - r is proportional to $(t - r)^4$. Hence we have

$$Flux \propto \frac{(t - r)^4}{r^2} \; . \tag{11}$$

Intrinsically similar galaxies, all observed at time t, will have flux values that depend on their distances r according to (11). The redshift z is given by

$$1 + z = \frac{m(t)}{m(t - r)} = \frac{t^2}{(t - r)^2} \; . \tag{12}$$

Eliminating r between (11) and (12) gives

$$Flux \propto \frac{1}{1 + z} \cdot \frac{1}{[\sqrt{1 + z} - 1]^2} \; , \tag{13}$$

which is Hubble's law for the Einstein de Sitter model.

The mass determination (9) is illustrated in Figure 1, except that twice the retarded part of $\tilde{G}$ has been taken. By taking $\frac{1}{2}\lambda^2$ instead of λ^2 the calculated mass remains unchanged, so that this issue is purely formal. The situation in Figure 1 agrees with our usual ideas, a definite origin at t = 0 for the whole Universe, and an <u>ad hoc</u> choice of the retarded potential.

4

However the mass computation is unchanged if we pass now to the scheme of Figure 2, and here we have a drastic change of concept. The situation is now symmetric with respect to advanced and retarded potentials, and the Universe does not begin at t = 0. Instead, we contemplate a switch of sign at t = 0, a switch in the sign of the contributions of particles to the mass field M. The significance of t = 0 is now that it represents a discontinuity, not an origin.

The emergence of a discontinuity suggests a quantum transition, not just of a single atom or small collection of atoms, but of the whole visible Universe. It would be better to replace Figure 2 by Figure 3, where the band around t = 0 denotes a quantum situation — i.e. a situation in which there is no unique spacetime metric . On either side of this band, for t < 0 as well as for t > 0, the Universe in the large is classical. Physical systems cease to be classical when the action becomes small, as it certainly does near t = 0, because all masses are zero at t = 0.

Our prodecure so far has been to begin with a mathematical structure identical in every respect with the usual Einstein de Sitter model (Fig. 1). We have then noticed that the new way of writing the theory is also equivalent to a situation in which the Universe undergoes a discontinuity, not an origin, at t = 0 (Fig 2) and that the discontinuity is very likely a quantum phenomenon. We also understand in a natural way why radiation generated near t = 0 is of very long wavelength, because the particle masses are small. It is this longwave radiation which we associate with the microwave background.

Can we connect these ideas with local phenomena in the modern Universe? In particular can the idea of a quantum transition, or a quantum state, b applied to a large aggregate of localised matter? Recently the words "black hole" have become an astronomical cliche. But is the conventional classical picture of a black hole at all correct? It seems to me that this classical picture is likely to be just as inapposite as was the classical picture of individual atoms in the first decade of the present century. The argument then was that the electron moving around a proton in the hydrogen atom will radiate electromagnetic waves, and will consequently collapse more and more on to the proton. In a similar way we argue today that an aggregate of matter with dimension close to the Schwarzschild radius will radiate, either gravitationally or electromagnetically, and will eventually collapse into a black hole. Former experience should warn us against this much too classical concept. The difficulty with attempts to formulate a quantum theory of large aggregates has hitherto been, however, that so long as the masses of particles are fixed and autonomous to the particles themselves the situation remains classical. This stranglehold on the problem is broken, at any rate in principle, when masses are taken to arise from the interactions of particles.

It is possible to prove an interesting result. Under cosmological conditions, or under conditions of weak local gravitational fields, the particle masses are dominated by interactions on the cosmological scale. Distant interactions overwhelm local interactions. This is no longer true for strong local gravitational fields, and it is here that the theory discribed above leads in a new direction. For a spherically symmetric object very close in radius to the Schwarzschild limit we have the following situation:

Distant Interactions tend to zero at a radius r given by

$$- \ln\left(1 - \frac{R}{r}\right) \approx 6 \quad , \tag{14}$$

where R is the usual Schwarzschild radius. Particle masses within the aggregate then arise only from internal interactions. With the local system no longer controlled by the external Universe there is at least a _prima facie_ case that the aggregate is able to find a multiplicity of different arrangements, but all with essentially the same total action, which is the condition needed for quantum considerations to become important.

For this reason I have little faith in the usual treatment of the "black hole" problem. I do not believe a classical treatment of systems at their Schwarzschild radii to be remotely adequate.

In some degree we are all used to the concept that the cosmological model affects the behavior of local systems. For example, the cosmological model affects the problem of the origin of galaxies — even within the range of conventional theory. The mass interaction discussed above represents a more far-reaching connection between the large and the small. Can we carry this connection still further, by requiring that the behavior of local systems affects the cosmological model? That is to say, could there be a feedback loop between the large and the small? I have a strong preference for an affirmative answer to this question but I have not yet seen any very strong argument to prove that it must be so. For such a relationship it is necessary that the energy interchange between local systems and the Universe as a whole, occurring in a characteristic time H^{-1}, shall be comparable to the whole energy of the Universe. The well-attested forms of energy output from localised objects — electromagnetic radiation of all kinds, high-energy particle emission — cannot comprise more than about one percent of the necessary amount. Even if cosmic rays are universal their energy density, $\sim 3 \times 10^{-12}$ erg per cm^3, falls more than two powers of ten short of the average mass energy density of the Universe. A more likely possibility comes from the possibility that all the apparently expanding clusters of galaxies really are expanding, that many such galaxies really are young and that the matter of which they are composed is of recent origin. I mean this in the same sense as we speak of the "origin" of the Universe as a whole. As I have emphasized above, this probably implies some kind of large scale discontinuity. There could be local discontinuities just as well as a single universal discontinuity. Again I have in mind the general concept of a quantum transition, but of large aggregates not just of individual atoms.

The alternative to supposing that apparently expanding clusters really are expanding is to assume the presence of "missing mass". However in many cases the amount of missing mass is very large compared to the mass of luminous material.

Another possibility which must be mentioned concerns the gravitational waves reported by Weber. Here the energy outflow is so very large that a feedback relationship between local systems and the Universe would be a distinct possibility. Quite apart from the importance of the phenomenon itself this would be a crucial implication of a general confirmation of Weber's results.

Although the representation of the masses of particles as functions of cosmic time can be related by strict mathematical transformations to the usual cosmological models this different way of writing cosmological theory prepares us for the possibility that masses could depend on spatial location as well as on time — if we have $m(t)$ then why not $m(t, \underset{\sim}{x})$, where $\underset{\sim}{x}$

8

refers to spatial position? This concept appears necessary if we are to
understand the result reported by Arp for the galaxy NGC 7603 and its
appendage. It also appears necessary if we are to understand the strange
situation which has gradually been uncovered for the quasi-stellar objects.

In 1965 Burbidge and I wrote a paper which has often been
misquoted as saying that QSO redshifts were not normal cosmological redshifts. What we actually said was that the possibility that QSO's were
not at cosmological distances deserved serious consideration. The idea
we sought to develop was that QSO's were systems expelled from comparatively
nearby galaxies. Our arguments were essentially physical. Little of the
empirical properties of QSO's were then known. Only about ten had had
their redshifts determined and these seemed to approximate to the Hubble
law. Also for the one known QSO of large measured redshift, 3C 9, there
was a suspicion that the continuum shortward of Ly α was partially absorbed
by intergalactic atomic hydrogen. Both these objections soon disappeared,
however,

This was before any QSO's near bright galaxies were known. The first
case to be discovered was the source 3C 275.1 which was found to lie about
3 arc minutes from the galaxy NGC 4651. The probability of finding one of
the forty or so QSO's in the 3C catalogue lying within 3 arc minutes of an
NGC galaxy is 1/10, if the association were due to chance. The next case
was a 20'th magnitude radioquiet QSO found by Stockton within about 1 arc
minute of the remarkable system of NGC 3561.

By early 1971 three further 3C QSO's had been found near bright galaxies.
Taking these later cases together with 3C 275.1, Burbidge, Burbidge, Solomon,
and Strittmatter have estimated that the probability of all these associations
being due to chance is not greater than 1/250. It is relevant as how one

estimates this situation that the proposition that QSO's might be associated with nearby galaxies was not a post hoc hypothesis. It was suggested in 1965 in my paper with Burbidge before the relevant data were known.

Commentators on the paper of Burbidge, Burbidge, Solomon, and Strittmatter draw attention to those 3C QSO's which do not lie near bright galaxies, apparently being under the impression that if the rest could be shown to have no correlation with nearby galaxies this would make the acceptance of a one-in-two hundred-and-fifty-coincidence somehow more palatable. No sooner were these comments in the press than a fifth 3C source, 3C 455, was found to lie near the galaxy NGC 7413. The source coincided within about 1 arc second (the error of measurement) with a blue stellar object which was found to be a QSO. The angular separation in this case was only 23", and the probability that the association was due to chance was now as small as 2.5×10^{-3}, for this one case alone. The probability that all five 3C cases were due to chance is as low as 10^{-5}.

Burbidge, O'Dell, and Strittmatter noticed that the five sources just mentioned have angular separations from the NGC galaxies that are inversely related to the redshifts of the galaxies, so that if the QSO's are associated with the galaxies, and if we accept the usual redshift-distance relation for the galaxies, then the distance separations are closely similar in all five cases.

Peterson and Bolton have further pointed out that the source PKS 2020 - 370 appears to be similar to 3C 455. It is a QSO with redshift 1.11 lying about 15" from a 16'th magnitude galaxy. The probability of this association being due to chance is more difficult to estimate than are probabilities for the 3C sample, but an order of magnitude estimate shows the probability to be small, probably $\sim 10^{-2}$. This sixth case lies close to 3C 455 in the plot of Burbidge, O'Dell, and Strittmatter.

I referred above to the early situation when the redshifts and magnitudes of the first handful of QSO's appeared to be following the Hubble law. The current situation is shown in Figure 4. Whatever the cause of redshift, quanta lose energy by the factor $(1 + z)^{-1}$. There must also be a second $(1 + z)^{-1}$ factor arising from a reduction in the quantum counting rate (the "number effect"), so that observed fluxes must be increased by $(1 + z)^2$ in order that QSO's can be compared in a redshift-magnitude plot. If there is no further K-term (as is usually supposed), then we have to plot the QSO's in a z, V - 5 log(1 + z) plane, as is done in Figure 4. If we make the further usual supposition that intrinsic luminosity is not correlated with z, then we have the following alternatives:

(1) If z is uncorrelated with distance, the QSO plot should form a scatter diagram, which essentially it does.

(2) If z is cosmological in origin, the QSO plot should form a Hubble diagram, which clearly it does not.

The cosmological hypothesis cannot survive unless:

(a) We suppose that a large number of QSO's of large redshift have been missed.

(b) The sample is too small to show up very bright QSO's of small redshift.

The objection to (a), so far as the 3C sample at any rate is concerned, is that many QSO's have <u>not</u> been missed. The sample is substantially complete. Other radio samples used in Figure 4 are less completely identified and for these samples one might seek to fall back on (a). However, Hazard and Murdoch have recently identified a substantial fraction of about 350 sources

observed with the Molonglo Cross . The sample is complete over two strips

of sky having $\triangle$R.A. = 6 hours, $\triangle\delta$ = 2° down to 0.5 flux units at 408 MHz.

With the use of computer overlays on to the Palomar sky charts it has been

possible to make identifications with blue stellar objects for about 1/8

of these sources. Since this is not much less than the fraction of QSO's

found in the 3C sample, it is believed that a substantial fraction of QSO's

in the new sample have been identified. The optical magnitudes are not

appreciably different from the 3C QSO's. Typically V runs from 17 to 18.

Although redshifts remain to be determined, the fact that the new QSO's

have been just as readily identified as before makes it likely that Figure

4 does represent a genuine departure from the Hubble law.

Let me return to $m(t,\underset{\sim}{x})$. By adjusting the spatial dependence on $\underset{\sim}{x}$

it is possible to account for all these strange observations. The theore-

tical difficulty lies in understanding the kind of dependence that the

observations seem to demand. The simple classical theory based on (6) is

not adequate. Something far more sophisticated would be needed. It is

just at this point that a crisis is reached. If the observations are judged

sufficient to force us still further along the road which (6) begins to

open up, then we shall soon find ourselves involved with intricate issues

of particle physics. The implication is that cosmology will have a great

deal to say about fundamental physics. This is a conclusion which many

physicists and indeed many astronomers are trying hard to resist. To what

point in the accumulation of data this is a reasonable position to maintain

is a matter of subjective judgment. For each of us there is a decision to be

made. Do we cross a bridge into wholly unfamiliar territory or do we try

12

to remain safely within well-known concepts?   This depends on how each of us sees the data.

For me, personally, the exact state of the data of any given moment is less important than the trend of the data.   There seems to me no doubt that the trend is towards forcing us, whether we like it or not, across this exceedingly important bridge.   Either the bridge must be crossed or one must judge the data of the past five years to be extremely freakish.

REFERENCES

Arp, H. C. 1971, Astr. Letters, 7, 221.

Arp, H. C., Burbidge, E. M., MacKay, C. D., and Strittmatter, P. A. 1972,
    Ap. J. (letters), 171, L41.

Burbidge, E. M., Burbidge, G. R., Solomon, P. M., and Strittmatter, P. A.
    1971, Ap. J., 170, 233.

Hoyle, F., and Burbidge, G. R. 1966, Ap. J., 144, 534.

Hoyle, F., and Narlikar, J. V. 1972, M.N.R.A.S., 155, 305.

Peterson, B. A., and Bolton, J. G. 1972, Astr. Letters, 10, 105.

Sandage, A., Véron, P., and Wyndham, J. D. 1965, Ap. J., 142, 1307.

Stockton, A. 1969, Ap. J. (Letters), 155, L141.

FIGURE CAPTIONS

Fig. 1.   Determination of the mass field using only retarded potentials.

Fig. 2.   The mass interaction is to be multiplied by ϵ, with $\epsilon = +1$ for
$t > 0$ and $\epsilon = -1$ for $t < 0$. This permits both advanced and
retarded potentials to be used.

Fig. 3.   Since masses are zero at $t = 0$, the situation near $t = 0$ is
quantum mechanical. This is indicated by the heavy band near
$t = 0$. The switch in the sign of ϵ at $t = 0$ may be thought of
as a quantum transition.

Fig. 4.   Plot of redshifts z against $V - 5 \log_{10} (1 + z)$ for quasi-stellar
radio sources. Here V is the visual magnitude. Subtraction of
$5 \log_{10} (1 + z)$ takes account of loss of quantum energies due to
redshift, and also of the 'number effect'. If z is uncorrelated
with distance this should be a scatter diagram, which it is.

15

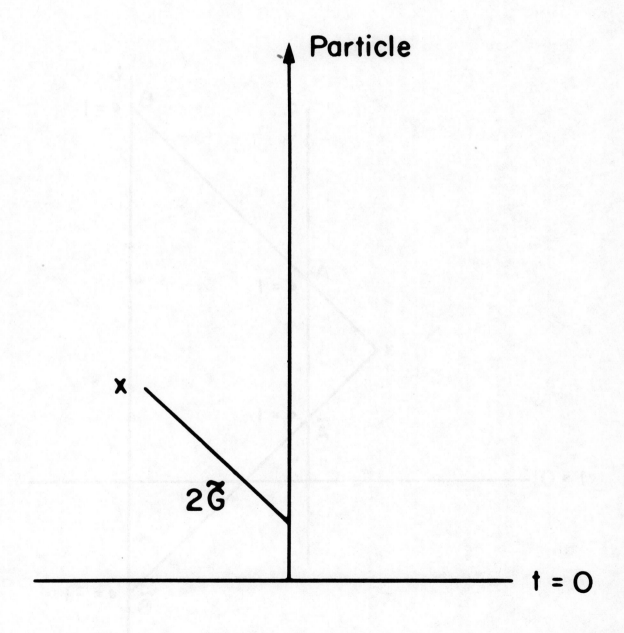

Fig. 1

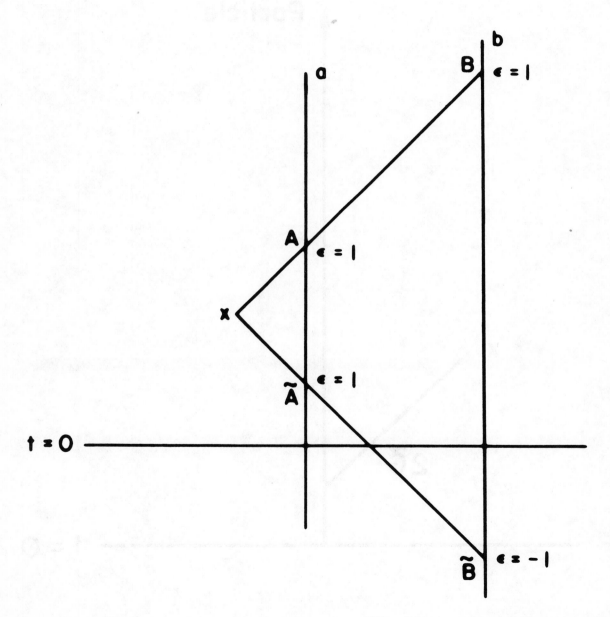

Fig. 2

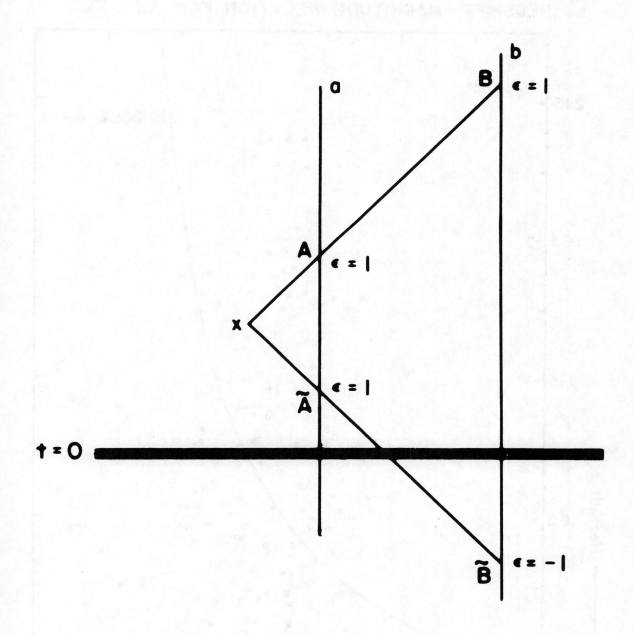

Fig. 3

REDSHIFT - MAGNITUDE RELATION FOR QSR

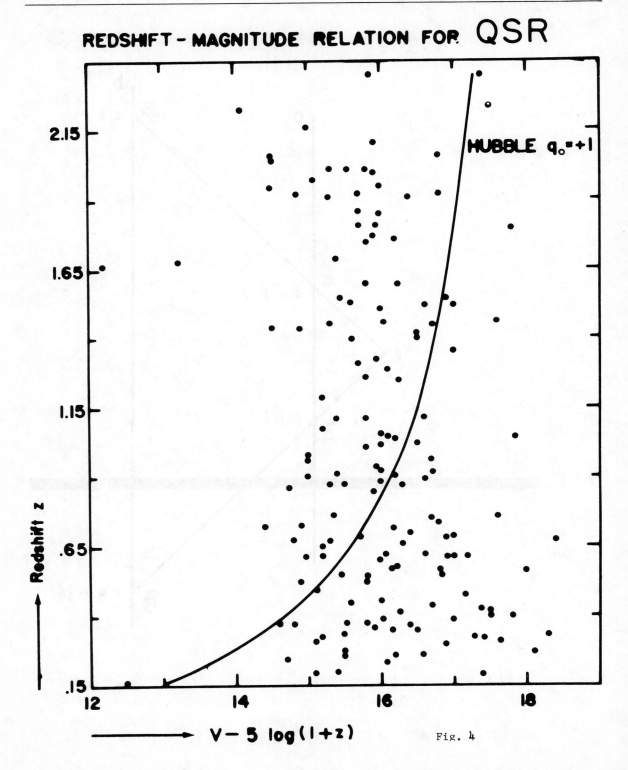

Fig. 4

INDEX

9383